Annals of Mathematics Studies

Number 151

The Geometry and Cohomology of Some Simple Shimura Varieties

by

Michael Harris

and

Richard Taylor

With an appendix by Vladimir G. Berkovich

PRINCETON UNIVERSITY PRESS

PRINCETON AND OXFORD
2001

Published by Princeton University Press, 41 William Street, Princeton, New Jersey 08540

In the United Kingdom: Princeton University Press, 3 Market Place, Woodstock, Oxfordshire OX20 1SY

The Annals of Mathematics Studies are edited by John N. Mather and Elias M. Stein

ISBN 0–691–09090–4 (cloth)
ISBN 0–691–09092–0 (pbk.)

The publisher would like to acknowledge the authors of this volume for providing the camera-ready copy from which this book was printed.

Printed on acid-free paper. ∞

www.pup.princeton.edu

Printed in the United States of America

10 9 8 7 6 5 4 3 2 1

10 9 8 7 6 5 4 3 2 1

(Pbk.)

To Béatrice and Christine

Contents

Introduction

This book has twin aims. On the one hand we prove the local Langlands conjecture for GL_n over a p-adic field. On the other hand in many cases we are able to identify the action of the decomposition group at a prime of bad reduction on the l-adic cohomology of the "simple" Shimura varieties studied by Kottwitz in [Ko4]. These two problems go hand in hand.

The local Langlands conjecture is one of those hydra-like conjectures which seems to grow as it gets proved. However the generally accepted formulation seems to be the following (see [He2]). Let K be a finite extension of $\mathbb{Q}_p$. Fix a non-trivial additive character $\Psi : K \to \mathbb{C}^\times$. We will denote the absolute value on K which takes uniformisers to the reciprocal of the number of elements in the residue field by $| \ |_K$. We will let W_K denote its Weil group. Recall that local class field theory gives us a canonical isomorphism

$$\mathrm{Art}_K : K^\times \to W_K^{\mathrm{ab}}.$$

(Normalised so that geometric Frobenius elements correspond to uniformisers.) The local Langlands conjecture provides some sort of description of the whole of W_K in the same spirit.

We will let $\mathrm{Irr}(GL_n(K))$ denote the set of isomorphism classes or irreducible admissible representations of $GL_n(K)$ over $\mathbb{C}$ (or what comes to the same thing: irreducible smooth representations). If $[\pi_1] \in \mathrm{Irr}(GL_{n_1}(K))$ and $[\pi_2] \in \mathrm{Irr}(GL_{n_2}(K))$ then there is an L-factor $L(\pi_1 \times \pi_2, s)$ and an epsilon factor $\epsilon(\pi_1 \times \pi_2, s, \Psi)$ associated to the pair π_1, π_2 (see for instance [JPSS]).

On the other hand let $\mathrm{WDRep}_n(W_K)$ denote the set of isomorphism classes of n-dimensional Frobenius semi-simple Weil-Deligne representations of the Weil group, W_K, of K over $\mathbb{C}$. By a Frobenius semi-simple Weil-Deligne representation of W_K over $\mathbb{C}$ we mean a pair (r, N) where r is a semi-simple representation of W_K on a finite dimensional complex vector space V, which is trivial on an open subgroup, and an element $N \in \mathrm{End}_{\mathbb{C}}(V)$ such that

$$r(\sigma) N r(\sigma)^{-1} = |\mathrm{Art}_K^{-1}(\sigma)|_K N$$

1

for all $\sigma \in W_K$. Again if $[(r, N)] \in \mathrm{WDRep}_{n_1}(W_K)$ then there is an L-factor $L((r, N), s)$ and an epsilon factor $\epsilon((r, N), s, \Psi)$ associated to (r, N) (see for instance [Tat2] and section VII.2 of this book for the precise normalisations we are using).

By a *local Langlands correspondence* for K we shall mean a collection of bijections

$$\mathrm{rec}_K : \mathrm{Irr}(GL_n(K)) \longrightarrow \mathrm{WDRep}_n(W_K)$$

for every $n \geq 1$ satisfying the following properties.

1. If $\pi \in \mathrm{Irr}(GL_1(K))$ then $\mathrm{rec}_K(\pi) = \pi \circ \mathrm{Art}_K^{-1}$.

2. If $[\pi_1] \in \mathrm{Irr}(GL_{n_1}(K))$ and $[\pi_2] \in \mathrm{Irr}(GL_{n_2}(K))$ then

$$L(\pi_1 \times \pi_2, s) = L(\mathrm{rec}_K(\pi_1) \otimes \mathrm{rec}_K(\pi_2), s)$$

   and

$$\epsilon(\pi_1 \times \pi_2, s, \Psi) = \epsilon(\mathrm{rec}_K(\pi_1) \otimes \mathrm{rec}_K(\pi_2), s, \Psi).$$

3. If $[\pi] \in \mathrm{Irr}(GL_n(K))$ and $\chi \in \mathrm{Irr}(GL_1(K))$ then

$$\mathrm{rec}_K(\pi \otimes (\chi \circ \det)) = \mathrm{rec}_K(\pi) \otimes \mathrm{rec}_K(\chi).$$

4. If $[\pi] \in \mathrm{Irr}(GL_n(K))$ and π has central character χ then

$$\det \mathrm{rec}_K(\pi) = \mathrm{rec}_K(\chi).$$

5. If $[\pi] \in \mathrm{Irr}(GL_n(K))$ then $\mathrm{rec}_K(\pi^\vee) = \mathrm{rec}_K(\pi)^\vee$ (where $\vee$ denotes contragredient).

Henniart showed (see [He5]) that there is at most one set of bijections rec_K with these properties. The commonest formulation of the local Langlands conjecture for GL_n is the following theorem.

Theorem A *A local Langlands correspondence* rec_K *exists for any finite extension* $K/\mathbb{Q}_p$.

However it seems to us that one would really like more than this simple existence theorem. On the one hand it would be very useful if one had some sort of explicit description of this map rec_K. Our methods shed no light on this. One might well hope that the methods of Bushnell, Henniart and Kutzko will lead to an explicit version of this theorem. On the other hand one would also like to know that the local reciprocity map rec_K is compatible with global reciprocity maps whenever the global map is known to exist. Our methods do not resolve this latter question but they do shed considerable light on it. For instance in the cases considered by Clozel in [Cl1] we settle this question affirmatively up to semisimplification. (We do not identify the two N's.)

Maybe a remark on the history of this problem is in order. The existence of $\mathrm{rec}_K|_{\mathrm{Irr}(GL_1(K))}$ with the desired properties follows from local class field theory (due originally to Hasse [Has]), but this preceded the general conjecture. The key generalisation to $n > 1$ is due to Langlands (see [Lan]), who formulated some much more wide ranging, if less precise, conjectures. The formulation in the form described here, with its emphasis on epsilon factors of pairs, seems to be due to Henniart (see [He2]). Henniart's formulation has the advantage that there is at most one such correspondence, but as remarked above it limits somewhat the scope of Langlands' original desiderata. The existence of $\mathrm{rec}_K|_{\mathrm{Irr}(GL_2(K))}$ with the desired properties was established by Kutzko ([Ku]), following earlier partial work by a number of people. The existence of $\mathrm{rec}_K|_{\mathrm{Irr}(GL_3(K))}$ with almost all the desired properties was established by Henniart ([He1]). In particular, his correspondence had enough of these properties to characterise it uniquely. Both the work of Kutzko and Henniart relied on a detailed classification of all elements of $\mathrm{Irr}(GL_n(K))$. These methods have since been pushed much further, but to date have not provided a construction of rec_K which demonstrably has the desired properties on $\mathrm{Irr}(GL_n(K))$ for any $n > 3$. In the case of completions of functions fields of transcendence degree 1 over finite fields, the corresponding theorem was proved by Laumon, Rapoport and Stuhler ([LRS]).

We will let $\mathrm{Cusp}\,(GL_n(K))$ denote the subset of $\mathrm{Irr}(GL_n(K))$ consisting of equivalence classes of supercuspidal representations. Let $\mathrm{Rep}_n(W_K)$ denote the subset of $\mathrm{WDRep}_n(W_K)$ consisting of equivalence classes of pairs (r, N) with $N = 0$. Also let $\mathrm{Irr}_n(W_K)$ denote the subset of $\mathrm{Rep}_n(W_K)$ consisting of equivalence classes of pairs $(r, 0)$ with r irreducible. It follows from important work of Zelevinsky [Ze] that it suffices to construct bijections

$$\mathrm{rec}_K : \mathrm{Cusp}\,(GL_n(K)) \longrightarrow \mathrm{Irr}_n(W_K)$$

with the properties listed above (see [He2].) In a key breakthrough, Henniart [He4] showed that there did exist bijections

$$\mathrm{rec}_K : \mathrm{Cusp}\,(GL_n(K)) \longrightarrow \mathrm{Irr}_n(W_K),$$

which preserved conductors and were compatible with twists by unramified characters. He was however unable to show that these bijections had enough of the other desired properties to characterise them uniquely. The usefulness of this result is that it allows one to use counting arguments, for instance any injection $\mathrm{Cusp}\,(GL_n(K)) \hookrightarrow \mathrm{Irr}_n(W_K)$ satisfying the desired properties must be a bijection. (This result is usually referred to as the numerical local Langlands theorem.)

We will give a natural construction of a map

$$\mathrm{rec}_K : \mathrm{Cusp}\,(GL_n(K)) \longrightarrow \mathrm{Irr}_n(W_K),$$

which we will show is compatible with the association of l-adic representations to many automorphic forms on certain unitary groups. Using this global compatibility and some instances of non-Galois automorphic induction discovered by one of us (M.H., see [Har3]) we will see that there is a subset $\mathrm{Cusp}\,(GL_n(K))' \subset \mathrm{Cusp}\,(GL_n(K))$ such that

$$\mathrm{rec}_K : \mathrm{Cusp}\,(GL_n(K))' \xrightarrow{\sim} \mathrm{Irr}_n(W_K)$$

and such that $\mathrm{rec}_K|_{\mathrm{Cusp}\,(GL_n(K))'}$ has all the desired properties. The subset $\mathrm{Cusp}\,(GL_n(K))'$ may be described as those elements of $\mathrm{Cusp}\,(GL_n(K))$ which become unramified after some series of cyclic base changes. Appealing to Henniart's numerical local Langlands theorem ([He4]) we can conclude that $\mathrm{Cusp}\,(GL_n(K)) = \mathrm{Cusp}\,(GL_n(K))'$ and so deduce theorem A.

One of us (M.H. see [Har2]) had previously given a different construction of a map

$$\mathrm{rec}'_K : \mathrm{Cusp}\,(GL_n(K)) \longrightarrow \mathrm{Irr}_n(W_K).$$

In some cases he was able to show its compatibility with the association of l-adic representations to certain classes of automorphic forms on unitary groups. As a result he could deduce the local Langlands conjecture only for $p > n$ (see [BHK] and [Har3]). A posteriori we can show that $\mathrm{rec}'_K = \mathrm{rec}_K$. Since the distribution of a preliminary version of this work, but before the distribution of the final version, Henniart [He6] has given a much simpler proof of theorem A by making much cleverer use of the non-Galois automorphic induction of [Har3] and of his own numerical local Langlands theorem [He4]. He does not need the a priori construction of a map rec_K compatible with some instances of the global correspondence, and thus he is able to by-pass all the main results in this book. For the reader interested only in theorem A his is clearly the better proof. None the less we believe the results of this book are still important as they establish many instances of compatibility between the global and local correspondences.

Let us now explain our construction of maps

$$\mathrm{rec}_K : \mathrm{Cusp}\,(GL_n(K)) \longrightarrow \mathrm{Rep}_n(W_K).$$

To this end choose a prime $l \neq p$ and fix an isomorphism $\mathbb{C} \cong \overline{\mathbb{Q}}_l^{\mathrm{ac}}$. Let k denote the residue field of K. For any $g \geq 1$ there is, up to isomorphism, a unique one-dimensional formal $\mathcal{O}_K$-module $\Sigma_{K,g}/k^{\mathrm{ac}}$ of $\mathcal{O}_K$-height g. Then $\mathrm{End}_{\mathcal{O}_K}(\Sigma_{K,g}) \otimes_{\mathbb{Z}} \mathbb{Q} \cong D_{K,g}$, the division algebra with centre K and Hasse invariant $1/g$. The functor which associates to any Artinian local $\mathcal{O}_K$-algebra A with residue field k^{ac} the set of isomorphism classes of deformations of $\Sigma_{K,g}$ to A is prorepresented by a complete noetherian local $\mathcal{O}_K$ algebra $R_{K,g}$ with residue field k^{ac}. (In fact $R_{K,g}$ is a formal power series ring in $g-1$ variables over the ring of integers of the completion of the maximal unramified extension of K.) We will let $\widetilde{\Sigma}_{K,g}$ denote the universal deformation of $\Sigma_{K,g}$ over $R_{K,g}$. (In the case $g = 1$ one just obtains the base change to the ring of integers of the completion of the maximal unramified extension of K of any Lubin-Tate formal $\mathcal{O}_K$ module over K.) Drinfeld showed that for any integer $m \geq 0$ there is a finite flat $R_{K,g}$-algebra $R_{K,g,m}$ over which $\widetilde{\Sigma}_{K,g}$ has a universal Drinfeld level p^m-structure. We will consider the direct limit over m of the formal vanishing cycle sheaves of $\mathrm{Spf}\, R_{K,g,m}$ with coefficients in $\mathbb{Q}_l^{\mathrm{ac}}$. This gives a collection $\{\Psi_{K,l,g}^i\}$ of infinite-dimensional $\mathbb{Q}_l^{\mathrm{ac}}$ vector spaces with natural admissible actions of the subgroup of $GL_g(K) \times D_{K,g}^{\times} \times W_K$ consisting of elements (γ, δ, σ) such that

$$|\det \delta||\det \gamma|^{-1}|\mathrm{Art}_K^{-1}\sigma| = 1.$$

For any irreducible representation ρ of $D_{K,g}^{\times}$ we set

$$\Psi_{K,l,g}^i(\rho) = \mathrm{Hom}_{\mathcal{O}_{D_{K,g}}^{\times}}(\rho, \Psi_{K,l,g}^i).$$

This becomes an admissible $GL_g(K) \times W_K$-module. In the case $g = 1$ we have $\Psi_{K,l,1}^i = (0)$ for $i > 0$, while it follows from the theory of Lubin-Tate formal groups (see [LT]) that $\Psi_{K,l,1}^0(\rho) = \mathbb{Q}_l^{\mathrm{ac}}$ with an action of $K^{\times} \times W_K$ via $\rho^{-1} \times (\rho \circ \mathrm{Art}_K^{-1})$ (see section 3.4 of [Car3]).

To describe $\Psi_{K,l,g}^i(\rho)$ in greater generality we must recall that Deligne, Kazhdan and Vigneras (see [DKV]) and Rogawski ([Rog2]) have given a bijection between irreducible representations of $D_{K,g}^{\times}$ and (quasi-)square integrable irreducible admissible representations of $GL_g(K)$ characterised by a natural character identity (see section I.3). This generalises work of Jacquet and Langlands in the case $g = 2$ so we will denote the correspondence $\rho \mapsto \mathrm{JL}(\rho)$. Carayol essentially conjectured ([Car3]) that if $\mathrm{JL}(\rho)$ is supercuspidal then

$$\Psi_{K,l,g}^{g-1}(\rho) \cong \mathrm{JL}(\rho)^{\vee} \times \mathrm{rec}_K(\mathrm{JL}(\rho) \otimes |\det|^{(1-g)/2}).$$

We do not quite prove this (though it may be possible by our methods to do so). However, motivated by Carayol's conjecture, our first main theorem is the following. To state it let $[\Psi_{K,g}(\rho)]$ denote the virtual representation $(-1)^{g-1} \sum_{i=0}^{g-1} (-1)^i [\Psi_{K,l,g}^i(\rho)]$.

Theorem B *If π is an irreducible supercuspidal representation of $GL_g(K)$ then there is a (true) representation*

$$r_l(\pi) : W_K \to GL_g(\mathbb{Q}_l^{ac}),$$

such that in the Grothendieck group

$$[\Psi_{K,l,g}(\mathrm{JL}\,(\pi)^\vee)] = [\pi \otimes r_l(\pi)].$$

In the case $n = 1$ Lubin-Tate theory allows one to identify

$$r_l(\pi) = \pi^{-1} \circ \mathrm{Art}\,_K^{-1}.$$

We use this theorem to define

$$\mathrm{rec}_K : \mathrm{Cusp}\,(GL_g(K)) \longrightarrow \mathrm{Rep}_g(W_K)$$

by the formula

$$\mathrm{rec}_K(\pi) = r_l(\pi^\vee \otimes (|\ |\circ \det)^{(1-g)/2}).$$

It will also be convenient for us to extend r_l to all irreducible admissible representations of $GL_g(K)$ as follows. If π is an irreducible admissible representation of $GL_g(K)$, then we can find positive integers $g_1, \ldots, g_t$ which sum to g and irreducible supercuspidal representations π_i of $GL_{g_i}(K)$ such that π is a subquotient of n-Ind$\,(\pi_1 \times \cdots \times \pi_t)$, where we are using the usual normalised induction (see section I.2). Then we set

$$r_l(\pi) = \bigoplus_{i=1}^t r_l(\pi_i) \otimes |\mathrm{Art}\,_K^{-1}|^{(g_i-g)/2}.$$

This is well defined and

$$\mathrm{rec}_K(\pi) = (r_l(\pi^\vee \otimes (|\ |\circ \det)^{(1-g)/2}), N),$$

for some N.

Our second key result is that r_l is compatible with many instances of the global Langlands correspondence. The following theorem strengthens a theorem of Clozel [Cl1] (in which he only identifies $[R(\Pi)|_{W_{F_y}}]$ for all but finitely many places y, and specifically for none of the bad places).

Theorem C *Suppose that L is a CM field and that Π is a cuspidal automorphic representation of $GL_g(\mathbb{A}_L)$ satisfying the following conditions:*

- $\Pi^\vee \cong \Pi^c$,

- Π_∞ *has the same infinitesimal character as some algebraic representation over $\mathbb{C}$ of the restriction of scalars from L to $\mathbb{Q}$ of GL_g,*

- *and for some finite place x of L the representation Π_x is square integrable.*

Then there is a continuous representation

$$R_l(\Pi) : \text{Gal}\,(L^{\text{ac}}/L) \longrightarrow GL_g(\mathbb{Q}_l^{\text{ac}})$$

such that for any finite place y of L not dividing l we have

$$[R(\Pi)|_{W_{L_y}}] = [r_l(\Pi_y)].$$

Moreover for each finite place y of F the representation Π_y is tempered.

In the case $n = 2$ and $K = \mathbb{Q}_p$ and $F^+ = \mathbb{Q}$ both theorems B and C were essentially proved by Deligne in his beautiful letter [De2]. (The argument was completed by Brylinski [Bry].) Carayol [Car2] generalised Deligne's method to essentially prove both theorems B and C in the general $n = 2$ case. We will simply generalise Deligne's approach to $n > 2$. The combination of theorems B and C, Henniart's numerical local Langlands theorem [He4], and the non-Galois automorphic induction of [Har3] suffice to prove theorem A.

Both theorems B and C follow without great difficulty from an analysis of the bad reduction of certain Shimura varieties. We will next explain this analysis. Unfortunately we must first establish some notation. Let E denote an imaginary quadratic field in which p splits: $p = uu^c$. Let F^+ denote a totally real field of degree d and set $F = EF^+$. Fix a place w of F above u. Let B be a division algebra with centre F such that

- the opposite algebra B^{op} is isomorphic to $B \otimes_{E,c} E$;

- B is split at w;

- at any place x of F which is not split over F^+, B_x is split;

- at any place x of F which is split over F^+ either B_x is split or B_x is a division algebra,

- if n is even then $1 + dn/2$ is congruent modulo 2 to the number of places of F^+ above which B is ramified.

Let n denote $[B : F]^{1/2}$. We can pick a positive involution of the second kind $*$ on B (i.e. $*|_F = c$ and $\text{tr}_{B/\mathbb{Q}}(xx^*) > 0$ for all nonzero $x \in B$). If $\beta \in B^{*=-1}$ then we will let

- G denote the algebraic group with $G(\mathbb{Q})$ the subgroup of elements $x \in (B^{\text{op}})^\times$ so that $x^*\beta x = \nu(x)\beta$ for some $\nu(x) \in \mathbb{Q}^\times$,

- $\nu : G \to \mathbb{G}_m$ the corresponding character,

- G_1 the kernel of ν,

- and (,) the pairing on B defined by

$$(x, y) = (\mathrm{tr}\,_{F/\mathbb{Q}} \circ \mathrm{tr}\,_{B/F})(x\beta y^*).$$

We can and will choose β such that

- G is quasi-split at all rational primes x which do not split in E

- and $G_1(\mathbb{R}) \cong U(1, n-1) \times U(0, n)^{[F^+ : \mathbb{Q}]-1}$.

If $U \subset G(\mathbb{A}^\infty)$ is an open compact subgroup we will consider the following moduli problem. If S is a connected F-scheme and s is a closed geometric point of S then we consider equivalence classes of quadruples $(A, \lambda, i, \overline{\eta})$ where

- A is an abelian scheme of dimension $[F^+ : \mathbb{Q}]n^2$;

- $\lambda : A \to A^\vee$ is a polarisation;

- $i : B \hookrightarrow \mathrm{End}\,(A) \otimes_{\mathbb{Z}} \mathbb{Q}$ such that $\lambda \circ i(b) = i(b^*)^\vee \circ \lambda$ for all $b \in B$ and such that

$$\mathrm{tr}\,(b|_{Lie(A)}) = (c \circ \mathrm{tr}\,_{F/E} \circ \mathrm{tr}\,_{B/F})(nb) + \mathrm{tr}\,_{B/F}(b) - (c \circ \mathrm{tr}\,_{B/F})(b)$$

  for all $b \in B$;

- $\overline{\eta}$ is a $\pi_1(S, s)$-invariant U-orbit of isomorphisms of $B \otimes_{\mathbb{Q}} \mathbb{A}^\infty$-modules $\eta : V \otimes_{\mathbb{Q}} \mathbb{A}^\infty \to VA_s$ which take the standard pairing (,) on V to a scalar multiple of the λ-Weil pairing on VA_s.

We consider two such quadruples equivalent if the abelian varieties are isogenous in a way that preserves the rest of the structure (but only need preserve the polarisation up to $\mathbb{Q}^\times$ multiples). The set of equivalence classes is canonically independent of the choice of s. If U is sufficiently small this moduli problem is represented by a smooth proper scheme of finite type X_U/F.

If ξ is a representation of the algebraic group G over $\mathbb{Q}_l^{ac}$ then we can define a lisse $\mathbb{Q}_l^{ac}$ sheaf $\mathcal{L}_\xi$ on X_U. Then we will consider the $\mathbb{Q}_l^{ac}$-vector spaces

$$H^i(X, \mathcal{L}_\xi) = \lim_{\substack{\longrightarrow \\ U}} H^i_{et}(X_U \times F^{ac}, \mathcal{L}_\xi).$$

This is naturally a $G(\mathbb{A}^\infty) \times \mathrm{Gal}\,(F^{ac}/F)$-module, admissible as a $G(\mathbb{A}^\infty)$-module. In fact we can write

$$H^i(X, \mathcal{L}_\xi) = \bigoplus_\pi \pi \otimes R^i_\xi(\pi),$$

where π runs over irreducible admissible representations of $G(\mathbb{A}^\infty)$ and $R^i_\xi(\pi)$ is a finite dimensional continuous representation of $\mathrm{Gal}\,(F^{\mathrm{ac}}/F)$. We will focus on the virtual representation

$$[R_\xi(\pi)] = (-1)^{n-1} \sum_i (-1)^i [R^i_\xi(\pi)].$$

Kottwitz (see [Ko4]) determined $\mathrm{tr}\,[R_\xi(\pi)](\mathrm{Frob}_x)$ in terms of π for all but finitely many places x of F. He thus completely determined the virtual representation $[R_\xi(\pi)]$. We will extend Kottwitz's description to all places $x \nmid l$ of F (see corollary VII.1.10).

We have an isomorphism

$$G(\mathbb{Q}_p) \cong E^\times_{u^c} \times \prod_{x|u}(B^{\mathrm{op}}_x)^\times$$

and hence a decomposition

$$G(\mathbb{A}^\infty) \cong G(\mathbb{A}^{\infty,p}) \times E^\times_{u^c} \times \prod_{x|u}(B^{\mathrm{op}}_x)^\times.$$

Thus we may decompose an irreducible admissible representation π of $G(\mathbb{A}^\infty)$ as

$$\pi \cong \pi^p \otimes \pi_{p,0} \otimes \bigotimes_{x|u} \pi_x.$$

For $h = 1,\ldots,n$ we will let P^{op}_h denote the parabolic subgroup of $GL_n(F_w)$ consisting of block lower triangular matrices with an $(n-h) \times (n-h)$-block in the top left and an $h \times h$ block in the bottom right. It has Levi component $GL_{n-h}(F_w) \times GL_h(F_w)$. Let N^{op}_h denote its unipotent radical. Suppose that ρ is an irreducible representation of $D^\times_{F_w,n-h}$. We will let $\varphi_{\mathrm{JL}\,(\rho)^\vee} \in C^\infty(GL_{n-h}(F_w))$ denote a pseudo-coefficient for $\mathrm{JL}\,(\rho)^\vee$ (so that $\varphi_{\mathrm{JL}\,(\rho)^\vee}$ is compactly supported mod centre, and for any irreducible tempered admissible representation α of $GL_{n-h}(F_w)$ with the same central character as $\mathrm{JL}\,(\rho)^\vee$ we have $\mathrm{tr}\,\alpha(\varphi_{\mathrm{JL}\,(\rho)^\vee}) = \mathrm{vol}\,(D^\times_{F_w,n-h}/F^\times_w)$ if $\alpha \cong \mathrm{JL}\,(\rho)^\vee$ and $= 0$ otherwise). Then we define a homomorphism

$$\text{n-red}^{(h)}_\rho : \mathrm{Groth}\,(GL_n(F_w)) \longrightarrow \mathrm{Groth}\,(GL_h(F_w))$$

as a composite

$$\mathrm{Groth}\,(GL_n(F_w)) \to \mathrm{Groth}\,(GL_{n-h}(F_w) \times GL_h(F_w))$$
$$\to \mathrm{Groth}\,(GL_h(F_w)),$$

where

- the first map takes the class, $[\pi]$, of an irreducible admissible representation, π, to the class, $[J_{N_h^{op}}(\pi)]$, of its normalised Jacquet module (see section I.2),

- and the second map takes the class, $[\alpha \otimes \beta]$, to

  - vol$(D_{F_w,n-h}^{\times}/F_w^{\times})^{-1}tr\,\alpha(\varphi_{\mathrm{JL}\,(\rho)^{\vee}})$ times $[\beta]$ if the central characters of α and JL$\,(\rho)^{\vee}$ are equal,

  - and to 0 otherwise.

Our key technical result is the following theorem relating $[R_\xi(\pi)]$ and $[\Psi_{F_w,l,g}]$ for $1 \leq g \leq n$. From it both theorem B and theorem C follow without undue difficulty. (As does corollary VII.1.10.)

Theorem D *Suppose that π is an irreducible admissible representation of $G(\mathbb{A}^\infty)$ such that $\pi_{p,0}|_{\mathbb{Z}_p^\times} = 1$. Then*

$$n[\pi_w][R_\xi(\pi)|_{W_{F_w}}] = (\dim[R_\xi(\pi)]) \sum_{h=0}^{n-1} \sum_\rho \text{n-Ind}_{P_h^{op}(F_w)}^{GL_n(F_w)}$$

$$((\text{n-red}_\rho^{(h)}[\pi_w])[\Psi_{F_w,l,n-h}(\rho) \otimes ((\pi_{p,0}^{-1} \otimes |\ |_p^{-h/2}) \circ \text{Art}_{\mathbb{Q}_p}^{-1})|_{W_{F_w}}])$$

where ρ runs over irreducible admissible representations of $D_{F_w,n-h}^{\times}$.

Almost all of this book is devoted to proving this theorem. In the rest of this introduction we will give a very brief sketch of the strategy. We caution the reader that in the rest of this introduction we will not make precise mathematical statements, but rather comments that we hope will convey an idea of our methods. We refer the reader to the body of our book for the accurate formulation of these ideas.

We compute the cohomology groups $H_{et}^i(X_U \times F_w^{ac}, \mathcal{L}_\xi)$ via vanishing cycle sheaves on the special fibre $\overline{X}_U$ of X_U. Thus we are led to try to compute the cohomology groups

$$H_{et}^i(\overline{X}_U \times k(w)^{ac}, R^j\Psi_\eta(\mathbb{Q}_l^{ac}) \otimes \mathcal{L}_\xi),$$

where $k(w)$ denotes the residue field of w. The first key idea is to introduce a certain stratification on $\overline{X}_U$ and compute stratum by stratum. Consider the w^∞ torsion points on the universal abelian variety over X_U. It has an action of $B_w \cong M_n(F_w)$. Applying the idempotent $(e_{ij}) \in M_n(F_w)$ with $e_{11} = 1$ and $e_{ij} = 0$ otherwise, we obtain a divisible $\mathcal{O}_{F_w}$-module $\mathcal{G}/X_U$ of $\mathcal{O}_{F_w}$-height n and of dimension 1. For $h = 0, \ldots, n-1$ we will let $\overline{X}_U^{(h)}$ denote the (h-dimensional) locally closed reduced subscheme of $\overline{X}_U$ where the maximal etale quotient of $\mathcal{G}$ has $\mathcal{O}_{F_w}$-height h. (Then the closure of $\overline{X}_U^{(h)}$ is the union of the $\overline{X}_U^{(h')}$ for $h' \leq h$.) It will suffice to compute

$$H_c^i(\overline{X}_U^{(h)} \times k(w)^{ac}, R^j\Psi_\eta(\mathbb{Q}_l^{ac}) \otimes \mathcal{L}_\xi).$$

Next we restrict to U of the form $U^w \times U_w$ where $\mathcal{O}_{E,u^c}^\times \subset U^w \subset$ $G(\mathbb{A}^{\infty,p}) \times E_{u^c}^\times \times \prod_{x|u, x \neq w}(B_x^{\mathrm{op}})^\times$ and $U_w \subset GL_n(F_w)$. If U_w is the group of matrices in $GL_n(\mathcal{O}_{F,w})$ congruent to 1 modulo w^m we will write $U = U^w(m)$. To analyse $\overline{X}_{U^w(m)}^{(h)}$ we introduce an analogue of Igusa curves in this setting: we call them *Igusa varieties of the first kind*. More precisely $I_{U^w,m}^{(h)}$ will denote the etale cover of $\overline{X}_{U^w(0)}^{(h)}$ which parametrises isomorphisms

$$(\mathcal{O}_{F_w}/w^m)^h \overset{\sim}{\to} \mathcal{G}^{\mathrm{et}}[w^m].$$

One can define this Igusa variety of the first kind not only as a variety in characteristic p but as a formal scheme. Thus we obtain formal schemes $(I_{U^w,m}^{(h)})^\wedge(t)$ with special fibre $I_{U^w,m}^{(h)}$ over which there is a universal deformation of the formal $\mathcal{O}_{F_w}$-module $\mathcal{G}^0$ together with its Drinfeld level w^t-structure.

One can show that $\overline{X}_{U(m)}^{(h)}$ is a disjoint union of copies of $I_{U^w,m}^{(h)}$ except that the structure map down to $\overline{X}_{U(0)}^{(h)}$ is twisted by a power of Frobenius. If P_h denotes the opposite parabolic to P_h^{op} then one can obtain an isomorphism

$$\lim_{\to m} H_c^i(\overline{X}_{U(m)}^{(h)} \times k(w)^{\mathrm{ac}}, R^j \Psi_\eta(\mathbb{Q}_l^{\mathrm{ac}}) \otimes \mathcal{L}_\xi) \cong$$
$$\mathrm{Ind}_{P_h}^{GL_n(F_w)} \lim_{\to m,t} H_c^i(I_{U^w,m}^{(h)} \times k(w)^{\mathrm{ac}}, R^j \Psi_\eta(\mathbb{Q}_l^{\mathrm{ac}})_{(I_{U^w,m}^{(h)})^\wedge(t)_\eta} \otimes \mathcal{L}_\xi).$$

Thus it will suffice to compute

$$H_c^i(I_{U^w,m}^{(h)} \times k(w)^{\mathrm{ac}}, R^j \Psi_\eta(\mathbb{Q}_l^{\mathrm{ac}})_{(I_{U^w,m}^{(h)})^\wedge(t)_\eta} \otimes \mathcal{L}_\xi).$$

The next step is to understand the vanishing cycle sheaves

$$R^j \Psi_\eta(\mathbb{Q}_l^{\mathrm{ac}})_{(I_{U^w,m}^{(h)})^\wedge(t)_\eta}.$$

To do so we introduce a second generalisation of Igusa varieties, which we will call *Igusa varieties of the second kind*. More specifically we let $J_{U^w,m,s}^{(h)}/I_{U^w,m}^{(h)} \times k(w)^{\mathrm{ac}}$ denote the moduli space for isomorphisms

$$\alpha : \Sigma_{F_w,n-h}[w^s] \overset{\sim}{\to} \mathcal{G}^0[w^s],$$

which for every $s' > s$ lift etale locally to isomorphisms of the $w^{s'}$-division schemes. As s varies we get a system of finite etale Galois covers and the system has Galois group $\mathcal{O}_{D_{F_w,n-h}}^\times$. It is perhaps worth noting three things. One is that we need now the technical condition that the isomorphism must lift locally - this is because $\mathrm{Aut}(\Sigma_{F_w,n-h}) \to \mathrm{Aut}(\Sigma_{F_w,n-h}[w^s])$ is not usually surjective. Secondly we remark that if one looked at a similar construction in the case of the ordinary locus of a modular curve one would just obtain the familiar Igusa curves. This is because in that case there is a duality between the connected and etale part of the analogue of $\mathcal{G}$ (i.e. the p-divisible group of the universal elliptic curve). To the best of our knowledge, for $n - h > 1$ the varieties $J_{U^w,m,s}^{(h)}$ do not occur in the reduction of any Shimura variety. They seem to naturally exist only in characteristic p.

The idea is now that over the "pro-object" $\varprojlim_s J_{U^w,m,s}^{(h)}$ we have an isomorphism $\mathcal{G}^0 \cong \Sigma_{F_w,n-h}$ and $R^j \Psi_\eta(\mathbb{Q}_l^{ac})_{(I_{U^w,m}^{(h)})^\wedge(e_{F_w}/\mathbb{Q}_p t)_\eta}$ becomes the constant sheaf $R^j \Psi_\eta(\mathbb{Q}_l^{ac})_{(\mathrm{Spf}\, R_{F_w,n-h,t})}$. If one descends this isomorphism back down to $I_{U^w,s}^{(h)}$ one obtains an isomorphism

$$\varinjlim_t R^j \Psi_\eta(\mathbb{Q}_l^{ac})_{(I_{U^w,m}^{(h)})^\wedge(t)_\eta} \cong \bigoplus_\rho \mathcal{F}(\Psi_{F_w,l,n-h}^j[\rho]),$$

where ρ runs over irreducible representations of $D_{F_w,n-h}^\times$ (up to unramified twist). If $\mathcal{F}_\rho$ is the lisse $\mathbb{Q}_l^{ac}$-sheaf on $I_{U^w,m}^{(h)}$ associated to the representation ρ of $\mathrm{Gal}\,(J_{U^w,m,\infty}^{(h)}/I_{U^w,m}^{(h)})$ then the sheaf $\mathcal{F}(\Psi_{F_w,l,n-h}^j[\rho])$ is closely related to

$$\mathcal{F}_\rho \otimes \Psi_{F_w,l,n-h}^j(\rho),$$

where now the rather mysterious action of

$$GL_{n-h}(\mathcal{O}_{F,w}) \times I_{F_w}$$

is concentrated on the constant sheaf $\Psi_{F_w,l,n-h}^j(\rho)$. At least in our unskilled hands it took some effort to make sense of the non-mathematical ideas of this paragraph. We are very grateful to Berkovich for providing a key step in the argument.

In this way we obtain an isomorphism

$$\varinjlim_m H_c^i(\overline{X}_{U(m)}^{(h)} \times k(w)^{ac}, R^j \Psi(\mathbb{Q}_l^{ac}) \otimes \mathcal{L}_\xi)^{\oplus(n-h)} \cong$$
$$\bigoplus_\rho \mathrm{Ind}_{P_h}^{GL_n(F_w)}(\varinjlim_m H_c^i(I_{U^w,m}^{(h)} \times k(w)^{ac}, \mathcal{L}_\xi \otimes \mathcal{F}_\rho))$$
$$\otimes \Psi_{F_w,l,n-h}^j(\rho)^{\oplus(n-h)/e[\rho]}$$

for some explicit integers $e[\rho]|(n - h)$ (see section II.2). (Here we are using unnormalised induction.) To complete the proof of theorem D it remains to compute

$$\varinjlim H_c^i(I_{U^w,m}^{(h)} \times k(w)^{ac}, \mathcal{L}_\xi \otimes \mathcal{F}_\rho)$$

as a $G(\mathbb{A}^{\infty,p}) \times E_{u^c}^\times \times \mathbb{Z} \times GL_h(F_w) \times \prod_{x|u,x\neq w}(B_x^{\mathrm{op}})^\times$-module.

At this point we return to Langlands' idea of using the Lefschetz trace formula to calculate the trace of the action of correspondences on Shimura varieties in characteristic p. In our case we use Fujiwara's "Deligne conjecture" to compute the trace of a Hecke operator acting on

$$\varinjlim H_c^i(I_{U^w,m}^{(h)} \times k(w)^{\mathrm{ac}}, \mathcal{L}_\xi \otimes \mathcal{F}_\rho)$$

in terms of data at fixed points. For this to be applicable there is a condition on the Hecke operator which corresponds to it being sufficiently twisted by Frobenius. Following Kottwitz we combine the results of Honda and Tate with some group theory (which we need in order to understand polarisations) to describe the points of $I_{U^w,m}^{(h)} \times k(w)^{\mathrm{ac}}$. We find an expression for the sum of the terms at fixed points in terms of orbital integrals in the group

$$G(\mathbb{A}^{\infty,p}) \times E_{u^c}^\times \times D_{F_w,n-h}^\times \times GL_h(F_w) \times \prod_{x|u,x\neq w}(B_x^{\mathrm{op}})^\times.$$

Unlike Kottwitz's work there is no distinguished Frobenius element, we use a classification of points over $k(w)^{\mathrm{ac}}$ rather than over finite extensions of $k(w)$ and we get only orbital integrals not twisted orbital integrals. We are able to manipulate this expression so that it becomes a sum of orbital integrals in $G(\mathbb{A})$. In doing so we use again the condition that the Hecke operator "was sufficiently twisted by Frobenius" (cf [Cas]). Next we apply the trace formula on $X_U \times F_u^{\mathrm{ac}}$ to relate traces of Hecke operators on

$$\sum_i (-1)^i [\varinjlim H_c^i(I_{U^w,m}^{(h)} \times k(w)^{\mathrm{ac}}, \mathcal{L}_\xi \otimes \mathcal{F}_\rho)]$$

with traces of related Hecke operators on

$$\sum_i (-1)^i [\varinjlim H_{\mathrm{et}}^i(X_U \times F_w^{\mathrm{ac}}, \mathcal{L}_\xi)].$$

From this comparison it is not hard to deduce theorem D.

We remark that we recover in this way some of Kottwitz's results from [Ko4]. Although we have borrowed many of Kottwitz's ideas our argument in the case of overlap does seem to be somewhat different. For instance we make no appeal to the fundamental lemma for stable base change.

The knowledgeable reader may prefer to skip some of the early sections of this book (in particular chapter I), which review more or less well known material. However we strongly recommend that the reader consult section I.7 where we establish certain important running assumptions that are specific to this book.

Acknowledgements

R.T. is very grateful to the Miller Institute at the University of California at Berkeley for its hospitality and support during the very lengthy revision of this book. He was also partially supported by NSF Grant DMS-9702885. M.H. is a member of Institut de Mathématiques de Jussieu, U.M.R. 7586 du CNRS. We are both very grateful to Matthew Leingang for resolving a number of tricky Latex problems.

We are also very grateful for the help (conscious and unconscious) of many mathematicians. We would like to thank Bultel, Clozel, Coleman, Brian Conrad, Fargues, Gabber, Gaitsgory, Gross, Henniart, R.Huber, de Jong, Katz, Kazhdan, Kottwitz, Labesse, Laumon, Mantovan, Mazur, Rapoport and Strauch for helpful conversations. We came at this work from different starting points. One of us (M.H.) was inspired by the book of Rapoport and Zink [RZ2] and by Boyer's Paris XI thesis [Bo]. He would like to thank Laumon for explaining Boyer's work. The other (R.T.) was inspired by teaching a class on Deligne's letter [De2] at Harvard; he would like to thank the students of math 253x for their help with understanding this letter.

Our debt to the work of Kottwitz (see eg [Ko3] and [Ko4]) on the l-adic cohomology of these same Shimura varieties will be clear to the reader. We have borrowed many of his ideas.

Our work would have been impossible without Berkovich's vanishing cycles for formal schemes. We are extremely grateful to him for a number of things. Firstly for the help he has given us in understanding his results and in writing section I.5. Secondly because he found a proof of a key result we needed, which he kindly wrote it up in an appendix to this book.

Finally we would like to acknowledge our debt to the work of Carayol, Deligne and Drinfeld; on whose ideas the present work is based. In particular we owe a very great debt to Deligne: our work is simply the natural generalisation of the arguments of his beautiful letter [De2] from modular curves to the unitary group Shimura varieties we consider.

15

Chapter I

Preliminaries

In this chapter we will establish some notation, recall some standard facts and prove a couple of technical lemmas. The reader may like to read only section I.7 (which establishes some very important assumptions which are particular to this book) and simply refer back to the rest of this chapter as the need arises.

I.1   General notation

In this section we will introduce some notation which we will use throughout this book.

We will let p and l denote rational primes. *We will often assume that $l \neq p$. In particular from the middle of section III.4 onwards this will always be the case.* Let $\mathbb{Z}_{(p)}$ denote the ring of elements of $\mathbb{Q}$ with denominator coprime to p, val_p the p-adic valuation (so that $\mathrm{val}_p(p) = 1$) and $| \ |_p$ the p-adic absolute value (so that $|p|_p = 1/p$).

If X is a scheme and x is a point of X we will let $k(x)$ denote the residue field at x. We will let $\mathcal{O}_{X,x}$ denote the local ring of X at x and we will let $\mathcal{O}_{X,x}^{\wedge}$ denote its completion at its maximal ideal. If $Y \subset X$ is a locally closed subscheme we will let $X_Y^{\wedge}$ denote the completion of X along Y. For instance $X_x^{\wedge} = \mathrm{Spf}\, \mathcal{O}_{X,x}^{\wedge}$. If $\mathcal{X}$ is a locally noetherian formal scheme then $\mathcal{X}$ has a unique largest ideal of definition $\mathcal{I}$. The formal scheme with the same underlying topological space as $\mathcal{X}$ and with structure sheaf $\mathcal{O}_{\mathcal{X}}/\mathcal{I}$ is in fact a scheme which we will refer to as the *reduced subscheme of $\mathcal{X}$*, and will denote $\mathcal{X}^{\mathrm{red}}$. (See section 10.5 of [EGAI].)

If L is a field we will let L^{ac} denote its algebraic closure. If L'/L is a finite field extension we will let $N_{L'/L}$ denote the norm from L' to L. If X/L' is a scheme we will write $\mathrm{RS}_L^{L'} X$ for the restriction of scalars of X to

17

L, i.e. $(\mathrm{RS}_L^{L'} X)(S) = X(S \times_{\mathrm{Spec}\,L} \mathrm{Spec}\,L')$ for any L-scheme S.

If k is a field and A/k is an abelian variety we will let TA denote the Tate module of A, i.e.

$$TA = \varprojlim_N A[N](k^{\mathrm{ac}}),$$

where the limit is over all positive integers N. We will also introduce the characteristic zero version of the Tate module

$$VA = TA \otimes_{\mathbb{Z}} \mathbb{Q}.$$

If S is a finite set of rational primes we will let $T^S A$ and $V^S A$ denote the "away from S" Tate modules, i.e.

$$T^S A = \varprojlim_N A[N](k^{\mathrm{ac}})$$

where the limit is over all positive integers coprime to S, and $V^S A = T^S A \otimes_{\mathbb{Z}} \mathbb{Q}$. We similarly define the "at S" Tate modules $T_S A$ and $V_S A$.

If A and A'/S are abelian schemes then by an isogeny $\alpha : A \to A'$ we shall mean an invertible element of $\mathrm{Hom}\,(A, A') \otimes_{\mathbb{Z}} \mathbb{Q}$. We will denote $\mathrm{Hom}\,(A, A) \otimes_{\mathbb{Z}} \mathbb{Q}$ by $\mathrm{End}^0(A)$. By a polarisation λ of A we shall mean a homomorphism $\lambda : A \to A^{\vee}$ such that for each geometric point s of S the homomorphism λ_s is a polarisation in the usual sense. If p is a rational prime then by a prime-to-p-isogeny we shall mean an invertible element of $\mathrm{Hom}\,(A, A') \otimes_{\mathbb{Z}} \mathbb{Z}_{(p)}$. By a prime-to-$p$-polarisation of A we shall mean a polarisation $\lambda : A \to A^{\vee}$ which is also a prime-to-p-isogeny.

If R is an $\mathbb{F}_p$-algebra we will let $\mathrm{Fr} : R \to R$ denote the Frobenius morphism which takes $x \in R$ to $x^p \in R$. If $X/\mathbb{F}_p$ is a scheme we will let $\mathrm{Fr}^* : X \to X$ denote the Frobenius morphism induced by Fr on structure sheaves. If $Y \to X$ is a morphism of schemes over $\mathbb{F}_p$ then we will let $Y_{/X}^{(p)}$ (or simply $Y^{(p)}$, if no confusion seems likely to arise) denote the pullback of Y by $\mathrm{Fr}^* : X \to X$. We will also let $F_{Y/X} : Y \to Y_{/X}^{(p)}$ (or simply $F : Y \to Y^{(p)}$ when no confusion seems likely to arise) denote the relative Frobenius, i.e. the morphism that arises from $\mathrm{Fr} : Y \to Y$ and the universal property of the pull back $Y_{/X}^{(p)}$. If Y/X is a finite flat group scheme then we will let $V : Y^{(p)} \to Y$ denote the dual of $F : Y^{\vee} \to Y^{\vee,(p)} = Y^{(p),\vee}$, where $Y^{\vee}$ is the Cartier dual of Y. This definition then extends to p-divisible groups Y/X. The morphism $V : Y^{(p)} \to Y$ induces a morphism of quasi-coherent sheaves of $\mathcal{O}_X$-modules

$$V_* : (\mathrm{Fr}^*)^* \mathrm{Lie}\,Y \cong \mathrm{Lie}\,Y^{(p)} \to \mathrm{Lie}\,Y.$$

Composing this with the natural map $\mathrm{Fr} : \mathrm{Lie}\,Y \to (\mathrm{Fr}^*)^* \mathrm{Lie}\,Y$ we get a map, which we will also denote V_*, from $\mathrm{Lie}\,Y$ to itself over X, which satisfies

$$V_*(xy) = x^p V_*(y)$$

for x a section of $\mathcal{O}_X$ and for y a section of $\mathrm{Lie}\, Y$.

If $k/\mathbb{F}_p$ is a finite extension we will let $\mathrm{Frob}_k \in \mathrm{Gal}\,(k^{\mathrm{ac}}/k)$ denote $\mathrm{Fr}^{-[k:\mathbb{F}_p]}$, i.e. it will denote a geometric Frobenius element.

Throughout this book K will denote a p-adic field, i.e. a finite extension of $\mathbb{Q}_p$. We will let $v_K : K^\times \twoheadrightarrow \mathbb{Z}$ denote its unique valuation which is normalised to send uniformisers to 1. We will let $\mathcal{O}_K$ denote its ring of integers, $\wp_K$ the unique maximal ideal of $\mathcal{O}_K$ and $k(v_K) = k(\wp_K) = \mathcal{O}_K/\wp_K$ its residue field. We will often use ϖ_K to denote a uniformiser in $\mathcal{O}_K$. We will define an absolute value $|\ |_K = |\ |_{v_K}$ on K by

$$|x|_K = (\#k(v_K))^{-v_K(x)},$$

for $x \in K^\times$. We will let K^{nr} denote the maximal unramified extension of K. We will let $\widehat{K}^{\mathrm{ac}}$ denote the completion of the algebraic closure of K and $\widehat{K}^{\mathrm{nr}}$ denote the completion of K^{nr}.

We will let $I_K \subset \mathrm{Gal}\,(K^{\mathrm{ac}}/K)$ denote the inertia subgroup, so that

$$\mathrm{Gal}\,(K^{\mathrm{ac}}/K)/I_K \overset{\sim}{\to} \mathrm{Gal}\,(K^{\mathrm{nr}}/K) \overset{\sim}{\to} \mathrm{Gal}\,(k(v_K)^{\mathrm{ac}}/k(v_K)).$$

We will let $W_K \subset \mathrm{Gal}\,(K^{\mathrm{ac}}/K)$ denote the Weil group, i.e. the inverse image in $\mathrm{Gal}\,(K^{\mathrm{ac}}/K)$ of $\mathrm{Frob}_{k(v_K)}^{\mathbb{Z}} \subset \mathrm{Gal}\,(k(v_K)^{\mathrm{ac}}/k(v_K))$. We will write Frob_{v_K} for $\mathrm{Frob}_{k(v_K)}$, and will without comment think of it as an element of W_K/I_K. If $\sigma \in W_K$ then we define $v_K(\sigma)$ by $\sigma I_K = \mathrm{Frob}_{v_K}^{v_K(\sigma)}$. We will let $f_K = f_{v_K} = [k(v_K) : \mathbb{F}_p]$ and $e_K = e_{v_K} = [K : \mathbb{Q}_p]/f_K$. If g is a positive integer we will let $D_{K,g}$ denote the division algebra with centre K and Hasse invariant $1/g$. The algebra $D_{K,g}$ has a unique maximal order, which we will denote $\mathcal{O}_{D_{K,g}}$. We will let $\Pi_{K,g}$ denote a uniformiser in $\mathcal{O}_{D_{K,g}}$. *We will let $A_{K,g}$ denote the subgroup of*

$$GL_g(K) \times D_{K,g}^\times \times W_K$$

consisting of elements (γ, δ, σ) such that the valuation of the reduced norm of δ equals $v_K(\det \gamma) - v_K(\sigma)$. This group will be of particular importance for us.

Local class field theory gives us a canonical isomorphism

$$\mathrm{Art}_K : K^\times \overset{\sim}{\longrightarrow} W_K^{\mathrm{ab}}.$$

There is a choice of sign in the definition of Art_K. We will choose a normalisation which makes uniformisers and geometric Frobenius elements correspond. If $\sigma \in W_K$ we will define

$$|\sigma|_K = |\mathrm{Art}_K^{-1}\sigma|_K = p^{-f_K v_K(\sigma)}.$$

We will let c denote the non-trivial element of $\mathrm{Gal}\,(\mathbb{C}/\mathbb{R})$. We take $\mathrm{Art}_{\mathbb{C}} : \mathbb{C}^{\times} \to \mathrm{Gal}\,(\mathbb{C}/\mathbb{C})$ to be the trivial homomorphism and we take

$$\mathrm{Art}_{\mathbb{R}} : \mathbb{R}^{\times}/\mathbb{R}^{\times}_{>0} \xrightarrow{\sim} \mathrm{Gal}\,(\mathbb{C}/\mathbb{R})$$

to be the unique isomorphism. We take $|\ |_{\mathbb{R}}$ to be the usual absolute value on $\mathbb{R}$ and $|\ |_{\mathbb{C}}$ to be the square of the usual absolute value $|\ |$ on $\mathbb{C}$, i.e. $|z|_{\mathbb{C}} = |z|^2 = |zz^c|_{\mathbb{R}}$.

If L is a number field we will let $\mathbb{A}_L$ denote the adeles of L. If S is a finite set of places of L we decompose $\mathbb{A}_L = \mathbb{A}_L^S \times L_S$ where $\mathbb{A}_L^S$ denotes the adeles away from S and where $L_S = \prod_{x \in S} L_x$. Also let $\overline{\mathbb{A}}_L^S$ denote

$$\varinjlim \mathbb{A}_{L'}^{S(L')},$$

where L' runs over finite extensions of L and where $S(L')$ denotes the set of places of L' above S. The product of the normalised absolute values gives a homomorphism

$$|\ | = \prod_x |\ |_{L_x} : L^{\times} \backslash \mathbb{A}_L^{\times} \longrightarrow \mathbb{R}^{\times}_{>0}.$$

Global class field theory tells us that the product of the local Artin maps gives an isomorphism

$$\mathrm{Art}_L : \mathbb{A}_L^{\times}/\overline{L^{\times}(L_{\infty}^{\times})^0} \xrightarrow{\sim} \mathrm{Gal}\,(L^{\mathrm{ac}}/L)^{\mathrm{ab}},$$

where $(L_{\infty}^{\times})^0$ denotes the connected component of the identity in $L_{\infty}^{\times}$.

Suppose that $\imath : \mathbb{Q}_l^{\mathrm{ac}} \xrightarrow{\sim} \mathbb{C}$. If

$$\psi : \mathbb{A}_L^{\times}/L^{\times} \longrightarrow \mathbb{C}^{\times}$$

is a continuous character, we will call ψ algebraic if we can find integers n_{σ} for each embedding $\sigma : L \hookrightarrow \mathbb{C}$ such that

$$\psi|_{((L \otimes_{\mathbb{Q}} \mathbb{R})^{\times})^0} = \prod_{\sigma} (\sigma \otimes 1)^{n_{\sigma}}.$$

In this case there is a unique continuous character

$$\mathrm{rec}_{l,\imath}(\psi) = \mathrm{rec}(\psi) : \mathrm{Gal}\,(L^{\mathrm{ac}}/L) \longrightarrow (\mathbb{Q}_l^{\mathrm{ac}})^{\times}$$

such that for any finite place $x \nmid l$ of L we have

$$\imath \circ \mathrm{rec}_{l,\imath}(\psi) \circ \mathrm{Art}_{L_x} = \psi|_{L_x^{\times}}.$$

More explicitly $\mathrm{rec}_{l,\imath}(\psi) = \psi' \circ \mathrm{Art}_L^{-1}$ where

$$\psi' : \mathbb{A}_L^{\times}/\overline{L^{\times}(L_{\infty}^{\times})^0} \longrightarrow (\mathbb{Q}_l^{\mathrm{ac}})^{\times}$$

is defined by

$$\psi'(x) = \imath(\psi(x) \prod_\sigma (\sigma \otimes 1)(x_\infty)^{-n_\sigma}) \prod_\sigma ((\imath \circ \sigma) \otimes 1)(x_l)^{n_\sigma}.$$

If M is any $\mathrm{Gal}\,(L^{\mathrm{ac}}/L)$-module we will let

$$\ker^1(L, M)$$

denote the subset of $H^1(L, M)$ consisting of elements which become trivial in $H^1(L_x, M)$ for every place x of L.

If L is a CM field we will let c denote complex conjugation on L, i.e. the unique automorphism of L which coincides with complex conjugation on $\mathbb{C}$ for any embedding $L \hookrightarrow \mathbb{C}$.

Let $\varepsilon \in M_n(\mathbb{Z})$ denote the idempotent

$$\begin{pmatrix} 1 & 0 & 0 & & 0 \\ 0 & 0 & 0 & \cdots & 0 \\ 0 & 0 & 0 & & 0 \\ & \vdots & & \ddots & \vdots \\ 0 & 0 & 0 & \cdots & 0 \end{pmatrix}.$$

Thus for any ring R we have isomorphisms

- $M_n(R)\varepsilon \cong R^n$ via the map sending x to its first column;

- $\varepsilon M_n(R) \cong (R^n)^\vee$ via the map sending x to its first row;

- the map $M_n(R)\varepsilon \otimes_R \varepsilon M_n(R) \to M_n(R)$ sending $x\varepsilon \otimes \varepsilon y$ to $x\varepsilon y$ is an isomorphism.

If $x \in M_n(R)$ then we will let x^t denote its transpose.

I.2 Generalities on representations

If G is a group and $g \in G$ we will let

- $Z(G)$ denote the centre of G,

- $Z_G(g)$ denote the centraliser of g in G,

- and $[g]$ the conjugacy class of g in G.

If π is a representation of G we will let W_π denote the vector space on which the image of π acts, and we will let ψ_π denote the central character of π (if it has one).

Now suppose that G is a topological group such that every neighbourhood of the identity contains a compact open subgroup. Suppose also that Ω is an algebraically closed field of characteristic 0. Then we will let

$$C_c^\infty(G)$$

denote the space of locally constant Ω-valued functions on G with compact support. If $\psi : Z(G) \to \Omega^\times$ is a smooth character of $Z(G)$ then we will let

$$C_c^\infty(G, \psi)$$

denote the space of locally constant Ω-valued functions φ on G such that

- $\varphi(zg) = \psi(z)\varphi(g)$ for all $z \in Z(G)$ and $g \in G$

- and the image of the support of φ in $G/Z(G)$ is compact.

We may choose a (left or right) Haar measure μ on G such that every compact subgroup of G has measure in $\mathbb{Q}$. Then we may speak of an Ω-valued Haar measure meaning a non-zero element of $\Omega\mu$. If $\varphi \in C_c^\infty(G)$, if π is an admissible representation G over Ω and if we fix an (Ω-valued) Haar measure on G then we have a well defined endomorphism $\pi(\varphi)$ of W_π. The endomorphism $\pi(\varphi)$ has finite rank and so $\operatorname{tr}\pi(\varphi)$ makes sense. Similarly if ψ is an admissible character of $Z(G)$, if $\varphi \in C_c^\infty(G, \psi^{-1})$, if π is an admissible representation G with central character ψ and if we fix Haar measures on G and $Z(G)$ then again we have a well defined endomorphism $\pi(\varphi)$ of W_π. Again the endomorphism $\pi(\varphi)$ has finite rank and so $\operatorname{tr}\pi(\varphi)$ makes sense. If $\varphi \in C_c^\infty(G)$ or $C_c^\infty(G, \psi)$ and if we fix Haar measures on G and $Z_G(g)$ then we will let

$$O_g^G(\varphi)$$

denote the integral

$$\int_{G/Z_G(g)} \varphi(xgx^{-1})dx,$$

if this integral converges. All these notations depend on a choice of Haar measure(s) which we are suppressing. We will try to make clear in the accompanying text which measures we have chosen. If the choice of Haar measure on a particular group enters twice into a particular formula we will always suppose that we make the same choice both times, unless there is an explicit statement to the contrary. In such cases it will often be irrelevant which choice we make, only that we make a consistent choice.

Now suppose that H/K is a reductive algebraic group. Let $P \subset H$ be a parabolic subgroup with unipotent radical $N \subset P$. If π is an admissible representation of $(P/N)(K)$ then we define an admissible representation $\operatorname{Ind}_P^H(\pi)$ of $H(K)$ as follows. The underlying space will be the set of functions $f : H(K) \to W_\pi$ such that

- $f(hg) = \pi(h)(f(g))$ for all $g \in H(K)$ and $h \in P(K)$;

- there exists an open subgroup $U \subset H(K)$ such that $f(gu) = f(g)$ for all $u \in U$ and $g \in H(K)$.

If f is such a function and $g \in H(K)$ then we set

$$(g(f))(h) = f(hg)$$

for all $h \in H(K)$. Conversely if π is an admissible representation of $G(K)$ then the space of $N(K)$-coinvariants $W_{\pi, N(K)}$ is naturally an admissible representation π_N of $(P/N)(K)$.

Often it is convenient to use instead an alternative normalisation. To describe this, choose a square root $|\ |_K^{1/2} : K^\times \to \Omega^\times$ of $|\ |_K$. If f_K is even we will suppose that $|\ |_K^{1/2}$ takes a uniformiser to $p^{-f_K/2}$. Then we will let n-Ind$_P^H(\pi)$ denote the normalised induction as in [BZ]. Thus n-Ind$_P^H(\pi) = $ Ind$_P^H(\pi \otimes \delta_P^{1/2})$ where $\delta_P^{1/2}(h) = |\det(\text{ad}(h)|_{\text{Lie }N})|_K^{1/2}$. Similarly if π is an admissible representation of $H(K)$, we will define the Jacquet module $J_N(\pi)$ to be the admissible representation $\pi_N \otimes \delta_P^{-1/2}$ of $(P/N)(K)$.

Now return to the general topological group G such that every neighbourhood of the identity in G contains a compact open subgroup. Let Irr(G) denote the set of isomorphism classes of irreducible admissible representations of G over Ω. Let Groth (G) denote the abelian group of formal (possibly infinite) sums

$$\sum_{\Pi \in \text{Irr}(G)} n_\Pi \Pi$$

where $n_\Pi \in \mathbb{Z}$ and where for any open compact subgroup $U \subset G$ there are only finitely many $\Pi \in \text{Irr}(G)$ with both $\Pi^U \neq (0)$ and $n_\Pi \neq 0$. If (π, V) is an admissible representation of G then we will define

$$[\pi] = \sum_{\Pi \in \text{Irr}(G)} n_\Pi(\pi)\Pi \in \text{Groth}(G)$$

as follows. Given $\Pi \in \text{Irr}(G)$ choose an open compact subgroup $U \subset G$ such that $\Pi^U \neq (0)$. Then Π^U is an irreducible $\mathcal{H}(U \backslash G/U)$-module. We let $n_\Pi(\pi)$ denote the number of $\mathcal{H}(U \backslash G/U)$-Jordan-Hölder factors of π^U isomorphic to Π^U. This is independent of the choice of U. (To see this suppose $U' \subset U$. Let F^i be a Jordan-Hölder filtration on π^U. Let $(F^i)' = \mathcal{H}(U' \backslash G/U)F^i$. It suffices to show that $(F^i)'/(F^{i+1})'$ contains $\Pi^{U'}$ once or not at all depending on whether F^i/F^{i+1} is or is not congruent to Π^U. This is easy to verify.) We list some basic properties of this construction.

1. [ ] is additive on short exact sequences.

2. Let K denote a p-adic field. Suppose that $G = G_1 \times GL_n(K)$ and that $H = G_1 \times P(K)$ where $P \subset GL_n$ is a parabolic subgroup. Then there is a unique homomorphism $\operatorname{Ind}_{P(K)}^{GL_n(K)} : \operatorname{Groth}(H) \to \operatorname{Groth}(G)$ such that for any admissible representation π of H we have $\operatorname{Ind}_{P(K)}^{GL_n(K)}[\pi] = [\operatorname{Ind}_{P(K)}^{GL_n(K)} \pi]$.

3. Suppose that $G = G_1 \times G_2$ and that π_i is an admissible representation G_i for $i = 1, 2$. Then $\pi_1 \otimes \pi_2$ is an admissible representation of $G_1 \times G_2$. If π_1 and π_2 are irreducible so is $\pi_1 \otimes \pi_2$. We can define a product $\operatorname{Groth}(G_1) \otimes \operatorname{Groth}(G_2) \to \operatorname{Groth}(G_1 \times G_2)$ which sends $[\Pi_1] \otimes [\Pi_2]$ to $[\Pi_1 \otimes \Pi_2]$ for any irreducible admissibles Π_1 and Π_2. Then for any admissible representations π_i of G_i for $i = 1, 2$ we have $[\pi_1 \otimes \pi_2] = [\pi_1][\pi_2]$.

4. More generally suppose that $G = G_1 \times G_2$ and that we have a continuous homomorphism $d : G_2 \to Z(G_1)$ with discrete image. Suppose that π_1 is an admissible representation of G_1 and that π_2 is an admissible representation of G_2. Then we define the representation $\pi_1 \otimes_d \pi_2$ of $G_1 \times G_2$ by $(\pi_1 \otimes_d \pi_2)(g_1, g_2) = \pi_1(g_1 d(g_2)) \otimes \pi_2(g_2)$. Then $\pi_1 \otimes_d \pi_2$ is admissible. If Π_1 and Π_2 are irreducible so is $\Pi_1 \otimes_d \Pi_2$ and so we can define a product $\operatorname{Groth}(G_1) \otimes \operatorname{Groth}(G_2) \to \operatorname{Groth}(G_1 \times G_2)$ which sends $[\Pi_1] \otimes [\Pi_2]$ to $[\Pi_1 \otimes_d \Pi_2]$. We will denote this product $*_d$. Then for any admissible representation π_1 of G_1 and π_2 of G_2 we have $[\pi_1 \otimes_d \pi_2] = [\pi_1] *_d [\pi_2]$.

It some special cases it will be convenient to introduce a slight variant of $\operatorname{Groth}(G)$. For this suppose that l is a prime number and $\Omega = \overline{\mathbb{Q}}_l^{ac}$. Suppose that G has an open subgroup $H \times \Gamma$. By an $H \times \Gamma$-smooth/continuous representation of G we shall mean a representation π of G such that

- $\pi|_H$ is smooth (i.e. the stabiliser of any element of W_π is open)

- we can write $W_\pi = \lim_{\to} W_i$ where W_i are finite-dimensional Γ-invariant subspaces of W_π such that the representation

$$\pi : \Gamma \longrightarrow \operatorname{Aut}(W_i)$$

is continuous with respect to the l-adic topology on W_i.

By an $H \times \Gamma$-admissible/continuous representation of G we shall mean a representation π of G such that

- $\pi|_H$ is smooth (i.e. the stabiliser of any element of W_π is open)

- and for any open subgroup $U \subset H$, the vector space W_π^U is finite dimensional and the representation

$$\pi : \Gamma \longrightarrow \mathrm{Aut}\,(W_\pi^U)$$

is continuous with respect to the l-adic topology on W_π^U.

We will let $\mathrm{Irr}_{H \times \Gamma, l}(G)$ denote the set of isomorphism classes of irreducible $H \times \Gamma$-admissible/continuous representations of G. We will also let $\mathrm{Groth}\,_{H \times \Gamma, l}(G)$ denote the abelian group of formal sums

$$\sum_{\Pi \in \mathrm{Irr}_{H \times \Gamma, l}(G)} n_\Pi \Pi$$

where $n_\Pi \in \mathbb{Z}$ and where for any open compact subgroup $U \subset H$ there are only finitely many $\Pi \in \mathrm{Irr}_{H \times \Gamma, l}(G)$ with both $\Pi^U \neq (0)$ and $n_\Pi \neq 0$. If (π, V) is an $H \times \Gamma$-admissible/continuous representation of G then we can define $[\pi] \in \mathrm{Groth}\,_{H \times \Gamma, l}(G)$ as before.

Our examples will all be of one of the following forms. Here K will denote a finite field extension of $\mathbb{Q}_p$.

1. H is a topological group such that every neighbourhood of the identity in H contains a compact open subgroup, Γ is a Galois group with the Krull topology and $G = H \times \Gamma$.

2. H is a topological group such that every neighbourhood of the identity in H contains a compact open subgroup, $\Gamma = I_K$ and $G = H \times W_K$.

3. $G = A_{K,g}$ (see section II.2), $H = GL_g(\mathcal{O}_K) \times \mathcal{O}_{D_{K,g}}^\times$ and $\Gamma = I_K$.

In the latter two cases it is a theorem of Grothendieck that $\mathrm{Irr}_l(G) = \mathrm{Irr}(G)$ and hence that $\mathrm{Groth}\,_l(G) \subset \mathrm{Groth}\,(G)$. In the rest of this book we will usually suppress the choice of H and Γ, it being always assumed to be chosen in accordance with the above list.

Now consider the following special situation. Let K be a p-adic field. For $h = 1, \ldots, n-1$ we will let P_h denote the parabolic subgroup of $GL_n(K)$ consisting of matrices (g_{ij}) with $g_{ij} = 0$ if $i > n - h$ and $j \leq n - h$. We will let N_h denote the unipotent radical of P_h and L_h the Levi component consisting of matrices $(g_{ij}) \in P_h$ with $g_{ij} = 0$ for $i \leq n - h$ and $j > n - h$. Also let Z_h denote the centre of L_h. Abusing notation we will write $N_h(\wp_K^m)$ for the elements in $N_h(F) \cap \wp_K^m M_n(\mathcal{O}_K)$. This is in fact a group.

Lemma I.2.1 *1. If V is an admissible $P_h(K)$-module and if for all $u \in N_h(K)$ we have $(u-1)^2 = 0$ on V then $N_h(K)$ acts trivially on V.*

2. *If V is a smooth $P_h(K)$-module which is admissible as a $L_h(K)$-module then $N_h(K)$ acts trivially on V.*

3. *If G_1 is any locally compact totally disconnected group and if V is a smooth $G_1 \times P_h(K)$-module which is admissible as a $G_1 \times L_h(K)$-module then $N_h(K)$ acts trivially on V.*

Proof: For the first part consider the open compact subgroups U_m consisting of elements of $P_h(\mathcal{O}_K)$ which reduce modulo $\wp_K^m$ to the identity and modulo $\wp_K^{2m}$ to an element of $N_h(\mathcal{O}_K/\wp_K^{2m})$. Note that U_m is normalised by $N_h(\wp_K^{-m})$. Thus V^{U_m} is a finite dimensional smooth $N_h(\wp_K^m)$-module, and as $N_h(\wp_K^{-m})$ is compact, V^{U_m} is semi-simple. Thus if $u \in N(\wp_K^{-m})$ we see that $u = 1$ on V^{U_m}. As $N_h(K) = \bigcup_m N_h(\wp_K^{-m})$ and $V = \bigcup_m V^{U_m}$, the first part of the lemma follows.

Now consider the second part of the lemma. If χ is a character of $Z_h(K)$ then set

$$V_i^\chi = \bigcap_{z \in Z_h(K)} \ker(z - \chi(z))^i$$

and

$$V_\infty^\chi = \bigcup_i V_i^\chi.$$

Because V is an admissible $L_h(K)$-module we have

$$V = \bigoplus_\chi V_\infty^\chi.$$

(This follows because for any open compact subgroup $U \subset L_h(K)$, V^U is a finite dimensional smooth $Z_h(K)$-module, and hence

$$V^U = \bigoplus_\chi (V_\chi^\infty \cap V^U).)$$

Thus it suffices to show for each χ and i that V_i^χ is a $P_h(K)$-module on which $N_h(K)$ acts trivially. We will do this by induction on i for fixed χ. For $i = 0$ there is nothing to prove. Thus assume the result is true for V_{i-1}^χ and we will prove it for V_i^χ. By the first part of this lemma it suffices to show that V_i^χ/V_{i-1}^χ is a $P_h(K)$-submodule of V/V_{i-1}^χ on which $N_h(K)$ acts trivially. Suppose that $v \in V/V_{i-1}^\chi$. By smoothness, v is invariant by $N_h(\wp_K^m)$ for some m. If u is any element of $N_h(K)$ we may choose $z \in Z_h(K)$ such that $zuz^{-1} \in N_h(\wp_K^m)$. Then we have

$$uv = uz^{-1}\chi(z)v = \chi(z)z^{-1}(zuz^{-1})v = \chi(z)z^{-1}v = v.$$

The second part of the lemma follows.

The third part of the lemma follows easily from the second. □

Finally we have the following lemma.

Lemma I.2.2 *Let Γ be a profinite group and let*

$$R : \Gamma \longrightarrow GL_{an}(\mathbb{Q}_l^{ac})$$

be a continuous representation. Suppose that

1. *Γ contains a dense subset Σ such that for each $\sigma \in \Sigma$ every eigenvalue of $R(\sigma)$ has multiplicity at least a; and*

2. *the $\widehat{\mathbb{Q}_l^{ac}}$-span of the Lie algebra of the image of R contains an element which has n distinct eigenvalues each with multiplicity a.*

Then there is a continuous representation

$$\widetilde{R} : \Gamma \longrightarrow GL_{an}(\mathbb{Q}_l^{ac})$$

such that

$$[R] = a[\widetilde{R}].$$

Proof: Let W_R denote the space underlying the representation R. We may assume that R is semi-simple. The condition on a monic polynomial of degree na that all its roots have multiplicity at least a is Zariski closed. (The polynomial $P(X)$ has all roots with multiplicity $\geq a$ if and only if $P(X)|((d/dX)^i P(X))^a$ for $i = 1, \ldots, a - 1$.) Let H denote the Zariski closure of the image of R in $GL_{an}/\mathbb{Q}_l^{ac}$ and let H^0 denote the connected component of the identity in H. Then W_R is semi-simple as a representation of H and hence of H^0. Thus H^0 is a reductive group (as it has a faithful semi-simple representation). Moreover every eigenvalue on W_R of every element of $H(\mathbb{Q}_l^{ac})$ has multiplicity at least a. In particular it has at most n eigenvalues. Finally $(\text{Lie } H^0)(\mathbb{Q}_l^{ac})$ has an element with n distinct eigenvalues on W_R each of multiplicity a.

Let $B \supset T$ denote a Borel and maximal torus in H^0. Also let N denote the stabiliser of the pair (B, T) in H. Thus $N \cap H^0 = T$ and $H = H^0.N$. We see that T has exactly n distinct weights on W_R and each weight space has dimension a. If λ is such a weight, let N_λ denote the stabiliser of λ in N, let $H_\lambda = H^0 N_\lambda$ and let $W_{R,\lambda}$ denote the λ-weight space. Thus $\dim W_{R,\lambda} = a$.

Suppose that $n \in N_\lambda$. Let $O_1 = \{\lambda\}, O_2, \ldots, O_s$ denote the orbits of weights of T on W_R under $\langle n \rangle$. Let $W_{R,i}$ denote the sum of the μ weight spaces for $\mu \in O_i$. If $x_1, \ldots, x_u$ are the eigenvalues of n on $W_{R,i}$ and if $t \in T$ then any eigenvalue x of nt on $W_{R,i}$ satisfies

$$x^b = x_j^b \prod_{\mu \in O_i} \mu(t)^{b(\mu)},$$

for some $j \in \{1, \ldots, u\}$. Here b is a positive integer (for example the order of n in N/T) and the $b(\mu)$ are non-negative integers with $\sum_\mu b(\mu) = b$. They depend on n but not on t. Suppose that for any $i > 1$ we have

$$\lambda \neq \left(\sum_{\mu \in O_i} b(\mu)\mu \right)/b.$$

(This would be true for instance if λ lies at a vertex of the convex hull of the set of weights of T on W_R.) In this case, if n had two distinct eigenvalues y_1, y_2 on $W_{R,\lambda}$, then for generic $t \in T$ the element nt would have eigenvalues $y_1\lambda(t)$ and $y_2\lambda(t)$ on $W_{R,\lambda}$ but on no other $W_{R,i}$ (with $i > 1$). That would imply that $a = \dim W_{R,\lambda} \geq 2a$, a contradiction. We conclude that in this case λ extends to a homomorphism $\lambda : N_\lambda \to \mathbb{G}_m$ such that N_λ acts on $W_{R,\lambda}$ by λ.

Now suppose that λ does lie at a vertex of the convex hull of the set of weights of T on W_R and that λ is minimal with respect to B amongst this set of weights. Let X (resp. Y) denote the sum of the H^0 isotypical components of W_R corresponding to irreducible representations of H^0 with minimal weight λ (resp. any conjugate under N of λ). Then $Y \cong \operatorname{Ind}_{H_\lambda}^H X$. On the other hand if $x_1, \ldots, x_a$ is a basis of $W_{R,\lambda}$ then $X \cong \bigoplus_{i=1}^a \langle Bx_i \rangle$ as an H_λ-module. We conclude that

$$W_R \cong (\operatorname{Ind}_{H_\lambda}^H \langle By_1 \rangle)^a \oplus Z$$

as H-modules. Arguing recursively we see that $W_R \cong U^a$ for some H-module U. The lemma follows. $\square$

I.3   Admissible representations of GL_g

Let Ω be an algebraically closed field of characteristic 0 and of cardinality equal to that of $\mathbb{C}$. Let K be a finite extension of $\mathbb{Q}_p$. Suppose that V/Ω is a vector space and that $\pi : GL_g(K) \to \operatorname{Aut}(V)$ is an irreducible admissible representation with central character ψ_π. We will call π supercuspidal if for any $v \in V$ and f in the smooth dual of V the function $GL_g(K) \to \Omega$ which sends

$$x \longmapsto f(xv)$$

is compactly supported modulo the centre $K^\times$ of $GL_g(K)$. Choose an embedding of fields $\imath : \Omega \hookrightarrow \mathbb{C}$. We will call π square integrable if for any $v \in V$ and f in the smooth dual of V the function $GL_g(K)/K^\times \to \mathbb{R}$ which sends

$$x \longmapsto |\imath(f(xv))|^2 |\imath(\psi_\pi(\det x))|^{-2/g}$$

is integrable. It follows from Zelevinsky's classification [Ze] that this definition is independent of the choice of $\imath$. We will say that a representation π is $\imath$-*preunitary* if there is a pairing $(\ ,\ )$ from $V \times V$ to $\mathbb{C}$ such that

- $(av_1 + v_2, v_3) = \imath(a)(v_1, v_3) + (v_2, v_3)$ for all $a \in \Omega$ and $v_1, v_2, v_3 \in V$,

- $(v_1, v_2) = c(v_2, v_1)$ for all $v_1, v_2 \in V$ (where c denotes complex conjugation),

- $(v, v) > 0$ for all non-zero $v \in V$,

- $(\pi(x)v_1, \pi(x)v_2) = |\imath(\psi_\pi(\det x))|^{2/g}(v_1, v_2)$ for all $v_1, v_2 \in V$ and $x \in GL_g(K)$.

Rogawski ([Rog2]) and Deligne, Kazhdan and Vigneras (see [DKV]) have shown the existence of a unique bijection, which we will denote JL, from irreducible admissible representations of $D_{K,g}^\times$ to square integrable irreducible admissible representations of $GL_g(K)$ such that if ρ is an irreducible admissible representation of $D_{K,g}^\times$ then the character $\chi_{\mathrm{JL}(\rho)}$ of JL (ρ) satisfies

- $\chi_{\mathrm{JL}(\rho)}(\gamma) = 0$ if $\gamma \in GL_g(K)$ is regular semi-simple but not elliptic,

- $\chi_{\mathrm{JL}(\rho)}(\gamma) = (-1)^{g-1}\mathrm{tr}\,\rho(\delta)$ if $\gamma \in GL_g(K)$ is regular semi-simple and elliptic and if δ is an element of $D_{K,g}^\times$ with the same characteristic polynomial as γ.

If π is a square integrable irreducible admissible representation of the group $GL_g(K)$ with central character ψ_π then Deligne, Kazhdan and Vigneras also show the existence of a function $\varphi_\pi \in C_c^\infty(GL_g(K), \psi_\pi^{-1})$, which we will call a pseudo-coefficient for π, with the following properties. (We always use associated measures on inner forms of the same group. See for instance page 631 in [Ko5], where they are called compatible measures.)

- $\mathrm{tr}\,\pi(\varphi_\pi) = \mathrm{vol}\,(D_{K,g}^\times/K^\times)$.

- Suppose that

$$GL_{g_1} \times \cdots \times GL_{g_s} = L \subset P \subset GL_g$$

is a Levi component of a parabolic subgroup of GL_g. Suppose also that for $i = 1, \ldots, s$ we are given a square integrable irreducible admissible representation π_i of $GL_{g_i}(K)$ such that $\pi_i \not\cong \pi$ and such that $\psi_{\pi_1} \ldots \psi_{\pi_s} = \psi_\pi$. Then

$$\operatorname{tr} \text{n-Ind}_{P(K)}^{GL_g(K)}(\pi_1 \times \cdots \times \pi_s)(\varphi_\pi) = 0.$$

- If $\gamma \in GL_g(K)$ is a non-elliptic regular semi-simple element then

$$O_\gamma^{GL_g(K)}(\varphi_\pi) = 0.$$

- If $\gamma \in GL_g(K)$ is an elliptic regular semi-simple element and if $\delta \in D_{K,g}^\times$ has the same characteristic polynomial as γ then

$$O_\gamma^{GL_g(K)}(\varphi_\pi) = (-1)^{g-1}\operatorname{vol}(D_{K,g}^\times/Z_{D_{K,g}^\times}(\delta))\operatorname{tr} JL^{-1}(\pi^\vee)(\delta).$$

(See section A.4 of [DKV], especially the introduction to that section and subsection A.4.1 .)

Lemma I.3.1 *Let π be a square integrable representation of $GL_g(K)$ and let φ_π be a pseudo-coefficient for π as above.*

1. If $\gamma \in GL_g(K)$ is a non-elliptic semi-simple element then

$$O_\gamma^{GL_g(K)}(\varphi_\pi) = 0.$$

2. If $\gamma \in GL_g(K)$ is an elliptic semi-simple element and if $\delta \in D_{K,g}^\times$ has the same characteristic polynomial as γ then

$$O_\gamma^{GL_g(K)}(\varphi_\pi) =$$
$$(-1)^{g(1-[K(\gamma):K]^{-1})}\operatorname{vol}(D_{K,g}^\times/Z_{D_{K,g}^\times}(\delta))\operatorname{tr} JL^{-1}(\pi^\vee)(\delta).$$

Proof: Consider the first part. Let T be a maximal torus containing γ. Then

$$0 = \sum_u \Gamma_u(t)O_{\gamma u}^{GL_g(K)}(\varphi_\pi)$$

where u runs over a set of representatives of the unipotent conjugacy classes in $Z_{GL_g}(\gamma)(K)$, where Γ_u denotes the Shalika germ associated to u and where t is any regular element of T sufficiently close to γ. Then homogeneity ([HC], theorem 14(1)) tells us that

$$0 = \Gamma_1(t)O_\gamma^{GL_g(K)}(\varphi_\pi)$$

for any regular $t \in T$ sufficiently close to γ. By [Rog1], $\Gamma_1(t)$ is not identically zero near γ and the first part of the lemma follows.

Consider now the second part. Let T be an elliptic maximal torus in $GL_g(K)$ containing γ. We can and will also think of $T \subset D_{K,g}^\times$. Then we can take δ to be $\gamma \in T \subset D_{K,g}^\times$. For t a regular element of T sufficiently close to γ we have

$$(-1)^{g-1}\mathrm{vol}\,(D_{K,g}^\times/T)\mathrm{tr}\,\mathrm{JL}^{-1}(\pi^\vee)(t) = \sum_u \Gamma_u(t)O_{\gamma u}^{GL_g(K)}(\varphi_\pi)$$

where u runs over a set of representatives of the unipotent conjugacy classes in $Z_{GL_g}(\gamma)(K)$ and where Γ_u denotes the Shalika germ associated to u. Again using homogeneity ([HC], theorem 14(1)) we see that

$$(-1)^{g-1}\mathrm{vol}\,(D_{K,g}^\times/T)\mathrm{tr}\,\mathrm{JL}^{-1}(\pi^\vee)(\delta) = O_\gamma^{GL_g(K)}(\varphi_\pi)\lim_{t\to\gamma}\Gamma_1(t).$$

Thus it suffices to check that

$$\lim_{t\to\gamma}\Gamma_1(t) = (-1)^{g/[K(\gamma):K]-1}\mathrm{vol}\,(Z_{D_{K,g}^\times}(\delta)/T).$$

This is independent of the choices of measures, as long as we choose associated measures on $Z_{GL_g}(\gamma)(K)$ and $Z_{D_{K,g}^\times}(\delta)$. (Choices of Haar measures on $Z_{GL_g}(\gamma)(K)$ and T are implicit in the definition of Γ_u.) Thus we may choose any measure on $K^\times$ and Euler-Poincaré measure on T and $Z_{GL_g}(\gamma)(K)$ (see section 1 of [Ko5]). Then we must use $(-1)^{g/[K(\gamma):K]-1}$ times Euler-Poincaré measure on $Z_{D_{K,g}^\times}(\delta)$ (by theorem 1 of [Ko5]). According to [Rog2] with these choices of measures $\Gamma_1(t) = 1$ for regular $t \in T$ sufficiently close to γ. On the other hand according to [Ser] with these measures $\mathrm{vol}\,(T/K^\times) = 1$ and $\mathrm{vol}\,(Z_{D_{K,g}^\times}(\delta)/K^\times) = 1$. The second part of the lemma follows. $\square$

Suppose that $s|g$ is a positive integer and that π is a supercuspidal representation of $GL_{g/s}(K)$. Let Q_s denote a parabolic subgroup of GL_g with Levi component $GL_{g/s}^s$. Zelevinsky ([Ze]) describes the irreducible subquotients of

$$\text{n-Ind}_{Q_s(K)}^{GL_g(K)}(\pi \times \pi \otimes |\det| \times \cdots \times \pi \otimes |\det|^{s-1})$$

as follows. Let $\Gamma(s,\pi)$ be the graph with vertices labelled $\pi \otimes |\det|^j$ for $j = 0, \ldots, s-1$ and with one edge between $\pi \otimes |\det|^j$ and $\pi \otimes |\det|^{j+1}$ for $j = 0, \ldots, s-2$ and no other edges. Zelevinsky shows that there is a bijection between directed graphs $\vec{\Gamma}$ with underlying undirected graph $\Gamma(s,\pi)$ and irreducible subquotients of

$$\text{n-Ind}_{Q_s(K)}^{GL_g(K)}(\pi \times \pi \otimes |\det| \times \cdots \times \pi \otimes |\det|^{s-1}),$$

which, following Zelevinsky's notation, we will denote $\vec{\Gamma} \mapsto \omega(\vec{\Gamma})$.

Some particular subquotients will be of special importance for us. So we will let $\vec{\Gamma}_{abc}(\pi)$ denote the directed graph with vertices labelled $\pi \otimes |\det|^j$ for $j = 0, \ldots, a + b + c - 1$ and

- a single edge from $\pi \otimes |\det|^j$ to $\pi \otimes |\det|^{j-1}$ for $j = 1, \ldots, a - 1$ and for $j = a + b + 1, \ldots, a + b + c - 1$,

- and a single edge from $\pi \otimes |\det|^{j-1}$ to $\pi \otimes |\det|^j$ for $j = a, \ldots, a + b$.

Similarly we will let $\vec{\Gamma}'_{abc}(\pi)$ denote the directed graph with vertices labelled $\pi \otimes |\det|^j$ for $j = 0, \ldots, a + b + c - 1$ and

- a single edge from $\pi \otimes |\det|^{j-1}$ to $\pi \otimes |\det|^j$ for $j = 1, \ldots, a$ and for $j = a + b, \ldots, a + b + c - 1$,

- and a single edge from $\pi \otimes |\det|^{j+1}$ to $\pi \otimes |\det|^j$ for $j = a, \ldots, a + b - 2$.

More over we will denote $\omega(\vec{\Gamma}_{t,s-1-t,1}) = \omega(\vec{\Gamma}'_{0,t,s-t})$ by

$$\mathrm{Sp}_t(\pi) \boxplus (\pi \otimes |\det|^t) \boxplus \cdots \boxplus (\pi \otimes |\det|^{s-1}),$$

for any $t = 0, \ldots, s$.

Zelevinsky ([Ze]) has proved the following results.

- $\mathrm{Sp}_s(\pi)$ is square integrable and any square integrable representation is of this form for a unique positive integer $s|g$ and a unique supercuspidal representation π of $GL_{g/s}(K)$.

- The only generic ("non-degenerate" in Zelevinsky's terminology) subquotient of $\text{n-Ind}_{Q_s(K)}^{GL_g(K)}(\pi \times \pi \otimes |\det| \times \cdots \times \pi \otimes |\det|^{s-1})$ is $\mathrm{Sp}_s(\pi)$.

Moreover Tadic ([Tad]) has shown that

- for any embedding $\imath : \Omega \hookrightarrow \mathbb{C}$ the only $\imath$-preunitary subquotients of $\text{n-Ind}_{Q_s(K)}^{GL_g(K)}(\pi \times \pi \otimes |\det| \times \cdots \times \pi \otimes |\det|^{s-1})$ are $\mathrm{Sp}_s(\pi)$ and $\pi \boxplus (\pi \otimes |\det|) \boxplus \cdots \boxplus (\pi \otimes |\det|^{s-1})$.

We also note that if ψ is a character of $K^\times$ then

$$\mathrm{JL}\,(\psi \circ \det) = \mathrm{Sp}_g(\psi| \; |^{(1-g)/2}).$$

Lemma I.3.2 *Suppose that $s_1 + s_2 = s|g$ are positive integers and that π is an irreducible supercuspidal representation of $GL_{g/s}(K)$. Set $g_i = s_i g/s$ and let P denote a parabolic subgroup of GL_g with Levi component $GL_{g_1} \times GL_{g_2}$. Let $\vec{\Gamma}_1$ (resp. $\vec{\Gamma}_2$) be an oriented graph with unoriented underlying graph $\Gamma(s_1, \pi)$ (resp. $\Gamma(s_2, \pi \otimes |\det|^{s_1})$). Let $\vec{\Gamma}$ and $\vec{\Gamma}'$ be* *the two (distinct) oriented graphs with underlying unoriented graph $\Gamma(s, \pi)$ which agree with $\vec{\Gamma}_1$ on $\Gamma(s_1, \pi)$ and with $\vec{\Gamma}_2$ on $\Gamma(s_2, \pi \otimes |\det|^{s_2})$. Then n-Ind$_{P(K)}^{GL_g(K)}(\omega(\vec{\Gamma}_1) \times \omega(\vec{\Gamma}_2))$ has a Jordan-Hölder series of length two and the two Jordan-Hölder factors are $\omega(\vec{\Gamma})$ and $\omega(\vec{\Gamma}')$.*

Proof: We may and will assume that $Q_s \subset P$. Let U_s denote the unipotent radical of Q_s. By section 1.6 and theorem 2.8 of [Ze] we can compute the Jordan-Hölder factors of $J_{U_s}(\text{n-Ind}_{P(K)}^{GL_g(K)}(\omega(\vec{\Gamma}_1) \times \omega(\vec{\Gamma}_2)))$. We find that they are all products (each taken with multiplicity one) of the form

$$(\rho \otimes |\det|^{j_1}) \times \cdots \times (\rho \otimes |\det|^{j_s}),$$

where $j_1, \ldots, j_s$ runs over all permutations of $0, \ldots, s-1$ such that $j_i < j_{i'}$ if there is an edge of either $\vec{\Gamma}_1$ or $\vec{\Gamma}_2$ running from i to i'. This is the same as the union (as sets with multiplicities) of the Jordan-Hölder factors of $\omega(\vec{\Gamma})$ and $\omega(\vec{\Gamma}')$ (see theorem 2.8 of [Ze]). The lemma then follows from theorem 2.2 of [Ze]. $\square$

Lemma I.3.3 *Suppose that $s|g$ are positive integers and that π is an irreducible supercuspidal representation of $GL_{g/s}(K)$. For $h = 0, \ldots, g-1$ let N_h^{op} be the unipotent subgroup of GL_g introduced at the start of section V.5.*

1. *If $g \nmid sh$ then $J_{N_h^{\text{op}}}(\text{Sp}_s(\pi)) = (0)$ and $J_{N_h^{\text{op}}}(\pi \boxplus \cdots \boxplus (\pi \otimes |\det|^{s-1})) = (0)$.*

2. *If $sh = gh'$ for some positive integer h' then*

$$J_{N_h^{\text{op}}}(\text{Sp}_s(\pi)) = \text{Sp}_{h'}(\pi \otimes |\det|^{s-h'}) \times \text{Sp}_{s-h'}(\pi),$$

and

$$J_{N_h^{\text{op}}}(\pi \boxplus \cdots \boxplus (\pi \otimes |\det|^{s-1})) = (\pi \boxplus \cdots \boxplus (\pi \otimes |\det|^{h'-1})) \times$$
$$\times ((\pi \otimes |\det|^{h'}) \boxplus \cdots \boxplus (\pi \otimes |\det|^{s-1})).$$

Proof: These results follow easily from theorem 2.2 of [Ze]. $\square$

Lemma I.3.4 *Suppose that $s|g$ are positive integers and that π is an irreducible supercuspidal representation of $GL_{g/s}(K)$.*

1. *If π' is an irreducible admissible representation of $GL_g(K)$ which is not a subquotient of n-Ind$_{Q_s(K)}^{GL_g(K)}(\pi \times \pi \otimes |\det| \times \cdots \times \pi \otimes |\det|^{s-1})$ then*

$$\text{tr } \pi'(\varphi_{\text{Sp}_s(\pi)}) = 0.$$

2. If P is a proper parabolic subgroup of GL_g with Levi component $GL_{g_1} \times GL_{g_2}$ and if π_1 (resp. π_2) is an irreducible admissible representation of $GL_{g_1}(K)$ (resp. $GL_{g_2}(K)$) then

$$\text{tr n-Ind}_{P(K)}^{GL_g(K)}(\pi_1 \times \pi_2)(\varphi_{\text{Sp}_s(\pi)}) = 0.$$

3. Suppose that $\vec{\Gamma}$ is an oriented graph with underlying unoriented graph $\Gamma(s, \pi)$ and that $\vec{\Gamma}$ has s edges oriented from $\pi \otimes |\det|^j$ to $\pi \otimes |\det|^{j+1}$. Then

$$\text{tr}\, \omega(\Gamma)(\varphi_{\text{Sp}_s(\pi)}) = (-1)^s \text{vol}\,(D_{K,g}^{\times}/K^{\times}).$$

Proof: For the first part note that the proof of lemma A.4.f of [DKV] shows that π' can be written in Groth $(GL_g(K))$ as an integral linear combination of n-Ind$_{P_i(K)}^{GL_g(K)}(\pi'_i)$ where P_i runs over parabolic subgroups of GL_g, where π'_i is an irreducible square integrable representation of the Levi component of $P_i(K)$ and where no (P_i, π'_i) is conjugate to $(GL_g, \text{Sp}_s(\pi))$. (In fact in the notation of [DKV] we have $r(\pi'_i) \neq r(\text{Sp}_s(\pi))$.)

For the second part note that it follows from lemma A.4.f of [DKV] that we may write $\pi_1 \in \text{Groth}\,(GL_{g_1}(K))$ as a finite sum

$$\pi_1 = \sum_i a_{1i} \text{n-Ind}_{P_{1i}(K)}^{GL_{g_1}(K)} \pi_{1i},$$

where $a_{1i} \in \mathbb{Z}$, $P_{1i} \subset GL_{g_1}$ is a parabolic subgroup and π_{1i} is an irreducible square integrable representation of the Levi component of $P_{1i}(K)$. Similarly we have

$$\pi_2 = \sum_i a_{2i} \text{n-Ind}_{P_{2i}(K)}^{GL_{g_2}(K)} \pi_{2i}.$$

For each pair of indices i, j choose a parabolic subgroup $P'_{ij} \subset P \subset GL_g$ such that $P'_{ij}(K) \cap GL_{g_1}(K) = P_{1i}(K)$ and $P'_{ij}(K) \cap GL_{g_2}(K) = P_{2i}(K)$. Note that for all i, j we have $P_{ij} \neq GL_g$. In Groth $(GL_g(K))$ we have the equality

$$\text{n-Ind}_{P(K)}^{GL_g(K)} = \sum_{ij} a_{1i} a_{2j} \text{n-Ind}_{P'_{ij}(K)}^{GL_g(K)} \pi_{1i} \times \pi_{2j}.$$

The second part of the lemma follows.

The third part follows by a simple recursion from the second part and lemma I.3.2. $\square$

Corollary I.3.5 *If $s|g$ are positive integers, if π is an irreducible supercuspidal representation of $GL_{g/s}(K)$ and if π' is an admissible representation of $GL_g(K)$ then*

$$\mathrm{vol}\,(D_{K,g}^\times / K^\times)^{-1} \mathrm{tr}\,\pi'(\varphi_{\mathrm{Sp}_s(\pi)}) \in \mathbb{Z}.$$

Corollary I.3.6 *If $s|g$ are positive integers, if π is an irreducible supercuspidal representation of $GL_{g/s}(K)$ and if π' is a generic irreducible admissible representation of $GL_g(K)$ such that*

$$\mathrm{tr}\,\pi'(\varphi_{\mathrm{Sp}_s(\pi)}) \neq 0$$

then $\pi' \cong \mathrm{Sp}_s(\pi)$.

Corollary I.3.7 *If $s|g$ are positive integers, if π is an irreducible supercuspidal representation of $GL_{g/s}(K)$, if $\imath : \Omega \hookrightarrow \mathbb{C}$ and if π' is an $\imath$-preunitary irreducible admissible representation of $GL_g(K)$ such that*

$$\mathrm{tr}\,\pi'(\varphi_{\mathrm{Sp}_s(\pi)}) \neq 0$$

then either $\pi' \cong \mathrm{Sp}_s(\pi)$ or $\pi' \cong \pi \boxplus \cdots \boxplus (\pi \otimes |\det|^{s-1})$.

Suppose that s_i and g_i are positive integers for $i = 1, \ldots, t$ such that $g = g_1 s_1 + \cdots + g_t s_t$. Suppose moreover that for $i = 1, \ldots, t$ we are given an irreducible supercuspidal representation π_i of $GL_{g_i}(K)$. Suppose first that

- if $\imath < \jmath$ then $\pi_j \not\cong \pi_i \otimes |\det|^a$ for any $a \in \mathbb{Z}_{\geq 1}$ with

$$1 + s_i - s_j \leq a \leq s_i.$$

Also let P denote the parabolic subgroup of GL_g consisting of block diagonal matrices with diagonal blocks of size $s_1 g_1 \times s_1 g_1, \ldots, s_t g_t \times s_t g_t$ from top left to bottom right. Then

$$\text{n-Ind}_{P(K)}^{GL_g(K)}(\mathrm{Sp}_{s_1}(\pi_1) \times \cdots \times \mathrm{Sp}_{s_t}(\pi_t))$$

has a unique irreducible quotient which we will denote

$$\mathrm{Sp}_{s_1}(\pi_1) \boxplus \cdots \boxplus \mathrm{Sp}_{s_t}(\pi_t).$$

If σ is any permutation of $\{1, 2, \ldots, t\}$ such that $(s_{\sigma 1}, \pi_{\sigma 1}), \ldots, (s_{\sigma t}, \pi_{\sigma t})$ still satisfies the above condition then

$$\mathrm{Sp}_{s_{\sigma 1}}(\pi_{\sigma 1}) \boxplus \cdots \boxplus \mathrm{Sp}_{s_{\sigma t}}(\pi_{\sigma t}) \cong \mathrm{Sp}_{s_1}(\pi_1) \boxplus \cdots \boxplus \mathrm{Sp}_{s_t}(\pi_t).$$

Thus whether or not $(s_1, \pi_1), \ldots, (s_t, \pi_t)$ satisfy the above condition we may define

$$\mathrm{Sp}_{s_1}(\pi_1) \boxplus \cdots \boxplus \mathrm{Sp}_{s_t}(\pi_t) = \mathrm{Sp}_{s_{\sigma 1}}(\pi_{\sigma 1}) \boxplus \cdots \boxplus \mathrm{Sp}_{s_{\sigma t}}(\pi_{\sigma t}),$$

for any permutation σ of $\{1, 2, \ldots, t\}$ such that $(s_{\sigma 1}, \pi_{\sigma 1}), \ldots, (s_{\sigma t}, \pi_{\sigma t})$ does satisfy the above condition. It follows from theorem 2.8 of [Ze] that this notation is compatible with our previous use of $\boxplus$. Moreover any irreducible admissible representation π of $GL_g(K)$ is of this form and, moreover, the multiset $\{(s_1, \pi_1), \ldots, (s_t, \pi_t)\}$ is uniquely determined by π. (For these results see section 4.3 of [Rod], and note that a sketch of the unpublished result of I.N.Bernstein (proposition 11 of [Rod]) can be found in [JS2].)

We will call the collection $\{(s_i, \pi_i)\}$ *unlinked* if for all $i \neq j$ the following condition is satisfied.

- If $\pi_j \cong \pi_i \otimes |\det|^a$ for a positive integer a then either $a > s_i$ or $a + s_j \leq s_i$.

Zelevinsky shows ([Ze], theorem 9.7) that if $\{(s_i, \pi_i)\}$ is unlinked then

$$\mathrm{Sp}_{s_1}(\pi_1) \boxplus \cdots \boxplus \mathrm{Sp}_{s_t}(\pi_t) = \text{n-Ind}_{P(K)}^{GL_g(K)}(\mathrm{Sp}_{s_1}(\pi_1) \times \cdots \times \mathrm{Sp}_{s_t}(\pi_t))$$

and that this representation is generic. Zelevinsky also shows ([Ze], theorem 9.7) that any irreducible generic admissible representation of $GL_g(K)$ arises in this way from some unlinked collection $\{(s_i, \pi_i)\}$.

Combining this with the results of Tadic [Tad] we get the following lemma.

Lemma I.3.8 *Suppose that π is an irreducible, generic, $\imath$-preunitary representation of $GL_g(K)$ with $|\imath\psi_\pi| \equiv 1$. Then π is isomorphic to*

$$\text{n-Ind}_{P(K)}^{GL_g(K)}(\pi_1 \times \cdots \times \pi_s \times \pi_1'|\det|^{a_1} \times \pi_1'|\det|^{-a_1} \times \ldots$$
$$\times \pi_t'|\det|^{a_t} \times \pi_t'|\det|^{-a_t}),$$

where

- *P is a parabolic subgroup of GL_g,*

- *$\pi_1, \ldots, \pi_s, \pi_1', \ldots, \pi_t'$ are square integrable representations of smaller general linear groups,*

- *$|\imath\psi_{\pi_i}| \equiv |\imath\psi_{\pi_j'}| \equiv 1$ for all i and j,*

- *and $0 < a_j < 1/2$ for $j = 1, \ldots, t$.*

Moreover, $\iota\pi$ is tempered if and only if $t = 0$.

The following result follows from lemma 2.12 of [BZ].

Lemma I.3.9 *Suppose that s_i and g_i for $i = 1, \ldots, t$ are positive integers such that $g = g_1 s_1 + \cdots + g_t s_t$. Suppose also that π_i is an irreducible supercuspidal representation of $GL_{g_i}(K)$ for $i = 1, \ldots, t$; and that $P \subset GL_g$ is a parabolic subgroup with Levi factor $GL_{s_1 g_1} \times \cdots \times GL_{s_t g_t}$.*

For $h = 0, \ldots, g - 1$ let $N_h^{\mathrm{op}} < GL_g$ be the unipotent subgroup defined at the start of section V.5. Then in $\mathrm{Groth}\,(GL_h(K) \times GL_{g-h}(K))$ we have an equality between

$$[J_{N_h^{\mathrm{op}}}(\text{n-Ind}_{P(K)}^{GL_g(K)}(\mathrm{Sp}_{s_1}(\pi_1) \times \cdots \times \mathrm{Sp}_{s_t}(\pi_t)))]$$

and

$$\sum_{h_i}[\text{n-Ind}_{P'(K)}^{GL_h(K)}(\mathrm{Sp}_{h_1}(\pi_1 \otimes |\det|^{s_1 - h_1}) \times \cdots \times \mathrm{Sp}_{h_t}(\pi_t \otimes |\det|^{s_t - h_t}))]$$

$$[\text{n-Ind}_{P''(K)}^{GL_{n-h}(K)}(\mathrm{Sp}_{s_1 - h_1}(\pi_1) \times \cdots \times \mathrm{Sp}_{s_t - h_t}(\pi_t))],$$

where the sum is over all positive integers $h_1, \ldots, h_t$ with $h_i \leq s_i$ and $h = h_1 g_1 + \cdots + h_t g_t$ and where

- *$P' \subset GL_h$ is a parabolic subgroup with Levi component $GL_{h_1 g_1} \times \cdots \times GL_{h_t g_t}$*

- *and $P'' \subset GL_{n-h}$ is a parabolic subgroup with Levi component $GL_{(s_1 - h_1)g_1} \times \cdots \times GL_{(s_t - h_t)g_t}$.*

I.4 Base change

Suppose that K'/K is a cyclic Galois extension of p-adic fields of prime degree q. If π is an irreducible admissible representation of $GL_g(K)$ then one can associate to π its base change lifting $\mathrm{Res}_{K'}^K(\pi)$ to K' (see theorem 6.2 of chapter 1 and the discussion on pages 59 and 60 of [AC]). Also if π' is a $\mathrm{Gal}\,(K'/K)$-regular (see section 2.4 of [HH] for the definition of this concept) generic irreducible admissible representation of $GL_g(K')$ one can associate to π' its automorphic induction $\mathrm{Ind}_{K'}^K(\pi')$ to K (see theorem 2.4 of [HH]). Then $\mathrm{Res}_{K'}^K(\pi)$ is an irreducible admissible representation of $GL_g(K')$ and $\mathrm{Ind}_{K'}^K(\pi')$ is an irreducible admissible representation of $GL_{qg}(K)$.

Now suppose that L'/L is a cyclic Galois extension of number fields of prime degree q. Let τ denote a generator of $\mathrm{Gal}\,(L'/L)$ and let η denote a non-trivial character of

$$\mathbb{A}_L^\times/L^\times(L_\infty^\times)^0(N_{L'/L}\mathbb{A}_{L'}^\times).$$

If Π is a cuspidal automorphic representation of $GL_g(\mathbb{A}_L)$ then we can associate to Π an "induced from cuspidal" (see definition 4.1 of chapter 3 of [AC]) representation $\operatorname{Res}^L_{L'}\Pi$ of $GL_g(\mathbb{A}_{L'})$ with the following properties.

1. $\operatorname{Res}^L_{L'}(\Pi)$ is cuspidal if and only if $\Pi = \Pi \otimes (\eta \circ \det)$.

2. A cuspidal automorphic representation Π' of $GL_g(\mathbb{A}_{L'})$ is of the form $\operatorname{Res}^L_{L'}\Pi$ for some cuspidal automorphic Π if and only if $\Pi' \cong \Pi' \circ \tau$.

3. If x is a place of L which splits in L' and $\tilde{x}$ is a place of L' above x then
$$\operatorname{Res}^L_{L'}(\Pi)_{\tilde{x}} \cong \Pi_x.$$

4. If x is a finite place of L which is inert in L' then
$$\operatorname{Res}^L_{L'}(\Pi)_x \cong \operatorname{Res}^{L_x}_{L'_x}(\Pi_x).$$

(See theorems 4.2 and 5.1 of [AC].) Now suppose that Π' is a cuspidal automorphic representation of $GL_g(\mathbb{A}_{L'})$. Then there is an "induced from cuspidal" representation $\operatorname{Ind}^L_{L'}\Pi'$ of $GL_{gq}(\mathbb{A}_L)$ with the following properties.

1. $\operatorname{Ind}^L_{L'}(\Pi') \otimes (\eta \circ \det) \cong \operatorname{Ind}^L_{L'}(\Pi')$.

2. If $\Pi' \not\cong \Pi' \circ \sigma$ then $\operatorname{Ind}^L_{L'}(\Pi')$ is cuspidal.

3. If x is a place L which splits as $x_1 \ldots x_q$ in L' then
$$\operatorname{Ind}^L_{L'}(\Pi')_x \cong \operatorname{n-Ind}^{GL_{qg}(L_x)}_{Q(L_x)}(\Pi'_{x_1} \times \cdots \times \Pi'_{x_q}),$$

   where Q is a parabolic subgroup of GL_{qg} with Levi component GL^q_g.

4. For all but finitely many places x of L which are inert in L' we have
$$\operatorname{Ind}^L_{L'}(\Pi')_x \cong \operatorname{Ind}^{L_x}_{L'_x}(\Pi'_x).$$

(See theorem 6.2, lemma 6.4 and corollary 6.5 of [AC].) The following lemma seems to be well known (see section 1.5 of [HH]), but for lack of an explicit reference we give the proof.

Lemma I.4.1 *Keep the above notation and suppose that x is a finite place of L which is inert in L'.*

1. *Π'_x is $\operatorname{Gal}(L'_x/L_x)$-regular.*

2. *The only generic, irreducible, admissible representation π of the group*
   $GL_{qg}(L_x)$ *such that*

   - $\pi \otimes (\eta_x \circ \det) \cong \pi$
   - *and*

   $$\mathrm{Res}_{L'_x}^{L_x}(\pi) \cong \text{n-Ind}_{Q(L'_x)}^{GL_{qg}(L'_x)}(\Pi'_x \times \cdots \times (\Pi'_x \circ \tau^{q-1}))$$

   (where $Q \subset GL_{qg}$ is the parabolic subgroup defined above)

   is $\mathrm{Ind}_{L'_x}^{L_x}(\Pi'_x)$.

3. $\mathrm{Ind}_{L'}^{L}(\Pi')_x \cong \mathrm{Ind}_{L'_x}^{L_x}(\Pi'_x)$.

Proof: Note that the first part follows from lemma 2.3 of [HH]. Also note that the third part follows from the second part and the definition of $\mathrm{Ind}_{L'}^{L}$ (see section 6 of chapter 3 of [AC]). Thus it remains to prove the second part.

We can write

$$\Pi'_x \cong \left(\underset{i \in I'}{\boxplus} \mathrm{Sp}_{s'_i}(\pi'_i) \right) \boxplus \left(\underset{i \in J'}{\boxplus} \mathrm{Sp}_{s'_i}(\pi'_i) \right)$$

where $\pi'_i \cong \pi'_i \circ \tau$ if $i \in I'$, but not if $i \in J'$. For $i \in I'$ choose an irreducible admissible representation $\widetilde{\pi}_i$ such that $\mathrm{Res}_{L'_x}^{L_x}\widetilde{\pi}_i \cong \pi'_i$. Then

$$\text{n-Ind}_{Q(L'_x)}^{GL_{qg}(L'_x)}(\Pi'_x \times \cdots \times (\Pi'_x \circ \tau^{q-1}))$$

$$\cong \left(\underset{i \in I'}{\boxplus} \mathrm{Sp}_{s'_i}(\pi'_i)^{\boxplus q} \right) \boxplus \left(\underset{i \in J'}{\boxplus} \underset{j=0}{\overset{q-1}{\boxplus}} \mathrm{Sp}_{s'_i}(\pi'_i \circ \tau^j) \right)$$

(use the fact that Π'_x is $\mathrm{Gal}(L'_x/L_x)$-regular generic). Moreover

$$\mathrm{Ind}_{L'_x}^{L_x}(\Pi'_x) \cong \left(\underset{i \in I'}{\boxplus} \underset{j=0}{\overset{q-1}{\boxplus}} \mathrm{Sp}_{s'_i}\left(\widetilde{\pi}_i \otimes (\eta_x^j \circ \det)\right) \right) \boxplus \left(\underset{i \in J'}{\boxplus} \mathrm{Sp}_{s'_i}\left(\mathrm{Ind}_{L'_x}^{L_x}(\pi'_i)\right) \right)$$

(see [HH] and assertion 2.6 (a) of [BHK]), and hence

$$\mathrm{Res}_{L'_x}^{L_x}\mathrm{Ind}_{L'_x}^{L_x}(\Pi'_x) \cong \left(\underset{i \in I'}{\boxplus} \mathrm{Sp}_{s'_i}(\pi'_i)^{\boxplus q} \right) \boxplus \left(\underset{i \in J'}{\boxplus} \underset{j=0}{\overset{q-1}{\boxplus}} \mathrm{Sp}_{s'_i}(\pi'_i \circ \tau^j) \right)$$

(see [AC] and assertion 2.6 (b) of [BHK]). In particular

$$\mathrm{Res}_{L'_x}^{L_x}\mathrm{Ind}_{L'_x}^{L_x}(\Pi'_x) \cong \text{n-Ind}_{Q(L'_x)}^{GL_{qg}(L'_x)}(\Pi'_x \times \cdots \times (\Pi'_x \circ \tau^{q-1})).$$

Now suppose that π is a generic, irreducible, admissible representation of $GL_{qg}(L_x)$ such that

- $\pi \otimes (\eta_x \circ \det) \cong \pi$

- and

$$\mathrm{Res}^{L_x}_{L'_x}(\pi) \cong \text{n-Ind}^{GL_{qg}(L'_x)}_{Q(L'_x)}(\Pi'_x \times \cdots \times (\Pi'_x \circ \tau^{q-1})).$$

We may write

$$\pi \cong \left(\underset{i \in I}{\boxplus} \, \overset{q-1}{\underset{j=0}{\boxplus}} \, \mathrm{Sp}_{\,s_i}(\pi_i \otimes (\eta_x^j \circ \det)) \right) \boxplus \left(\underset{i \in J}{\boxplus} \, \mathrm{Sp}_{\,s_i}(\pi_i) \right)$$

where $\pi_i \cong \pi_i \otimes (\eta_x \circ \det)$ if $i \in J$, but not if $i \in I$. If $i \in J$ then we can write $\pi_i = \mathrm{Ind}^{L_x}_{L'_x}\widetilde{\pi}'_i$, where $\widetilde{\pi}'_i$ is an irreducible supercuspidal representation. Then

$$\mathrm{Res}^{L_x}_{L'_x}\pi \cong \left(\underset{i \in I}{\boxplus} \, \mathrm{Sp}_{\,s_i}(\mathrm{Res}^{L_x}_{L'_x}\pi_i)^{\boxplus q} \right) \boxplus \left(\underset{i \in J}{\boxplus} \, \overset{q-1}{\underset{j=0}{\boxplus}} \, \mathrm{Sp}_{\,s_i}(\widetilde{\pi}'_i \circ \tau^j) \right)$$

(see [AC] and assertion 2.6 (b) of [BHK]). Note that for $i \in I$ $\mathrm{Res}^{L_x}_{L'_x}\pi_i \cong (\mathrm{Res}^{L_x}_{L'_x}\pi_i) \circ \tau$, while for $i \in J$ we have $\widetilde{\pi}'_i \not\cong \widetilde{\pi}'_i \circ \tau$. Thus we may identify I with I' and J with J' so that

- for $i \in I$ we have $\mathrm{Res}^{L_x}_{L'_x}\pi_i \cong \pi'_i$

- and for $i \in J$ we have $\widetilde{\pi}'_i \cong \pi'_i \circ \tau^{j(i)}$ for some $j(i)$.

Then

- for $i \in I$ we have $\pi_i \cong \pi'_i \otimes (\eta_x \circ \det)^{j(i)}$, for some $j(i)$

- and for $i \in J$ we have $\pi_i \cong \widetilde{\pi}_i$.

Thus

$$\pi \cong \mathrm{Ind}^{L_x}_{L'_x}\Pi'_x,$$

as desired. $\square$

I.5   Vanishing cycles and formal schemes

In this section let K be a p-adic field and l be a prime integer different from p. We will also let $\mathcal{O}$ denote the ring of integers of $\widehat{K}^{\mathrm{nr}}$ and k ($\cong \mathbb{F}^{\mathrm{ac}}_p$) its residue field.

Suppose that $X/\mathcal{O}$ is a proper scheme of finite type. We will let X_η denote the generic fibre $X \times_\mathcal{O} \widehat{K}^{\mathrm{nr}}$. Suppose also that $\mathcal{L}/X_\eta$ is a constructible

$\mathbb{Q}_l^{\mathrm{ac}}$-sheaf. Then one can form the vanishing cycle sheaves $R^i \Psi_\eta(\mathcal{L})$, which are constructible $\mathbb{Q}_l^{\mathrm{ac}}$-sheaves on $X_s = X \times_{\mathcal{O}} k$ (combine theorem 3.2 of section "théorème de finitude" in [SGA4$\frac{1}{2}$] with proposition 5.3.1 of exposé V in [SGA5]). If $\sigma \in W_K$, if

$$
\begin{array}{ccc}
X & \xrightarrow{f} & Y \\
\downarrow & & \downarrow \\
\operatorname{Spec} \mathcal{O} & \xrightarrow{\sigma^*} & \operatorname{Spec} \mathcal{O}
\end{array}
$$

and if $g : f^* \mathcal{M} \to \mathcal{L}$ then there is a natural map

$$(\sigma, f^*, g)_* : f^* R^i \Psi_\eta(\mathcal{M}) \longrightarrow R^i \Psi_\eta(\mathcal{L}).$$

There is also a spectral sequence

$$H^i(X \times_{\mathcal{O}} k, R^j \Psi_\eta(\mathcal{L})) \Rightarrow H^{i+j}(X \times_{\mathcal{O}} (\widehat{K}^{\mathrm{nr}})^{\mathrm{ac}}, \mathcal{L}),$$

in which the action of $(f, (\sigma, f^*, g)_*)$ on the left corresponds to that of $(f \times \sigma^*, g)$ on the right.

Lemma I.5.1 *If x is a closed point of X then $R^i \Psi_\eta(\mathbb{Q}_l^{\mathrm{ac}})_x$ is a finite dimensional $\mathbb{Q}_l^{\mathrm{ac}}$-vector space with a continuous action of I_K.*

This follows at once from the constructibility of $R^i \Psi_\eta(\mathbb{Q}_l^{\mathrm{ac}})$.

Lemma I.5.2 *If $\mathcal{L}/X$ is a lisse $\mathbb{Q}_l^{\mathrm{ac}}$-sheaf then*

$$R^i \Psi_\eta(\mathcal{L}) \cong (R^i \Psi_\eta(\mathbb{Q}_l^{\mathrm{ac}})) \otimes \mathcal{L}_s,$$

where $\mathcal{L}_s$ is the restriction of $\mathcal{L}$ to the special fibre of X_s.

To prove this one can reduce to the case of a locally free $\mathbb{Z}/l^s\mathbb{Z}$-sheaf and then work etale locally so that $\mathcal{L}$ becomes free.

Lemma I.5.3 *Suppose that $X/\mathcal{O}$ is smooth. Then for $i > 0$ we have*

$$R^i \Psi_\eta(\mathbb{Q}_l^{\mathrm{ac}}) = (0).$$

Moreover the canonical map

$$\mathbb{Q}_l^{\mathrm{ac}} \to R^0 \Psi_\eta(\mathbb{Q}_l^{\mathrm{ac}})$$

is an isomorphism.

To prove this one can reduce to the case of coefficients $\mathbb{Z}/l^s\mathbb{Z}$ and then appeal to corollary 2.4 of exposé I of [SGA7].

Lemma I.5.4 *Suppose that Y/X is a finite cover with an action of a finite group G. Suppose that the generic fibres are a Galois etale cover with group G. Suppose that x is a closed point of X which is totally ramified in Y, and let y be its preimage in Y. Then*

$$R^i \Psi_\eta (\mathbb{Q}_l^{\mathrm{ac}})_x \xrightarrow{\sim} R^i \Psi_\eta (\mathbb{Q}_l^{\mathrm{ac}})_y^G.$$

To prove this one can use proposition 2.3 of exposé I of [SGA7] and the Hochschild-Serre spectral sequence to get a spectral sequence

$$H^i(G, R^j \Psi_\eta (\mathbb{Z}/l^s\mathbb{Z})_y) \Rightarrow R^{i+j} \Psi_\eta (\mathbb{Z}/l^s\mathbb{Z})_x.$$

Then taking a limit over s and tensoring with $\mathbb{Q}_l^{\mathrm{ac}}$ gives the result.

In [Berk3], a vanishing cycles functor is constructed for a certain class of formal schemes over $\mathcal{O}$. The comparison theorem 3.1 of [Berk3] implies that if $\mathcal{X}$ is a formal scheme over $\mathcal{O}$, which is isomorphic to the formal completion of a proper scheme of finite type $X/\mathcal{O}$ along a subscheme Y of X_s, then there is a canonical isomorphism of sheaves

$$R^i \Psi_\eta (\mathbb{Z}/l^m\mathbb{Z})_{\mathcal{X}_\eta} \cong R^i \Psi_\eta (\mathbb{Z}/l^m\mathbb{Z})_{X_\eta}|_Y.$$

It follows that the projective system of constructible sheaves

$$\varprojlim_m R^i \Psi_\eta (\mathbb{Z}/l^m\mathbb{Z})_{\mathcal{X}_\eta}$$

forms a $\mathbb{Z}_l$-sheaf and, hence, there is a well defined $\mathbb{Q}_l^{\mathrm{ac}}$-sheaf $R^i \Psi_\eta (\mathbb{Q}_l^{\mathrm{ac}})_{\mathcal{X}_\eta}$ on Y. It is canonically isomorphic to $R^i \Psi_\eta (\mathbb{Q}_l^{\mathrm{ac}})_X|_Y$. Now suppose only that $\mathcal{X}$ is a special formal scheme which has a Zariski open cover $\{\mathcal{U}_i\}$ such that each $\mathcal{U}_i$ is etale over the formal completion of a proper scheme of finite type $X_i/\mathcal{O}$ along a subscheme Y_i of $X_{i,s}$. Again the projective system of constructible sheaves $R^i \Psi_\eta (\mathbb{Z}/l^m\mathbb{Z})_{\mathcal{X}_\eta}$ form a $\mathbb{Z}_l$-sheaf and, therefore, there is a well defined $\mathbb{Q}_l^{\mathrm{ac}}$-sheaf $R^i \Psi_\eta (\mathbb{Q}_l^{\mathrm{ac}})_{\mathcal{X}_\eta}$ on $\mathcal{X}^{\mathrm{red}}$. (Use corollary 2.3 of [Berk3].) If $\mathcal{Y}$ is another such formal scheme, if $\sigma \in W_K$ and if

$$
\begin{array}{ccc}
\mathcal{X} & \xrightarrow{f} & \mathcal{Y} \\
\downarrow & & \downarrow \\
\mathrm{Spf}\,\mathcal{O} & \xrightarrow{\sigma^*} & \mathrm{Spf}\,\mathcal{O},
\end{array}
$$

then there is a natural map

$$(f^*, \sigma)_* : (f^{\mathrm{red}})^* R^i \Psi_\eta (\mathbb{Q}_l^{\mathrm{ac}})_{\mathcal{Y}_\eta} \longrightarrow R^i \Psi_\eta (\mathbb{Q}_l^{\mathrm{ac}})_{\mathcal{X}_\eta}.$$

The action of I_K on $R^i \Psi_\eta (\mathbb{Q}_l^{\mathrm{ac}})_{\mathcal{X}_\eta}$ is continuous (by comparison with the algebraic theory). Also the continuity theorem (corollary 4.5) from [Berk3] implies that there exists an ideal of definition of $\mathcal{X}$ such that any automorphism of $\mathcal{X}$, trivial modulo this ideal, acts trivially on the $\mathbb{Q}_l^{\mathrm{ac}}$-sheaves $R^i \Psi_\eta (\mathbb{Q}_l^{\mathrm{ac}})_{\mathcal{X}_\eta}$.

Lemma I.5.5 *Suppose that* $\mathcal{X} = \mathrm{Spf}\,\mathcal{O}[[T_1, \ldots, T_s]]$. *Then for* $i > 0$

$$R^i \Psi_\eta(\mathbb{Q}_l^{\mathrm{ac}})_{\mathcal{X}_{\bar\eta}} = (0),$$

while the canonical map

$$\mathbb{Q}_l^{\mathrm{ac}} \longrightarrow R^0 \Psi_\eta(\mathbb{Q}_l^{\mathrm{ac}})_{\mathcal{X}_{\bar\eta}}$$

is an isomorphism. In particular, if $\sigma \in W_K$ *and if* f *is a continuous automorphism of* $\mathcal{X}$ *such that*

$$
\begin{array}{ccc}
\mathcal{X} & \xrightarrow{f} & \mathcal{X} \\
\downarrow & & \downarrow \\
\mathrm{Spf}\,\mathcal{O} & \xrightarrow{\sigma^*} & \mathrm{Spf}\,\mathcal{O},
\end{array}
$$

commutes, then $(f^*, \sigma)_*$ *is the identity map on*

$$R^0 \Psi_\eta(\mathbb{Q}_l^{\mathrm{ac}})_{\mathcal{X}_{\bar\eta}} \cong \mathbb{Q}_l^{\mathrm{ac}}.$$

This can be proved by using theorem 3.1 of [Berk3] to reduce to lemma I.5.3.

Lemma I.5.6 *Suppose that* R *is a complete noetherian local* $\mathcal{O}$-*algebra. The natural map*

$$R^i \Psi_\eta(\mathbb{Q}_l^{\mathrm{ac}})_{\mathrm{Spf}\,R} \longrightarrow R^i \Psi_\eta(\mathbb{Q}_l^{\mathrm{ac}})_{\mathrm{Spf}\,R[[T_1, \ldots, T_s]]}$$

is an isomorphism.

This can be proved by combining the last lemma with theorem 7.8.1 of [Berk1].

The next lemma is well known to the experts. We are grateful to de Jong for explaining the proof to us. The case that X has semistable reduction follows from the weight spectral sequence of Rapoport and Zink [RZ1] (see also equations (3.8.2) and (3.8.3) in [I2]). The general case follows from this by de Jong's theory of alterations (as in the second paragraph of the proof of proposition 6.3.2 of [Bert]).

Lemma I.5.7 *Suppose that* X/K *is a smooth proper variety, that* $\sigma \in W_K$ *and that* α *is an eigenvalue of* σ *on* $H^i(X \times \mathrm{Spec}\,K^{\mathrm{ac}}, \mathbb{Q}_l^{\mathrm{ac}})$. *Then* $\alpha \in \mathbb{Q}^{\mathrm{ac}}$ *and for any embedding* $\mathbb{Q}^{\mathrm{ac}} \hookrightarrow \mathbb{C}$ *we have*

$$|\alpha|^2 \in (\#k(v_K))^{\mathbb{Z}}.$$

The next lemma is also a standard result, which follows easily from, for instance, proposition 1.1 of section I of [Arti]).

Lemma I.5.8 *Suppose that $\mathcal{Z} \to \mathcal{Y}$ is a finite etale morphism of locally noetherian formal schemes. Suppose that $\mathcal{X}$ is also a locally noetherian formal scheme and that we have a commutative diagram*

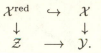

Then there is a unique diagonal morphism $\mathcal{X} \to \mathcal{Z}$ making the diagram still commute.

We end this section with a lemma of Berkovich's (see the lemma in his appendix to this book).

Lemma I.5.9 (Berkovich) *Let $\mathcal{X} = \operatorname{Spf} A$ and $\mathcal{Y} = \operatorname{Spf} B$ be special affine formal schemes. Let $J \subset B$ be the maximal ideal of definition for $\mathcal{Y}$, and set $\mathcal{Y}\{N\} = \operatorname{Spec}(B/J^{N+1})$. Assume that $\mathcal{X}$ is isomorphic to the formal completion of an affine scheme of finite type $X = \operatorname{Spec} A'/\operatorname{Spec} \mathcal{O}$ along a closed subscheme of its special fibre. Let $I' \subset A'$ be the maximal ideal of definition of this subscheme. Furthermore assume that we are given projective systems $\{\mathcal{X}_n\}_{n \geq 0}$ and $\{\mathcal{Y}_n\}_{n \geq 0}$ of finite etale coverings of $\mathcal{X}$ and $\mathcal{Y}$ respectively. Let $\mathcal{Y}_n\{N\}$ denote the pullback of $\mathcal{Y}\{N\}$ to $\mathcal{Y}_n$. Suppose finally that we are given compatible morphisms*

$$\varphi_n : \mathcal{Y}_n\{n\} \longrightarrow \mathcal{X}_n.$$

Then given any positive integer N for $m >> 0$ (depending on N) we can find a morphism

$$\varphi : \mathcal{Y}_m \longrightarrow \mathcal{X}_m$$

such that

$$\varphi|_{\mathcal{Y}_m\{N\}} = \varphi_m|_{\mathcal{Y}_m\{N\}}.$$

Proof: As in the proof of the lemma in the appendix we see that we have $\mathcal{X}_n = \operatorname{Spf} A_n$ and $\mathcal{Y}_n = \operatorname{Spf} B_n$, where A_n/A and B_n/B are finite etale. We set $B_\infty = \lim_{\to} B_n$ and $J_\infty = JB_\infty$. Again as in the proof of the lemma of the appendix we see that (B_∞, J_∞) is a Henselian pair. The map φ_n induces a homomorphism

$$\varphi'_n : A' \longrightarrow A_n \longrightarrow B_n/J^{n+1}B_n \longrightarrow B_\infty/J_\infty^{n+1}.$$

By corollary 1 on page 567 and remark 2 on page 587 of [El] we see that there exists an integer t such that for all $n >> 0$ there exists a homomorphism

$$\widetilde{\varphi}_n : A' \longrightarrow B_\infty$$

with

$$\widetilde{\varphi}_n \equiv \varphi'_n \bmod J_\infty^{n+1-t}.$$

Fix such an n, which we also suppose greater than $N + t$.

Note that as $A'/\mathcal{O}$ is finitely generated $\widetilde{\varphi}_n$ is in fact valued in some B_m for $m >> 0$. As $J_\infty^a \cap B_m = J^a B_m$ (use (4.C) of [Ma] and the faithful flatness of $B_{m'}$ over B_m) we see that for $m >> 0$

$$\widetilde{\varphi}_n \equiv \varphi_m \bmod J^{N+1} B_m.$$

Thus $\widetilde{\varphi}_n(I') \subset JB_m$ and so we may extend $\widetilde{\varphi}_n$ to a continuous homomorphism $A \to B_m$ such that

$$
\begin{array}{ccc}
A & \overset{\widetilde{\varphi}_n}{\longrightarrow} & B_m \\
\downarrow & & \downarrow \\
A_m & \overset{\varphi_m}{\longrightarrow} & B_m/J^{N+1} B_m
\end{array}
$$

commutes. By lemma I.5.8 we see that we get a morphism

$$\varphi : A_m \longrightarrow B_m$$

with $\varphi \equiv \varphi_m \bmod J^{N+1} B_m$, as desired. $\square$

I.6 Involutions and unitary groups

Suppose that L is a field of characteristic 0 and that C is a finite dimensional semi-simple L-algebra. We will let C^{op} denote the opposite algebra of C, and we will let $\det_{C/L}$ (resp. $\operatorname{tr}_{C/L}$) denote the composite of the reduced norm (resp. reduced trace) from C to its centre $Z(C)$ with the usual norm (resp. trace) from $Z(C)$ to L. We recall the following lemma.

Lemma I.6.1 *Keep the above notation and assumptions. Let $S/\operatorname{Spec} L$ be a scheme and $\mathcal{L}_1$ and $\mathcal{L}_2$ be locally free coherent $\mathcal{O}_S$-modules with an action of C. Then the following are equivalent.*

1. *Locally on S we have an isomorphism of $\mathcal{O}_S \otimes_L C$-modules $\mathcal{L}_1 \cong \mathcal{L}_2$.*

2. *Locally on $S \times \operatorname{Spec} L^{\mathrm{ac}}$ we have an isomorphism of $\mathcal{O}_{S \times \operatorname{Spec} L^{\mathrm{ac}}} \otimes_L C$-modules $\mathcal{L}_1 \cong \mathcal{L}_2$.*

3. $\operatorname{tr}_{/\mathcal{O}_S}(b|_{\mathcal{L}_1}) = \operatorname{tr}_{/\mathcal{O}_S}(b|_{\mathcal{L}_2})$ *for all $b \in C$.*

4. *Locally on S we have an isomorphism of $\mathcal{O}_S \otimes_L Z(C)$-modules $\mathcal{L}_1 \cong \mathcal{L}_2$.*

5. *For all closed points $s \in S$ there is an isomorphism of $k(s) \otimes_L Z(C)$-modules $\mathcal{L}_1 \otimes k(s) \cong \mathcal{L}_2 \otimes k(s)$.*

Proof: The equivalence of the first two conditions is proved by a standard argument using the facts that C-homomorphisms between $\mathcal{L}_1$ and $\mathcal{L}_2$ form a coherent sheaf $\mathcal{H}/S$ and that locally the condition for such a C-homomorphism to be an isomorphism is given by a certain polynomial function $\det : \mathcal{H} \to \mathcal{O}_S$ having invertible image.

Once the equivalence of the first two conditions is established we may reduce the rest of the lemma to the case that L is algebraically closed. In this case C has the form

$$\bigoplus_{i=1}^{s} M_{n_i}(L),$$

and locally each $\mathcal{L}_i$ has the form

$$\bigoplus_{i=1}^{s} \mathcal{O}_S^{m_i} \otimes_L L^{n_i}$$

with the obvious action of $\mathcal{O}_S \otimes_L C$. The local isomorphism class is determined by the non-negative integers $(m_1, \ldots, m_s)$. The lemma now follows easily. $\square$

Suppose that $*$ is an involution on C over L (i.e. $* : C \to C$ takes L to itself and satisfies $(x + y)^* = x^* + y^*$, $(xy)^* = y^* x^*$ and $*^2 = 1$) such that $*|_L \neq 1$. Set $L^+ = L^{*=1}$. We will call two such involutions $*$ and $*'$ equivalent if there exists $\gamma \in C^{\times}$ such that

$$\gamma x^{*'} \gamma^{-1} = (\gamma x \gamma^{-1})^*$$

for all $x \in C$.

Suppose that W is a C-module which is finite dimensional over L. We will call a non-degenerate L^+-alternating pairing

$$\langle \ , \ \rangle : W \times W \longrightarrow L^+$$

$*$-Hermitian if

$$\langle \gamma x, y \rangle = \langle x, \gamma^* y \rangle$$

for all $x, y \in W$ and all $\gamma \in C$. We will call two such pairings $\langle \ , \ \rangle$ and $\langle \ , \ \rangle_1$ equivalent if we can find $\delta \in \operatorname{End}_C(W)^{\times}$ and $\mu \in (L^+)^{\times}$ such that

$$\langle x, y \rangle_1 = \mu \langle \delta x, \delta y \rangle$$

for all $x, y \in W$. We may classify equivalence classes of non-degenerate L^+-alternating $*$-Hermitian pairings $W \times W \to L^+$ as follows.

Fix one such pairing $\langle \; , \; \rangle_0$ (if one exists). We will denote by $*_0$ the involution on $\operatorname{End}_{L^+}(W)$ such that

$$\langle \delta x, y \rangle_0 = \langle x, \delta^{*_0} y \rangle_0$$

for all $x, y \in W$ and all $\delta \in \operatorname{End}_{L^+}(W)$. Note that $*_0|_C = *$ and that $*_0$ preserves $\operatorname{End}_C(W)$. Define a reductive algebraic group H/L^+ by setting, for any L^+-algebra R, $H(R)$ equal to the set of $\delta \in (\operatorname{End}_C(W) \otimes_{L^+} R)^\times$ such that

$$\delta \delta^{*_0} \in R^\times.$$

Note that if R is an L-algebra then

$$\operatorname{End}_C(W) \otimes_{L^+} R \cong (\operatorname{End}_C(W) \otimes_{L^+} L) \otimes_L R \cong$$
$$\operatorname{End}_C(W) \otimes_L R \oplus \operatorname{End}_C(W) \otimes_{L, *|_L} R$$

and $*_0 \; (= *_0 \otimes 1)$ interchanges the two factors. Then $H(R)$ consists of the set of pairs $(x, \lambda x^{-*_0})$, where $x \in (\operatorname{End}_C(W) \otimes_L R)^\times$ and $\lambda \in R^\times$, i.e.

$$H(R) \cong (\operatorname{End}_C(W) \otimes_L R)^\times \times R^\times.$$

By Hilbert 90 we see that $H^1(L, H) = (0)$ and so

$$H^1(\operatorname{Gal}(L/L^+), H(L)) \xrightarrow{\sim} H^1(L^+, H).$$

We can describe $H^1(\operatorname{Gal}(L/L^+), H(L))$ as the set of equivalence classes of pairs $(\lambda, \gamma) \in L^\times \times \operatorname{End}_C(W)^\times$ such that $N_{L/L^+} \lambda = 1$ and

$$\gamma^{*_0} = \lambda^* \gamma.$$

We consider (λ, γ) and (λ', γ') equivalent if there exists $\mu \in L^\times$ and $\delta \in \operatorname{End}_C(W)^\times$ such that

$$(\lambda', \gamma') = (\lambda \mu / \mu^*, \mu^{-*} \delta \gamma \delta^{*_0}).$$

Applying Hilbert 90 to L/L^+ we see that $H^1(\operatorname{Gal}(L/L^+), H(L))$ is also in bijection with equivalence classes of

$$\gamma \in \operatorname{End}_C(W)^\times \cap \operatorname{End}_C(W)^{*_0 = 1},$$

where we consider γ and γ' equivalent if there exists $\mu \in (L^+)^\times$ and $\delta \in \operatorname{End}_C(W)^\times$ such that

$$\gamma' = \mu \delta \gamma \delta^{*_0}.$$

Any non-degenerate, L^+-alternating, $*$-Hermitian form $W \times W \to L^+$ is of the form

$$\langle x, y \rangle_\delta = \langle \delta x, \delta y \rangle_0$$

for some $\delta \in \operatorname{End}_{L^+}(W)^\times$ with

$$\delta^{*_0} \delta \in \operatorname{End}_C(W)^{*_0 = 1}.$$

Moreover $\langle \ , \ \rangle_\delta$ and $\langle \ , \ \rangle_{\delta'}$ are equivalent if and only if there exists $\gamma \in \operatorname{End}_C(W)^\times$ and $\lambda \in (L^+)^\times$ such that

$$(\delta')^{*_0} \delta' = \lambda \gamma^{*_0} (\delta^{*_0} \delta) \gamma.$$

Note that any element $\gamma \in \operatorname{End}_C(W)^{*_0 = 1}$ can be written $\delta^{*_0} \delta$ for some $\delta \in \operatorname{End}_{L^+}(W)$. (If we choose an L^+-basis of W we get an isomorphism of $\operatorname{End}_{L^+}(W)$ with $M_{2N}(L^+)$ for some integer N. Moreover $\langle \ , \ \rangle$ is represented by an anti-symmetric matrix $J \in GL_{2N}(L^+)$ and if $\delta \in M_{2N}(L^+)$ then $\delta^{*_0} = J^{-1} \delta^t J$ (where t denotes the transpose). Thus if $\delta^{*_0} = \delta$ we see that $J\delta$ is antisymmetric and hence that

$$J\delta = (\delta')^t J \delta'$$

for some $\delta' \in GL_{2N}(L^+)$. Thus $\delta = (\delta')^{*_0} \delta'$.) We deduce that the correspondence which associates $\langle \ , \ \rangle_\delta$ with $\delta\delta^*$ sets up a bijection between

- the set of equivalence classes of non-degenerate L^+-alternating $*$-Hermitian forms on W

- and $H^1(L^+, H)$.

Suppose now that L is a number field and that two classes ψ_1, ψ_2 in $H^1(L^+, H)$ correspond to non-degenerate L^+-alternating $*$-Hermitian forms $\langle \ , \ \rangle_1$ and $\langle \ , \ \rangle_2$. Then the same arguments show that $\langle \ , \ \rangle_1$ and $\langle \ , \ \rangle_2$ become equivalent over $\mathbb{A}_{L^+}^S$ (with the same definition of equivalence as over a field) if and only if ψ_1 and ψ_2 have the same image in $H^1(L^+, H(\overline{\mathbb{A}}_{L^+}^S))$.

We will call an L^+-bilinear pairing

$$(\ , \ ) : W \times W \longrightarrow L$$

$*$-symmetric if $(\ , \ )$ is L-linear in the first variable and satisfies

$$(y, x) = (x, y)^*$$

for all $x, y \in W$. We will call this pairing non-degenerate if $(x, y) = 0$ for all $y \in W$ implies that $x = 0$. We will call it $*$-Hermitian if

$$(\gamma x, y) = (x, \gamma^* y)$$

for all $x, y \in W$ and all $\gamma \in C$. We call two $*$-Hermitian $*$-symmetric pairings $(\ ,\ )_1$ and $(\ ,\ )_2$ equivalent if there exists $\delta \in \operatorname{End}_C(W)$ and $\lambda \in (L^+)^\times$ such that

$$(x, y)_2 = \lambda(\delta x, \delta y)_1$$

for all $x, y \in W$.

Suppose that $L = L^+(\sqrt{a})$ where $\sqrt{a}^2 = a \in L^+$. Then there is a bijection between equivalence classes of non-degenerate, $*$-Hermitian, $*$-symmetric pairings $W \times W \to L$ and equivalence classes of non-degenerate, $*$-Hermitian, L^+-alternating pairings $W \times W \to L^+$ given as follows. If $\langle\ ,\ \rangle$ is a non-degenerate, $*$-Hermitian, L^+-alternating pairing $W \times W \to L^+$ then associate to the equivalence class of $\langle\ ,\ \rangle$ the equivalence class of the non-degenerate, $*$-Hermitian, $*$-symmetric pairing given by

$$(x, y) = \langle \sqrt{a}x, y \rangle + \sqrt{a}\langle x, y \rangle.$$

Conversely if $(\ ,\ )$ is a non-degenerate, $*$-Hermitian, $*$-symmetric pairing $W \times W \to L^+$ then associate to the equivalence class of $(\ ,\ )$ the equivalence class of the non-degenerate, $*$-Hermitian, L^+-alternating pairing given by

$$\langle x, y \rangle = \operatorname{tr}_{L/L^+} \sqrt{a}(x, y).$$

This bijection is independent of the choice of $\sqrt{a} \in L$.

Suppose that $L = \mathbb{C}$, $L^+ = \mathbb{R}$, $C = \mathbb{C}^I$, $* = c^I$ and $W = (\mathbb{C}^n)^I$. Consider I-tuples $((a_i, b_i))_{i \in I}$ of pairs of non-negative integers (a_i, b_i) with $a_i + b_i = n$. We call two such I-tuples, $((a_i, b_i))_{i \in I}$ and $((a_i', b_i'))_{i \in I}$, equivalent either if they are equal or if for all $i \in I$ we have $(a_i', b_i') = (b_i, a_i)$. To such an I-tuple, $((a_i, b_i))_{i \in I}$, we associate the equivalence classe of the non-degenerate, $*$-Hermitian, $*$-symmetric pairing $W \times W \to \mathbb{C}$ given by

$$(\vec{x}_i)_{i \in I} \times (\vec{y}_i)_{i \in I} \longmapsto \sum_{i \in I} \vec{x}_i^t J_i \vec{y}_i^c,$$

where J_i is the diagonal $n \times n$-matrix with 1 on the diagonal a_i times followed by -1 on the diagonal $b_i = n - a_i$ times. This establishes a bijection between equivalence classes of non-degenerate, $*$-Hermitian, $*$-symmetric pairings $W \times W \to \mathbb{C}$ and equivalence classes of such I-tuples $((a_i, b_i))_{i \in I}$. We deduce that equivalence classes of non-degenerate, $*$-Hermitian, $\mathbb{R}$-alternating pairings $W \times W \to \mathbb{R}$ are also parametrised by equivalence classes of such I-tuples.

Now suppose that $L = \mathbb{C}$, $L^+ = \mathbb{R}$, $C = M_n(\mathbb{C})^I$, $(\gamma_i)^* = (\gamma_i^{c,t})$ and $W = C$. The equivalence classes of non-degenerate, $*$-Hermitian, $\mathbb{R}$-alternating pairings $W \times W \to \mathbb{R}$ are still parametrised by equivalence classes of such I-tuples $((a_i, b_i))_{i \in I}$. To see this one can note that to give a non-degenerate, $*$-Hermitian, $\mathbb{R}$-alternating form $W \times W \to \mathbb{R}$ is the same as giving a non-degenerate, $*$-Hermitian for $\mathbb{C}^I$, $\mathbb{R}$-alternating form $\varepsilon W \times \varepsilon W \to \mathbb{R}$. The equivalence sends $\langle \ , \ \rangle$ to the I-tuple parametrising

$$\langle \ , \ \rangle|_{\varepsilon W \times \varepsilon W}.$$

Now suppose that L is an imaginary quadratic field, that M is a totally real field, that C is a central simple LM-algebra with $\dim_{LM} C = n^2$, that $*$ is an involution of the second kind on C (i.e. $*|_{LM} = c$), that $*$ is positive (i.e. $\operatorname{tr}_{C/\mathbb{Q}}(\gamma\gamma^*) > 0$ for all non-zero $\gamma \in C$) and that $W = C$. Then the 5-tuple $(L_\infty, L_\infty^+, C_\infty, *, W_\infty)$ is isomorphic to the 5-tuple

$$(\mathbb{C}, \mathbb{R}, M_n(\mathbb{C})^{\operatorname{Hom}(M,\mathbb{R})}, (\gamma_\tau) \mapsto (\gamma_\tau^{c,t}), M_n(\mathbb{C})^{\operatorname{Hom}(M,\mathbb{R})}).$$

Hence equivalence classes of non-degenerate, $*$-Hermitian, $\mathbb{R}$-alternating bilinear forms $W_\infty \times W_\infty \to \mathbb{R}$ are parametrised by equivalence classes of $\operatorname{Hom}(M, \mathbb{R})$-tuples as above.

Again suppose that L is an imaginary quadratic field. Suppose that C and C' are semi-simple L-algebras with involutions $*$ and $*'$ with $*|_L = c = *'|_L$. Define algebraic groups H and H' over $\mathbb{Q}$ by setting

$$H(R) = \{g \in C \otimes_{\mathbb{Q}} R : gg^* \in R^\times\}$$

and

$$H'(R) = \{g \in C' \otimes_{\mathbb{Q}} R : gg^{*'} \in R^\times\}$$

for any $\mathbb{Q}$-algebra R. Let S be a finite set of places of $\mathbb{Q}$ and

$$i : C \otimes_{\mathbb{Q}} \mathbb{A}^S \xrightarrow{\sim} C' \otimes_{\mathbb{Q}} \mathbb{A}^S$$

an isomorphism of algebras taking $*$ to $*'$. Then i induces an isomorphism

$$i : H \times \mathbb{A}^S \xrightarrow{\sim} H \times \mathbb{A}^S.$$

Lemma I.6.2 *Keep the above notation and assumptions. Then we can find an isomorphism*

$$\psi : H \times \mathbb{Q}^{\operatorname{ac}} \xrightarrow{\sim} H' \times \mathbb{Q}^{\operatorname{ac}}$$

such that

- *for all $\sigma \in \operatorname{Gal}(\mathbb{Q}^{\operatorname{ac}}/\mathbb{Q})$ the automorphism $\psi^{-1}\sigma(\psi)$ of $H \times \mathbb{Q}^{\operatorname{ac}}$ is inner, and*

- *the automorphism $\psi^{-1}i$ of $H \times \overline{\mathbb{A}}^S$ is inner.*

Proof: From the existence of i we deduce the existence of an isomorphism

$$\psi : C \otimes_{\mathbb{Q}} \mathbb{Q}^{\mathrm{ac}} \xrightarrow{\sim} C' \otimes_{\mathbb{Q}} \mathbb{Q}^{\mathrm{ac}}.$$

Moreover we may adjust ψ so that it takes $*$ to $*'$. The result now follows on noting that any automorphism of $C \otimes_{\mathbb{Q}} \mathbb{Q}^{\mathrm{ac}}$ (resp. $C \otimes_{\mathbb{Q}} \overline{\mathbb{A}}^S$) which takes $*$ to itself is conjugation by an element of $H(\mathbb{Q}^{\mathrm{ac}})$ (resp. $H(\overline{\mathbb{A}}^S)$). $\square$

I.7   Notation and running assumptions

In this section we will establish notation and assumptions which will be used, often without comment, throughout the rest of the book.

We will let E denote an imaginary quadratic field in which p splits. We will let c denote complex conjugation in $\mathrm{Gal}\,(E/\mathbb{Q})$. *We will choose a prime u of E above p.* *We will also let $F^+/\mathbb{Q}$ denote a totally real field of degree d.* We will set $F = E.F^+$ so that F is a CM-field with maximal totally real subfield F^+. Let $w = w_1, w_2, \ldots, w_r$ denote the places of F above u and let $v = v_1, \ldots, v_r$ denote their restrictions to F^+. We will denote $[k(w_i) : \mathbb{F}_p]$ by f_i. *We will let B/F denote a division algebra of dimension n^2 such that*

- *F is the centre of B;*

- *the opposite algebra B^{op} is isomorphic to $B \otimes_{E,c} E$;*

- *B is split at w;*

- *at any place x of F which is not split over F^+, B_x is split;*

- *at any place x of F which is split over F^+ either B_x is split or B_x is a division algebra,*

- *if n is even then $1 + dn/2$ is congruent modulo 2 to the number of places of F^+ above which B is ramified.*

When no confusion seems likely to arise we may write det (resp. tr) for the reduced norm $\det_{B/F}$ (resp. reduced trace $\mathrm{tr}_{B/F}$). Define $n \in \mathbb{Z}_{>0}$ by $[B : F] = n^2$.

We may pick an involution of the second kind $*$ on B. (That we may chose such an involution follows from the second and fourth of the above assumptions on B. More precisely lemma 8.1 of [Sc] defines a homomorphism $\mathrm{Br}\,(F)^{\mathrm{op}=c} \to (F^+)^{\times}/N(F^{\times})$ and shows that B has an involution of the second kind if and only if $[B]$ is in the kernel of this homomorphism. But $[B]$ is in the kernel if and only if $[B_x]$ is in the kernel for all places x of F which are non split over F^+.) *We may and will further assume that*

* *is positive, i.e. for all nonzero* $x \in B$ *we have* $\mathrm{tr}_{B/\mathbb{Q}}(xx^*) > 0$. (To see that we may suppose that $*$ is positive one may argue as follows. The involutions of the second kind on B are exactly the maps of the form

$$x \longmapsto bx^*b^{-1},$$

where $b \in B^\times$ and $b^*b^{-1} \in F$. By Hilbert's theorem 90 we may alter any such b by an element of $F^\times$ so that $b^* = b$. Thus we may suppose that $b^* = b$. By lemma 2.8 of [Ko3] the set of invertible $b \in (B^{*=1} \otimes_{\mathbb{Q}} \mathbb{R})$ such that $x \mapsto bx^*b^{-1}$ is positive is a non-empty open set. Thus we can find an invertible $b \in B^{*=1}$ such that $x \mapsto bx^*b^{-1}$ is positive.)

We will let V denote the $B \otimes_F B^{\mathrm{op}}$ module B. We will be interested in alternating pairings $V \times V \to \mathbb{Q}$ (resp. $(V \otimes \mathbb{A}^\infty) \times (V \otimes \mathbb{A}^\infty) \to \mathbb{A}^\infty$) which are $*$-Hermitian for the action of B on V. Any such pairing is of the form

$$(x_1, x_2)_\beta = \mathrm{tr}_{B/\mathbb{Q}}(x_1\beta x_2^*),$$

for some $\beta \in B^{*=-1}$ (resp. $B^{*=-1} \otimes \mathbb{A}^\infty$). Define an involution of the second kind $\#_\beta$ on B (resp. $B \otimes \mathbb{A}^\infty$) by $x^{\#_\beta} = \beta x^* \beta^{-1}$. Then we have that

$$((b_1 \otimes b_2)x_1, x_2)_\beta = (x_1, (b_1^* \otimes b_2^{\#_\beta})x_2)_\beta$$

for all $x_1, x_2 \in V$ (resp. $V \otimes \mathbb{A}^\infty$), $b_1 \in B$ (resp. $B \otimes \mathbb{A}^\infty$) and $b_2 \in B^{\mathrm{op}}$ (resp $B^{\mathrm{op}} \otimes \mathbb{A}^\infty$). Also let $G_\beta/\mathbb{Q}$ (resp. $G_\beta/\mathbb{A}^\infty$) be the algebraic group whose R-points, for any $\mathbb{Q}$-algebra (resp. $\mathbb{A}^\infty$-algebra) R, are the set of pairs

$$(\lambda, g) \in R^\times \times (B^{\mathrm{op}} \otimes_{\mathbb{Q}} R)^\times$$

such that

$$gg^{\#_\beta} = \lambda.$$

This comes with a homomorphism $\nu : G_\beta \to \mathbb{G}_m$ which sends (λ, g) to λ. We will let $G_{\beta,1}$ denote the kernel of ν. Note that the structure map $G_{\beta,1} \to \mathrm{Spec}\,\mathbb{Q}$ (resp. $\mathbb{A}^\infty$) factors through $\mathrm{Spec}\,F^+$ (resp. $\mathrm{Spec}\,(F^+ \otimes \mathbb{A}^\infty)$) so we may also consider $G_{\beta,1}$ as an algebraic group over F^+ (resp. $F^+ \otimes \mathbb{A}^\infty$). We will let $PG_{\beta,1}$ denote the adjoint group of $G_{\beta,1}$.

Lemma I.7.1 *For any embedding* $\tau : F^+ \hookrightarrow \mathbb{R}$ *we may choose* $0 \neq \beta \in B^{*=-1}$ *such that*

 1. *if* x *is a rational prime which is not split in* E *then* $G_{\beta,1}$, *and hence* G_β, *is quasisplit at* x,

2. *and the pairing* $(\ ,\ )_\beta$ *on* $V \otimes_{\mathbb{Q}} \mathbb{R}$ *has invariants* $(1, n - 1)$ *at* τ *and* $(0, n)$ *at any* τ' *with* $\tau \neq \tau' : F^+ \hookrightarrow \mathbb{R}$.

Proof: Choose $0 \neq \beta_0 \in B^{*=-1}$ and suppose that if x is an infinite place of F^+ then $G_{\beta_0,1}(F_x^+) \cong U(p_{0,x}, q_{0,x})$. We will look for an element $\alpha \in B$, such that the element $\beta = \alpha\beta_0$ satisfies the conditions of the lemma. Firstly we require that

$$\alpha^{\#\beta_0} = \alpha.$$

Thus α defines a class in $H^1(F/F^+, PG_{\beta_0,1})$. Moreover every class in $H^1(F/F^+, PG_{\beta_0,1})$ arises in this way. (By definition such a class is represented by $\alpha \in (B^{\mathrm{op}})^{\times}$ such that $\alpha\alpha^{-\#\beta_0} = \lambda \in F^{\times}$. Then λ has norm 1 in F^+ and so by Hilbert's theorem 90 can be written as μ^c/μ. Then $\mu\alpha$ represents the same class as α and $\mu\alpha = (\mu\alpha)^{\#\beta_0}$.) Moreover $G_{\alpha\beta_0,1}$ is the inner form of $G_{\beta_0,1}$ classified by

$$[\alpha] \in H^1(F^+, PG_{\beta_0,1}).$$

If x is a place of F^+ which splits in F and if y is a place of F above x then we have natural maps

$$H^1(F_x^+, PG_{\beta_0,1}) \cong H^1(F_y, PG_{\beta_0,1}) \cong H^2(F_y, \mu_n) \cong \mathbb{Z}/n\mathbb{Z}.$$

If x is a finite place of F^+ which does not split in F then according to section 2 of [Cl1] we have

$$H^1(F_x^+, PG_{\beta_0,1}) \cong \mathbb{Z}/2\mathbb{Z}.$$

Moreover if x is an infinite place then $H^1(F_x^+, PG_{\beta_0,1})$ is in bijection with the set of unordered pairs of non-negative integers $\{p_x, q_x\}$ with $p_x + q_x = n$. In both these cases the map

$$H^1(F_x^+, PG_{\beta_0,1}) \longrightarrow H^1(F_x, PG_{\beta_0,1})$$

is trivial. Clozel also shows (lemma 2.1 of [Cl1]) that if n is odd then the map

$$H^1(F^+, PG_{\beta_0,1}) \longrightarrow \bigoplus_x H^1(F_x^+, PG_{\beta_0,1})$$

is surjective. If on the other hand n is even he shows there is a map

$$\bigoplus_x H^1(F_x^+, PG_{\beta_0,1}) \longrightarrow \mathbb{Z}/2\mathbb{Z}$$

whose kernel coincides with the image of $H^1(F^+, PG_{\beta_0,1})$. Clozel describes this map as the sum of the natural maps

$$H^1(F_x^+, PG_{\beta_0,1}) \twoheadrightarrow \mathbb{Z}/2\mathbb{Z}$$

if x is finite; and the map

$$H^1(F_x^+, PG_{\beta_0,1}) \twoheadrightarrow \mathbb{Z}/2\mathbb{Z}$$

which sends $\{p_x, q_x\}$ to $p_{x,0} - p_x \bmod 2$, if x is infinite. (See section 2 of [Cl1], particularly lemma 2.2.)

Suppose that B is ramified above s places of F^+ which split in F. If x is a place of F^+ which does not split in F let $u_x \in H^1(F_x^+, PG_{\beta_0,1})$ denote the class of the quasi-split inner form of $PG_{\beta_0,1}$ over F_x^+. If n is even then we see that

$$s + nd/2 + \sum_{x|\infty} p_{x,0} + \sum_{x \nmid \infty} u_x \equiv 0 \bmod 2.$$

Combining this with our assumptions on B we see that (again for n even)

$$\sum_{x|\infty} p_{x,0} + \sum_{x \nmid \infty} u_x \equiv 1 \bmod 2.$$

Let $A = \mathbb{Z}/\mathbb{Z}$ if n is odd and $A = \mathbb{Z}/2\mathbb{Z}$ if n is even. Also suppose that B is ramified above s places of F^+ which split in F. Then we see that we get maps

$$H^1(F^+, PG_{\beta_0,1}) \longrightarrow \bigoplus_x H^1(F_x^+, PG_{\beta_0,1}) \longrightarrow A,$$

where x runs over places of F^+ which do not split in F, where the second map is as described above and where the sequence is exact in the middle. Thus there is a class $\phi \in H^1(F^+, PG_{\beta_0,1})$ which maps to the following classes in $H^1(F_x^+, PG_{\beta_0,1})$:

- 0 if x splits in F,

- u_x if x is a finite place of F^+ which does not split in F,

- $\{1, n-1\}$ if x is an infinite place corresponding to $\tau : F^+ \hookrightarrow \mathbb{R}$,

- and $\{0, n\}$ if x is any other infinite place of F^+.

Then ϕ maps to zero in $H^1(F, PG_{\beta_0,1})$, because

$$H^1(F, PG_{\beta_0,1}) \hookrightarrow \bigoplus_y H^1(F_y, PG_{\beta_0,1}).$$

(We remark that these sets can be identified with the n-torsion in the Brauer group of F and F_y respectively.) Thus ϕ is the image of some class $[\alpha] \in H^1(F/F^+, PG_{\beta_0,1})$. Then

- $G_{\alpha\beta_0,1}$ is quasisplit at all places x of F^+ which are inert in F,

- $G_{\alpha\beta_0,1} \times_{F^+,\tau} \mathbb{R} \cong U(1, n-1)$, and

- for all $\tau' : F^+ \hookrightarrow \mathbb{R}$ other than τ we have $G_{\alpha\beta_0,1} \times_{F^+,\tau'} \mathbb{R} \cong U(0, n)$.

Now take $\beta = \lambda\alpha\beta_0$ for a suitable $\lambda \in (F^+)^\times$. (Note that $G_{\lambda\alpha\beta_0} = G_{\alpha\beta_0}$.) $\square$

Now fix

$$(\, , \,) : (V \otimes \mathbb{A}^\infty) \times (V \otimes \mathbb{A}^\infty) \longrightarrow \mathbb{A}^\infty$$

arising from some β as in the lemma (for some choice of τ). We will write simply $\#$ for the corresponding involution on $B^{\mathrm{op}} \otimes \mathbb{A}^\infty$, and G and G_1 for the corresponding groups over $\mathbb{A}^\infty$. If $\tau : F^+ \hookrightarrow \mathbb{R}$ then $(\, , \,)$ has a well-defined extension

$$(\, , \,)_\tau : (V \otimes \mathbb{A}) \times (V \otimes \mathbb{A}) \longrightarrow \mathbb{A}$$

with invariants $(1, n-1)$ at τ and $(0, n)$ at all other infinite places. Thus we get an involution $\#_\tau$ on $B^{\mathrm{op}} \otimes \mathbb{A}$ (up to equivalence) and groups G_τ and $G_{\tau,1}$ over $\mathbb{A}$. By the above lemma we see that $(\, , \,)_\tau$ can be taken to arise from some (non-canonical) $\beta_\tau \in B^{*=-1}$. Even the pairing $(\, , \,)_{\beta_\tau}$ need not be canonical, there are $\# \ker^1(\mathbb{Q}, G_{\beta_\tau})$ choices for it. If n is even then $\ker^1(\mathbb{Q}, G_{\beta_\tau}) = (0)$, while if n is odd then

$$\ker^1(\mathbb{Q}, G_{\beta_\tau}) = \ker((F^+)^\times/\mathbb{Q}^\times N_{F/F^+}(F^\times) \longrightarrow \mathbb{A}_{F^+}^\times/\mathbb{A}^\times N_{F^+/F}(\mathbb{A}_F^\times)).$$

In either case the natural map $\ker^1(\mathbb{Q}, G_{\beta_\tau}) \to H^1(\mathbb{Q}, PG_{\beta_\tau})$ is trivial. (See page 400 of [Ko3] for both these assertions.) Thus the involution $\#_{\beta_\tau}$ on B^{op} up to equivalence depends only on τ and the groups G_{β_τ} and $G_{\beta_\tau,1}$ also depend only on τ and not on the choice of β_τ. Thus we will denote them $\#_\tau$, G_τ and $G_{\tau,1}$.

If R is an E-algebra then $G_\tau(R)$ can be identified with the set of pairs

$$(g_1, g_2) \in (B^{\mathrm{op}} \otimes_E R) \times (B^{\mathrm{op}} \otimes_{E,c} R)$$

such that

$$(g_1 g_2^{\#_\tau}, g_2 g_1^{\#_\tau}) \in R^\times.$$

Thus we have

$$G_\tau(R) \cong (B^{\mathrm{op}} \otimes_E R)^\times \times R^\times$$

where

$$(g_1, g_2) \longmapsto (g_1, g_1 g_2^{\#\tau})$$

and inversely

$$(g, \nu) \longmapsto (g, \nu g^{-\#\tau}).$$

In particular we get an isomorphism

$$\mathrm{RS}_{\mathbb{Q}}^E(G_\tau \times_{\mathbb{Q}} E) \cong \mathrm{RS}_{\mathbb{Q}}^E(\mathbb{G}_m) \times H_{B^{\mathrm{op}}},$$

where $H_{B^{\mathrm{op}}}/\mathbb{Q}$ is the algebraic group defined by

$$H_{B^{\mathrm{op}}}(R) = (B^{\mathrm{op}} \otimes_{\mathbb{Q}} R)^\times.$$

Suppose that x is a place of $\mathbb{Q}$ which splits as $x = yy^c$ in E. Then the choice of a place $y|x$ allows us to consider $\mathbb{Q}_x \overset{\sim}{\to} E_y$ as an E-algebra and hence to identify

$$G(\mathbb{Q}_x) \cong (B_y^{\mathrm{op}})^\times \times \mathbb{Q}_x^\times.$$

In particular, we get an isomorphism

$$G(\mathbb{Q}_p) \overset{\sim}{\to} \mathbb{Q}_p^\times \times \prod_{i=1}^r (B_{w_i}^{\mathrm{op}})^\times,$$

which sends g to $(\nu(g), g_1, \ldots, g_r)$. We will often let

$$(g_0, g_1, \ldots, g_r) \in \mathbb{Q}_p^\times \times \prod_{i=1}^r (B_{w_i}^{\mathrm{op}})^\times$$

denote a typical element of $G(\mathbb{Q}_p)$. Similarly we will decompose a typical element $g \in G(\mathbb{A}^\infty)$ as $(g_x)_{x \neq p} \times (g_{p,0}, g_{w_1}, \ldots, g_{w_r})$ with $g_x \in G(\mathbb{Q}_x)$, $g_{p,0} \in \mathbb{Q}_p^\times$ and $g_{w_i} \in B_{w_i}^\times$; or as $g^p \times g_{p,0} \times g_w \times g_p^w$, where $g^p = (g_x)_{x \neq p}$, $g_w = g_{w_1}$ and $g_p^w = (g_{w_2}, \ldots, g_{w_r})$. We will let $G(\mathbb{A}^{\infty,w})$ denote the subgroup of $G(\mathbb{A}^\infty)$ consisting of elements with $g_{p,0} = 1$ and $g_w = 1$. Similarly if π is an irreducible admissible representation of $G(\mathbb{A}^\infty)$ over an algebraically closed field of characteristic 0 we may decompose it $\pi \cong \pi^p \otimes \pi_p \cong \pi^p \otimes \pi_{p,0} \otimes \pi_{p,1} \otimes \cdots \otimes \pi_{p,r} \cong \pi^p \otimes \pi_{p,0} \otimes \pi_w \otimes \pi_p^w \cong \pi^w \otimes \pi_{p,0} \otimes \pi_w$. Note that $\pi_{p,0} = \psi_\pi|_{E_{u^c}^\times}$ (where we recall that ψ_π is the character of $(\mathbb{A}_E^\infty)^\times$ which is the central character of π).

Fix a maximal order $\Lambda_i = \mathcal{O}_{B_{w_i}}$ in B_{w_i} for each $i = 1, \ldots, r$. Our pairing $(\ ,\ )$ gives a perfect duality between V_{w_i} and $V_{w_i^c}$. Let $\Lambda_i^\vee \subset V_{w_i^c}$ denote the dual of $\Lambda_i \subset V_{w_i}$. Then if

$$\Lambda = \bigoplus_{i=1}^r \Lambda_i \oplus \bigoplus_{i=1}^r \Lambda_i^\vee \subset V \otimes_{\mathbb{Q}} \mathbb{Q}_p,$$

we see that Λ is a $\mathbb{Z}_p$-lattice in $V \otimes_{\mathbb{Q}} \mathbb{Q}_p$ and that the pairing (,) on V restricts to give a perfect pairing $\Lambda \times \Lambda \to \mathbb{Z}_p$.

There is a unique maximal $\mathbb{Z}_{(p)}$-order $\mathcal{O}_B \subset B$ such that $\mathcal{O}_B^* = \mathcal{O}_B$ and $\mathcal{O}_{B,w_i} = \mathcal{O}_{B_{w_i}}$ for $i = 1, \ldots, r$. Then $\mathcal{O}_{B,p}$ equals the set of elements of B_p which carry Λ into itself. On the other hand the stabiliser of Λ in $G(\mathbb{Q}_p)$ is $\mathbb{Z}_p^{\times} \times \prod_{i=1}^r \mathcal{O}_{B_{w_i}}^{\times}$.

Fix an isomorphism $\mathcal{O}_{B_w} \cong M_n(\mathcal{O}_{F,w})$. Composing this with the transpose map t we also get an isomorphism $\mathcal{O}_{B_w}^{\mathrm{op}} \cong M_n(\mathcal{O}_{F,w})$. Moreover we get an isomorphism

$$\varepsilon \Lambda_1 \cong (\mathcal{O}_{F,w}^n)^{\vee}.$$

(See the end of section I.1 for the definition of ε.) The action of an element $g \in M_n(\mathcal{O}_{F,w}) \cong (\mathcal{O}_{B_w}^{\mathrm{op}})$ on this module is via right multiplication by g^t. We will write Λ_{11} as an abbreviation for $\varepsilon \Lambda_1$. We get an identification

$$\Lambda \cong ((\mathcal{O}_{F,w}^n \otimes \Lambda_{11}) \oplus (\mathcal{O}_{F,w}^n \otimes \Lambda_{11})^{\vee}) \oplus \bigoplus_{i=2}^r (\Lambda_i \oplus \Lambda_i^{\vee}).$$

Under this identification $(g_0, g_1, \ldots, g_r) \in G(\mathbb{Q}_p)$ acts as

$$((1 \otimes g_1) \oplus g_0(1 \otimes g_1^{-1})^{\vee}) \oplus \bigoplus_{i=2}^r (g_i \oplus g_0(g_i^{-1})^{\vee}).$$

Let ξ denote a representation of G on a $\mathbb{Q}_l^{\mathrm{ac}}$-vector space W_{ξ}. We will often assume that ξ is irreducible. This will always be the case from section III.3 onwards.

Fix a square root

$$\mid \ \mid_K^{1/2} : K^{\times} \longrightarrow (\mathbb{Q}_l^{\mathrm{ac}})^{\times}$$

of $\mid \ \mid_K : K^{\times} \to \mathbb{Q}_l^{\times}$, i.e. fix a square root of p^{f_K} in $\mathbb{Q}_l^{\mathrm{ac}}$. If f_K is even we assume that this square root is chosen to be $p^{f_K/2}$. Also choose $\imath : \mathbb{Q}_l^{\mathrm{ac}} \cong \mathbb{C}$, such that $\imath \circ \mid \ \mid^{1/2}$ is valued in $\mathbb{R}_{>0}^{\times}$. We apologise for making such an ugly choice. The reader will see that all our main results are independent of the choice of $\imath$, but it would require a lot of extra notation to make the proofs free of such a choice. Some of our main results do involve the choice of $\mid \ \mid^{1/2}$, but in each case this choice is involved in more than one place and all that matters is that the same choice is made at each place.

Chapter II

Barsotti-Tate groups

II.1 Barsotti-Tate groups

For the definition of a Barsotti-Tate group over a scheme S we refer the reader to section 2 of chapter I of [Me]. Suppose that S is a $\mathcal{O}_K$ scheme, then by a Barsotti-Tate $\mathcal{O}_K$-module H/S we shall mean a Barsotti-Tate group H/S together with an embedding $\mathcal{O}_K \hookrightarrow \mathrm{End}\,(H)$. (We remark that ring morphisms are assumed to send the multiplicative identity to itself.) We call a Barsotti-Tate $\mathcal{O}_K$-module H ind-etale if the underlying Barsotti-Tate group is ind-etale (see example 3.7 of chapter I of [Me]). There is an equivalence of categories between ind-etale Barsotti-Tate $\mathcal{O}_K$-modules and finite, torsion free lisse etale $\mathcal{O}_K$-sheaves on S (see example 3.7 of chapter I of [Me]). If S is connected we define the height of a Barsotti-Tate $\mathcal{O}_K$-module H to be the unique integer $h(H)$ such that $H[\wp_K^n]$ has rank $q_K^{nh(H)}$ for all $n \geq 1$. The usual height of H as a Barsotti-Tate group is $h(H)[K : \mathbb{Q}_p]$. In general we will let $H^\vee$ denote the unique Barsotti-Tate $\mathcal{O}_K$-module such that $H^\vee[p^r]$ is the Cartier dual of $H[p^r]$ for all p and such that the inclusions $H^\vee[p^r] \hookrightarrow H^\vee[p^s]$ for $s \geq r$ are the Cartier duals of $H[p^s] \xrightarrow{p^{s-r}} H[p^r]$. We will refer to $H^\vee$ as the Cartier dual of H.

Now suppose that p is locally nilpotent on S and that H/S is a Barsotti-Tate $\mathcal{O}_K$-module. We define $\mathrm{Lie}\,H$ locally to be $\mathrm{Lie}\,H[p^m]$ for any $m \gg 0$. It is a locally free sheaf on S. We refer to the rank of $\mathrm{Lie}\,H$ as the dimension of H and we will call H *compatible* if the two actions of $\mathcal{O}_K$ on $\mathrm{Lie}\,H$ (one from $\mathcal{O}_K \to \mathrm{End}\,(H)$ and the other from $\mathcal{O}_K \to \mathcal{O}_S$) coincide. We will call a H formal if the p-torsion $H[p]$ in H is radical. There is an equivalence of categories between the category of formal Barsotti-Tate $\mathcal{O}_K$-modules and the category of formal Lie groups Θ/S together with a morphism $\mathcal{O}_K \to \mathrm{End}\,(\Theta)$ such that

- $\Theta[p]$ is finite and locally free;

- $p : \Theta \to \Theta$ is an epimorphism;

(This follows from corollary 4.5 of chapter II of [Me].)

Let $\mathcal{X}$ be a locally noetherian formal scheme with ideal of definition $\mathcal{I}$ containing p. We will let $\mathcal{X}_n$ denote the scheme with underlying topological space $\mathcal{X}$ and structure sheaf $\mathcal{O}_\mathcal{X}/\mathcal{I}^n$. By a Barsotti-Tate $\mathcal{O}_K$-module $\mathcal{H}$ over the locally noetherian formal scheme $\mathcal{X}$ we shall mean a system of Barsotti-Tate $\mathcal{O}_K$-modules $\mathcal{H}_n$ over the schemes $\mathcal{X}_n$ together with compatible isomorphisms

$$\mathcal{H}_n \times_{\mathcal{X}_n} \mathcal{X}_m \cong \mathcal{H}_m$$

whenever $m \leq n$. This definition is easily checked to be canonically independent of the choice of ideal of definition $\mathcal{I}$. We will call $\mathcal{H}$ ind-etale (resp. formal) if each $\mathcal{H}_n$ is ind-etale (resp. formal). Note that in fact $\mathcal{H}$ is ind-etale (resp. formal) if and only if $\mathcal{H}_1$ is ind-etale (resp. formal, see paragraph 3.2 of chapter II of [Me]). We define $\mathrm{Lie}\,\mathcal{H}$ to be $\lim_{\leftarrow n} \mathrm{Lie}\,\mathcal{H}_n$. It is locally free on $\mathcal{X}$ and we refer to its rank as the dimension of $\mathcal{H}$. We call $\mathcal{H}$ *compatible* if the two actions of $\mathcal{O}_K$ on $\mathrm{Lie}\,\mathcal{H}$ coincide. If A is a noetherian ring complete with respect to the I-adic topology for some ideal I, then there is a natural functor from Barsotti-Tate $\mathcal{O}_K$-modules $H/\mathrm{Spec}\,A$ to Barsotti-Tate $\mathcal{O}_K$-modules $\mathcal{H}/\mathrm{Spf}\,A$. It follows from lemma 4.16 of chapter II of [Me] that if I contains some power of p then this is in fact an equivalence of categories which preserves exact sequences. We remark that $\mathcal{H}/\mathrm{Spf}\,A$ may be formal while the corresponding Barsotti-Tate $\mathcal{O}_K$-module $H/\mathrm{Spec}\,A$ is not.

Lemma II.1.1 *Suppose that p is locally nilpotent on a locally noetherian scheme S and that H/S is a Barsotti-Tate $\mathcal{O}_K$-module. Then for $h \in Z_{\geq 0}$ we can find reduced closed subschemes $S^{[h]} \subset S$ such that*

1. *$S^{[h]} \supset S^{[h-1]}$;*

2. *the codimension of any component of $S^{[h-1]}$ in any component of $S^{[h]}$ which contains it is at most 1;*

3. *for any geometric point s of S we have that s lies in $S^{[h]}$ if and only if $\#H[p](k(s)) \leq p^{[K:\mathbb{Q}_p]h}$;*

4. *on $S^{(h)} = S^{[h]} - S^{[h-1]}$ there is a short exact sequence of Barsotti-Tate $\mathcal{O}_K$-module*

$$(0) \longrightarrow H^0 \longrightarrow H \longrightarrow H^{\mathrm{et}} \longrightarrow (0)$$

where H^0 is a formal Barsotti-Tate $\mathcal{O}_K$-module and where H^{et} is an ind-etale Barsotti-Tate $\mathcal{O}_K$-module of height h.

Proof: By proposition 4.9 of [Me] it suffices to show that for $g \in \mathbb{Z}^{\geq 0}$ we can find closed subschemes $S'_g \subset S$ such that

1. $S'_g \supset S'_{g-1}$;

2. if s is a geometric point of S then s lies in S'_g if and only if $\#H[p](k(s))$ is less than or equal to p^g;

3. the codimension of any component of S'_{g-1} in any component of S'_g which contains it is at most 1.

The question is local on S so we may assume that $S = \operatorname{Spec} R$ for a noetherian ring R. We may further assume that S is reduced. By a simple inductive argument it suffices in fact to show that if for any geometric point s of S we have $\#H[p](k(s)) \leq p^g$ then we can find a reduced closed subscheme $S' \subset S$ such that

1. a geometric point s of S lies in S' if and only if $\#h[p](k(s)) < p^g$;

2. any irreducible component of S' has codimension at most one in any irreducible component of S containing it.

Finally we may assume that S is in fact integral.

We now follow the arguments of page 97 of [O]. Let $\mathcal{H}/S$ denote the locally free sheaf $\operatorname{Lie}(H[p]^\vee) = \operatorname{Lie} H^\vee$ (the equality here follows from remark 3.3.20 of chapter II of [Me] because $p = 0$ on S). For any geometric point s of S there is a canonical perfect pairing between $\mathcal{H}_s^{V_*=1}$ and $H_s[p](k(s))$. (This seems to be well known, but we know of no reference for the statement in exactly this form, so we will sketch the proof. On page 138 of [Mu1] we see that we can identify $\mathcal{H}_s$ with $\operatorname{Hom}(H_s[p], \mathbb{G}_a)$ and that V_* then becomes identified with the map $\phi \mapsto \phi \circ \operatorname{Fr}^*$. We get a pairing

$$
\begin{array}{ccccc}
H_s[p](k(s)) & \times & \mathcal{H}_s & \longrightarrow & k(s) \\
x & \times & \phi & \longmapsto & \phi \circ x,
\end{array}
$$

where $\phi \in \operatorname{Hom}(H_s[p], \mathbb{G}_a)$ and $\phi \circ x \in \mathbb{G}_a(k(s)) = k(s)$. We see that it restricts to a pairing

$$
H_s[p](k(s)) \times \mathcal{H}_s^{V_*=1} \longrightarrow \mathbb{F}_p.
$$

If $\phi \circ x = 0$ for all $x \in H_s[p](k(s))$ then ϕ factors through the local ring of $\mathbb{G}_a$ at 0. If moreover $\phi \circ \operatorname{Fr}^* = \phi$ then we see that $\phi = 0$. Thus our pairing gives an injection

$$
\mathcal{H}_s^{V_*=1} \hookrightarrow \operatorname{Hom}(H_s[p](k(s)), \mathbb{F}_p).
$$

To show this is in fact an isomorphism one can count orders. Suppose that $\#H_s[p](k(s)) = p^h$. Then we have an embedding $\mu_p^h \hookrightarrow H_s[p]^\vee$ and so an embedding $\operatorname{Lie} \mu_p^h \hookrightarrow \mathcal{H}_s$. But $(\operatorname{Lie} \mu_p)^{V_*=1}$ has order p (as follows easily from the results on page 143 of [Mu1]), and so

$$\#\mathcal{H}_s^{V_*=1} \geq p^h = \#H_s[p](k(s)).$$

The fact that our pairing is perfect follows at once.)

Again shrinking S we may assume that in fact $\mathcal{H}$ is free. Choose a basis $e_1, \dots, e_m$ and suppose that $V_* e_i = \sum_j v_{i,j} e_j$. Then

$$V_* \sum_i x_i e_i = \sum_{i,j} x_i^p v_{i,j} e_j.$$

Let $\mathcal{H}^{V_*=1}$ denote the subscheme of Aff_R^m defined by the equations

$$x_j = \sum_i v_{i,j} x_i^p$$

for $j = 1, \dots, m$. Then $\mathcal{H}^{V_*=1}/S$ is quasi-finite and etale (as the Jacobian is the identity matrix). Generically $\mathcal{H}^{V_*=1}/S$ has degree $\leq p^g$. We may suppose that in fact generically the degree equals p^g. The locus where the degree drops is closed (as the degree is locally constant). We must show that it has codimension 1. Let T denote the normalisation of S in a finite separable extension of the fraction field of R over which $\mathcal{H}^{V_*=1}$ has p^g points. As T/S is finite, it suffices to prove the result for T. Let $x^{(1)}, \dots, x^{(p^g)}$ denote the sections of $\mathcal{H}^{V_*=1}$ over the generic point of T. Then T' is simply the locus where some $x_i^{(j)}$ is not regular. But the locus where any given $x_i^{(j)}$ is not regular has codimension 1 because T is normal. $\square$

Corollary II.1.2 *Suppose that $\mathcal{H}/\mathcal{X}$ is a Barsotti-Tate $\mathcal{O}_K$-module over a locally noetherian formal scheme. Suppose also that $p = 0$ on $\mathcal{X}^{\mathrm{red}}$ and that the function from geometric points of $\mathcal{X}^{\mathrm{red}}$ to integers*

$$s \mapsto \#\mathcal{H}[p](k(s))$$

is constant. Then there is an exact sequence of Barsotti-Tate $\mathcal{O}_K$-modules

$$(0) \longrightarrow \mathcal{H}^0 \longrightarrow \mathcal{H} \longrightarrow \mathcal{H}^{\mathrm{et}} \longrightarrow (0)$$

over $\mathcal{X}$ with $\mathcal{H}^0$ formal and $\mathcal{H}^{\mathrm{et}}$ ind-etale.

For $g \geq 1$ there is a unique one-dimensional compatible formal Barsotti-Tate $\mathcal{O}_K$-module $\Sigma_{K,g}$ over $k(v_K)^{\mathrm{ac}}$ of height g. In fact if k is any separably closed field containing $k(v_K)$ then any one-dimensional compatible Barsotti-Tate $\mathcal{O}_K$-module over k is of the form $\Sigma_{K,g} \times (K/\mathcal{O}_K)^h$ for some

g and h. (If $H/k(v_K)^{\mathrm{ac}}$ is a one-dimensional compatible Barsotti-Tate $\mathcal{O}_K$-module over k we have an exact sequence

$$(0) \longrightarrow H^0 \longrightarrow H \longrightarrow H^{\mathrm{et}} \longrightarrow (0),$$

where H^0 is formal and H^{et} is ind-etale. By proposition 1.7 of [Dr] we see that $H^0 \cong \Sigma_{K,g}$. It only remains to find a splitting $H^{\mathrm{et}} \to H$, but this is the same as finding a splitting $H^{\mathrm{et}}(k) \to H(k)$. Finally note that $H(k) \to H^{\mathrm{et}}(k)$ is an isomorphism.) Moreover

$$\operatorname{End}(\Sigma_{K,g}/k) = \operatorname{End}(\Sigma_{K,g}/k(v_K)^{\mathrm{ac}}) \cong \mathcal{O}_{D_{K,g}}$$

(proposition 1.7 of [Dr]). We can extend the (left)-action of $\mathcal{O}_{D_{K,g}}^{\times}$ on $\Sigma_{K,g}/k(v_K)^{\mathrm{ac}}$ to an action of $D_{K,g}^{\times}$ on $\Sigma_{K,g}/k(v_K)$, such that for $\delta \in D_{K,g}^{\times}$

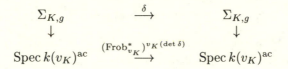

commutes. To see this one need only consider the case $v_K(\det \delta) \geq 0$: for $v_K(\det \delta) < 0$ we define the action of δ to be the inverse of the action of δ^{-1}. If $v_K(\det \delta) \geq 0$ then the kernel of $\delta \in \operatorname{End}(\Sigma_{K,g})$ is the same as the kernel of $F^{f_K v_K(\det \delta)} : \Sigma_{K,g} \to \Sigma_{K,g}^{(f_K v_K(\det \delta))}$. Thus δ induces a map $\delta F^{-f_K v_K(\det \delta)} : \Sigma_{K,g}^{(f_K v_K(\det \delta))} \to \Sigma_{K,g}$. We define our semi-linear action of δ as the composite of the pullback map

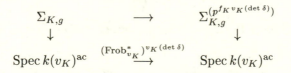

with the $\operatorname{Spec} k(v_K)^{\mathrm{ac}}$-linear map

$$\delta F^{-f_K v_K(\det \delta)} : \Sigma_{K,g}^{(p^{f_K v_K(\det \delta)})} \longrightarrow \Sigma_{K,g}.$$

Consider the functor from Artinian local $\mathcal{O}_K$-algebras with residue field $k(v_K)^{\mathrm{ac}}$ to sets which sends A to the set of isomorphism classes of pairs (H, j) where $H/\operatorname{Spec} A$ is a compatible Barsotti-Tate $\mathcal{O}_K$-module and $j : \Sigma_{K,g} \xrightarrow{\sim} H \times_A k(\wp_K)^{\mathrm{ac}}$. This functor is pro-represented by a complete noetherian local ring $R_{K,g}$ with residue field $k(\wp_K)^{\mathrm{ac}}$ and in fact $R_{K,g} \cong \mathcal{O}_{\widehat{K}^{\mathrm{nr}}}[[T_2, \ldots, T_g]]$ (proposition 4.2 of [Dr]). The universal deformation exists over $\operatorname{Spec} R_{K,g}$ (not just over $\operatorname{Spf} R_{K,g}$, see lemma 4.16 of chapter II of [Me]). We will denote this universal deformation by $(\widetilde{\Sigma}_{K,g}, \widetilde{j})/\operatorname{Spec} R_{K,g}$.

Note that $R_{K,g}$ has a continuous left action of $\mathcal{O}_{D_g}^{\times}$. (If $\delta \in \mathcal{O}_{D_{K,g}}$ then the push forward of $(\widetilde{\Sigma}_{K,g}, \widetilde{j})$ along $\delta : R_{K,g} \to R_{K,g}$ is $(\widetilde{\Sigma}_{K,g}, \widetilde{j} \circ \delta)$.) We will let $\overline{R}_{K,g}$ denote $R_{K,g} \times_{W(k(\wp_K))} k(\wp_K)^{\mathrm{ac}}$.

Set $H_0 = \Sigma_{k,g} \times (K/\mathcal{O}_K)^h$ a compatible Barsotti-Tate $\mathcal{O}_K$-module over $k(v_K)^{\mathrm{ac}}$. Let TH_0 denote its Tate module, i.e.

$$TH_0 = \mathrm{Hom}_{\mathcal{O}_K}(K/\mathcal{O}_K, H_0(k(v_K)^{\mathrm{ac}})) \cong \mathcal{O}_K^h.$$

Now consider the functor from Artinian local $\mathcal{O}_K$-algebras with residue field $k(v_K)^{\mathrm{ac}}$ to sets which sends A to the set of isomorphism classes of pairs (H, j) where $H/\mathrm{Spec}\, A$ is a compatible Barsotti-Tate $\mathcal{O}_K$-module and $j : H_0 \overset{\sim}{\to} H \times_A k(\wp_K)^{\mathrm{ac}}$. This functor is again pro-represented, this time by $\mathrm{Hom}\,(TH_0, \widetilde{\Sigma}_{K,g})$. By $\mathrm{Hom}\,(TH_0, \widetilde{\Sigma}_{K,g})$ we mean the $R_{K,g}$-formal scheme such that for any Artinian local $R_{K,g}$ algebra S we have

$$\mathrm{Hom}\,(TH_0, \widetilde{\Sigma}_{K,g})(S) = \mathrm{Hom}_{\mathcal{O}_K}(TH_0, \widetilde{\Sigma}_{K,g}(S)).$$

Noncanonically we have $\mathrm{Hom}\,(TH_0, \widetilde{\Sigma}_{K,g}) \cong \widetilde{\Sigma}_{K,g}^h$, where the fibre product is taken over $\mathrm{Spf}\, R_{K,g}$. We also have, again noncanonically,

$$\mathrm{Hom}\,(TH_0, \widetilde{\Sigma}_{K,g}) \cong \mathrm{Spf}\, \mathcal{O}_{\widehat{K}^{\mathrm{nr}}}[[T_2, \ldots, T_{g+h}]].$$

The universal deformation of H_0 over $\mathrm{Hom}\,(TH_0, \widetilde{\Sigma}_{K,g})$ is then the extension of $\widetilde{\Sigma}_{K,g}$ by $(K/\mathcal{O}_K)^h$ classified by the tautological class in

$$\mathrm{Hom}\,(TH_0, \widetilde{\Sigma}_{K,g})(S) = \mathrm{Hom}\,(TH_0, \widetilde{\Sigma}_{K,g}(S)) \cong$$
$$\mathrm{Ext}^1(TH_0 \otimes_{\mathcal{O}_K} (K/\mathcal{O}_K), \widetilde{\Sigma}_{K,g}(S)) \cong \mathrm{Ext}^1_S(TH_0 \otimes_{\mathcal{O}_K} (K/\mathcal{O}_K), \widetilde{\Sigma}_{K,g}),$$

where $S = \mathrm{Hom}\,(TH_0, \widetilde{\Sigma}_{K,g})$. (See proposition 4.5 of [Dr] and its proof.)

Lemma II.1.3 *Suppose that $S/k(v_K)^{\mathrm{ac}}$ is reduced of finite type. Suppose also that H/S is a one-dimensional compatible Barsotti-Tate $\mathcal{O}_K$-module. Suppose moreover that over S there is an exact sequence of Barsotti-Tate $\mathcal{O}_K$-modules*

$$(0) \longrightarrow H^0 \longrightarrow H \longrightarrow H^{\mathrm{et}} \longrightarrow (0),$$

where H^{et} has constant height h and H^0 has constant height g. Let s be a closed point of S and choose an isomorphism $j : \Sigma_{K,g} \overset{\sim}{\to} H_s^0$.

1. *Then $H^0/S_s^{\wedge}$ gives rise to a morphism $S_s^{\wedge} \to \mathrm{Spf}\, R_{K,g}$ which in fact factors through $\mathrm{Spec}\, k(v_K)^{\mathrm{ac}}$.*

2. *$H/S_s^{\wedge}$ gives rise to a morphism $S_s^{\wedge} \to \mathrm{Hom}\,(TH_s, \widetilde{\Sigma}_{K,g})$ which in fact factors through $\mathrm{Hom}\,(TH_s, \Sigma_{K,g}) \subset \mathrm{Hom}\,(TH_s, \widetilde{\Sigma}_{K,g})$.*

Proof: The statements are easily seen to be equivalent. We will prove the first one. Write $R_{K,g} = \mathcal{O}_{\widehat{K}^{\mathrm{nr}}}[[T_2, \ldots, T_g]]$, let P be a minimal prime of $\mathcal{O}_{S,s}^{\wedge}$ and let k denote the field of fractions of the image $R_{K,g} \to \mathcal{O}_{S,s}^{\wedge}/P$. As S is reduced, it suffices to show that $T_2, \ldots, T_g$ map to 0 in k. Suppose not.

For the rest of this proof we will use without comment the notation of [Dr]. We can arrange that $\widetilde{\Sigma}_{K,g}$ corresponds to a morphism $\Lambda_{\mathcal{O}_K} = \mathcal{O}_K[g_1, g_2, \ldots] \to R_{K,g}$ which

- sends $g_{p^{f_K i}-1}$ to T_{i+1} for $1 \leq i \leq g-1$;

- and sends g_j to zero for $1 \leq j < p^{f_K g} - 1$ and $j \neq p^{f_K i} - 1$ for some i in the above range.

(See the proof of proposition 4.2 of [Dr].) Choose i minimal such that T_i does not map to zero in k. Then $H^0 \times_S k$ corresponds to a morphism $\Lambda_{\mathcal{O}_K} \to k$ which sends g_j to 0 for $j = 1, 2, \ldots, p^{f_K (i-1)} - 2$ and sends $g_{p^{f_K(i-1)}-1}$ to something nonzero. Thus $H^0 \times_S k$ has height $i - 1 < g$ (see the proof of proposition 1.6 of [Dr]). This contradicts the fact that $H \times_S \operatorname{Spec} k$ is a compatible formal Barsotti-Tate $\mathcal{O}_K$-module of height g (because H/S is a compatible formal Barsotti-Tate $\mathcal{O}_K$-module of height g). $\square$

Corollary II.1.4 *Suppose that $S/k(v_K)^{\mathrm{ac}}$ is a smooth scheme of finite type. Suppose that H/S is a one-dimensional compatible Barsotti-Tate $\mathcal{O}_K$-module of constant height g. Suppose moreover that for each closed point s of S the formal completion $S_s^{\wedge}$ is isomorphic to the equidimensional universal formal deformation space of H_s. Then for $h = 0, \ldots, g-1$ the locally closed subscheme $S^{(h)} = S^{[h]} - S^{[h-1]}$ of S is either empty or smooth of dimension h. If s is a closed point of $S^{(h)}$ and if $j : \widetilde{\Sigma}_{K,g-h} \xrightarrow{\sim} H_s^0$ then we get an identification $S_s^{\wedge} \cong \operatorname{Hom}(TH_s, \widetilde{\Sigma}_{K,g-h}) \times_{\mathcal{O}_K} k(v_K)$ and under this identification $(S^{(h)})_s^{\wedge} \subset S_s^{\wedge}$ corresponds to $\operatorname{Hom}(TH_s, \Sigma_{K,g-h}) \subset \operatorname{Hom}(TH_s, \widetilde{\Sigma}_{K,g-h}) \times_{\mathcal{O}_K} k(v_K)$.*

Proof: Because the formal completion of S at any closed point is isomorphic to $k(v_K)^{\mathrm{ac}}[[T_2, \ldots, T_g]]$, every component of S has dimension $g - 1$. We must have $S = S^{[g-1]}$. Hence by lemma II.1.1 every irreducible component of $S^{[h]}$ has dimension at least h. Thus the same is true for $S^{(h)}$. On the other hand by the previous lemma if s is any closed point of $S^{(h)}$ then the formal completion $(S^{(h)})_s^{\wedge}$ corresponds to a sub-formal scheme of $\operatorname{Hom}(TH_s, \Sigma_{K,g-h}) \subset \operatorname{Hom}(TH_s, \widetilde{\Sigma}_{K,g-h})$. Thus we must have $(S^{(h)})_s^{\wedge} \cong \operatorname{Hom}(TH_s, \Sigma_{K,g-h})$ and, assuming such a closed point exists, we have that $S^{(h)}$ is smooth at s of dimension h. $\square$

The functor on schemes $S/k(v_K)^{\mathrm{ac}}$ which sends S to $\mathrm{Aut}\,(\Sigma_{K,g}[\wp_K^m]/S)$ is represented by a scheme $\mathrm{Aut}\,(\Sigma_{K,g}[\wp_K^m])$ of finite type over $k(v_K)^{\mathrm{ac}}$. (To see this simply think of these automorphisms as maps on sheaves of Hopf algebras.) If $m_1 > m_2$ there is a natural morphism

$$\mathrm{Aut}\,(\Sigma_{K,g}[\wp_K^{m_1}]) \longrightarrow \mathrm{Aut}\,(\Sigma_{K,g}[\wp_K^{m_2}]).$$

We will let $\mathrm{Aut}^{\,1}(\Sigma_{K,g}[\wp_K^m])$ denote the intersection of the scheme theoretic images of $\mathrm{Aut}\,(\Sigma_{K,g}[\wp_K^{m'}])$ in $\mathrm{Aut}\,(\Sigma_{K,g}[\wp_K^m])$ as m' varies over integers greater than or equal to m. We see that the scheme theoretic image of the morphism

$$\mathrm{Aut}^{\,1}(\Sigma_{K,g}[\wp_K^{m+1}]) \longrightarrow \mathrm{Aut}\,(\Sigma_{K,g}[\wp_K^m])$$

is just $\mathrm{Aut}^{\,1}(\Sigma_{K,g}[\wp_K^m])$.

Lemma II.1.5 $\mathrm{Aut}^{\,1}(\Sigma_{K,g}[\wp_K^m])/k(v_K)^{\mathrm{ac}}$ *is finite and*

$$\mathrm{Aut}^{\,1}(\Sigma_{K,g}[\wp_K^m])^{\mathrm{red}} \cong (\mathcal{O}_{D_{K,g}}/\wp_K^m \mathcal{O}_{D_{K,g}})^{\times}.$$

Proof: Suppose first that $\mathrm{Aut}^{\,1}(\Sigma_{K,g}[\wp_K^m])$ has an irreducible component V_m of dimension > 0. Then we can find irreducible components $V_{m'}$ of $\mathrm{Aut}^{\,1}(\Sigma_{K,g}[\wp_K^{m'}])$ for $m' > m$ such that whenever $m'' \geq m' \geq m$ then $V_{m''}$ maps to $V_{m'}$ and is dominating. Let $k(V_{m'})$ denote the function field of $V_{m'}^{\mathrm{red}}$, so that whenever $m'' \geq m' \geq m$ we have $k(V_{m'}) \hookrightarrow k(V_{m''})$. Let k be an algebraically closed extension field of $k(v_K)^{\mathrm{ac}}$ of uncountable transcendence degree. Then there are uncountably many maps $k(V_m) \hookrightarrow k$ and each can be extended into a compatible series of injections $k(V_{m'}) \hookrightarrow k$ for $m' > m$. Thus $\mathrm{Aut}^{\,1}(\Sigma_{K,g}[\wp_K^m])(k)$ has uncountably many points which can be lifted compatibly to each $\mathrm{Aut}^{\,1}(\Sigma_{K,g}[\wp_K^{m'}])(k)$ with $m' > m$. This implies that the image of

$$\mathrm{Aut}\,(\Sigma_{K,g}/k) \longrightarrow \mathrm{Aut}\,(\Sigma_{K,g}[\wp_K^m]/k)$$

is uncountably infinite. On the other hand it follows from proposition 1.7 of [Dr] that this image is just $(\mathcal{O}_{D_{K,g}}/\wp_K^m \mathcal{O}_{D_{K,g}})^{\times}$, which is finite. This contradiction shows that $\mathrm{Aut}^{\,1}(\Sigma_{K,g}[\wp_K^m])$ is zero dimensional.

As each $\mathrm{Aut}^{\,1}(\Sigma_{K,g}[\wp_K^m])$ is zero dimensional and as for $m' > m$ the morphism

$$\mathrm{Aut}^{\,1}(\Sigma_{K,g}[\wp_K^{m'}]) \longrightarrow \mathrm{Aut}^{\,1}(\Sigma_{K,g}[\wp_K^m])$$

is dominating we see that for $m' \geq m$

$$\mathrm{Aut}^{\,1}(\Sigma_{K,g}[\wp_K^{m'}])(k(v_K)^{\mathrm{ac}}) \twoheadrightarrow \mathrm{Aut}^{\,1}(\Sigma_{K,g}[\wp_K^m])(k(v_K)^{\mathrm{ac}}).$$

It follows that $\mathrm{Aut}^1(\Sigma_{K,g}[\wp_K^m])(k(v_K)^{\mathrm{ac}})$ equals the image of

$$\mathrm{Aut}\,(\Sigma_{K,g}/k(v_K)^{\mathrm{ac}}) \longrightarrow \mathrm{Aut}\,(\Sigma_{K,g}[\wp_K^m]/k(v_K)^{\mathrm{ac}}).$$

Again by proposition 1.7 of [Dr] this is just $(\mathcal{O}_{D_{K,g}}/\wp_K^m\mathcal{O}_{D_{K,g}})^\times$ and so the lemma follows. $\square$

We remark that for $m > 1$ the scheme $\mathrm{Aut}\,(\Sigma_{K,g}[\wp_K^m])$ has dimension > 0. By an explicit calculation with Dieudonne modules we checked in an earlier version of this work that $\mathrm{Aut}^1(\Sigma_{K,g}[\wp_K^m])^{\mathrm{red}}$ coincides with the reduced subscheme of the image of $\mathrm{Aut}\,(\Sigma_{K,g}[\wp_K^{m+1}]) \to \mathrm{Aut}\,(\Sigma_{K,g}[\wp_K^m])$. However we will not actually need that stronger result here, so we do not reproduce the argument.

Now suppose that S is a reduced $k(v_K)^{\mathrm{ac}}$-scheme and that H/S is a one-dimensional compatible formal Barsotti-Tate $\mathcal{O}_K$-module of constant height g. We want to investigate how far H differs from $\Sigma_{K,g} \times_{\mathrm{Spec}\,k(v_K)^{\mathrm{ac}}} S$. Consider the functor on S-schemes which sends T/S to the set of isomorphisms (over T)

$$j : \Sigma_{k,g}[\wp_K^m] \times_{\mathrm{Spec}\,k(v_K)^{\mathrm{ac}}} T \longrightarrow H[\wp_K^m] \times_S T.$$

It is easy to see that this functor is represented by a scheme $X_m(H/S)$ of finite type over S. (Think about j as a map of sheaves of Hopf algebras on T.) Then we define $Y_m(H/S)$ to be the intersection of the scheme theoretic images of the

$$X_{m'}(H/S) \longrightarrow X_m(H/S)$$

for $m' \geq m$. Finally we set $J^{(m)}(H/S) = Y_m(H/S)^{\mathrm{red}}$. We will also let j^{univ} denote the universal isomorphism

$$j^{\mathrm{univ}} : \Sigma_{K,g}[\wp_K^m] \overset{\sim}{\longrightarrow} H[\wp_K^m]$$

over $J^{(m)}(H/S)$.

Thus $Y_m(H/S)$ and $J_m(H/S)$ are finite type over S. If T/S is any scheme then $X_m(H/T) = X_m(H/S) \times_S T$. If T/S is flat then

$$Y_m(H/T) = Y_m(H/S) \times_S T$$

(because the formation of scheme theoretic image commutes with flat base change). We see that

$$J^{(m)}(\Sigma_{K,g}/k(v_K)^{\mathrm{ac}}) = \mathrm{Aut}^1(\Sigma_{K,g}[\wp_K^m])^{\mathrm{red}} \cong (\mathcal{O}_{D_{K,g}}/\wp_K^m\mathcal{O}_{D_{K,g}})^\times.$$

In fact if $S/k(v_K)^{\mathrm{ac}}$ is any reduced scheme then

$$J^{(m)}(\Sigma_{K,g}/S) = (J^{(m)}(\Sigma_{K,g}/k(v_K)^{\mathrm{ac}}) \times S)^{\mathrm{red}}$$
$$= ((\mathcal{O}_{D_{K,g}}/\wp_K^m\mathcal{O}_{D_{K,g}})_S^\times)^{\mathrm{red}} = (\mathcal{O}_{D_{K,g}}/\wp_K^m\mathcal{O}_{D_{K,g}})_S^\times.$$

Each of the schemes $X_m(H/S)$, $Y_m(H/S)$ and $J^{(m)}(H/S)$ has a natural right action of $(\mathcal{O}_{D_{K,g}}/\wp_K^m \mathcal{O}_{D_{K,g}})^\times$. ($\delta \in \mathcal{O}_{D,g}^\times$ takes j to $j \circ \delta$.) If $S = T \times_{\operatorname{Spec} k(v_K)} \operatorname{Spec} k(v_K)^{\mathrm{ac}}$ for a reduced scheme $T/k(v_K)$ and if $H = H_0 \times_T S$ for a compatible formal Barsotti-Tate $\mathcal{O}_K$-module H_0/T, then this action extends to one of $D_{K,g}^\times/(1 + \wp_K^m \mathcal{O}_{D_{K,g}})$ on each of $X_m(H/S)$, $Y_m(H/S)$ and $J^{(m)}(H/S)$ thought of as T-schemes. More precisely if $\delta \in D_{K,g}^\times$ we get a commutative diagram

$$
\begin{array}{ccc}
J^{(m)}(H/S) & \xrightarrow{\ \delta\ } & J^{(m)}(H/S) \\
\downarrow & & \downarrow \\
T \times \operatorname{Spec} k(v_K)^{\mathrm{ac}} & \xrightarrow{(1 \times \operatorname{Frob}_{v_K}^*)^{-v_K(\det \delta)}} & T \times \operatorname{Spec} k(v_K)^{\mathrm{ac}}.
\end{array}
$$

commutes. (Let X/S denote the pullback of $X_m(H/S)$ by

$$
T \times \operatorname{Spec} k(v_K)^{\mathrm{ac}} \xrightarrow{(1 \times \operatorname{Frob}_{v_K}^*)^{v_K(\det \delta)}} T \times \operatorname{Spec} k(v_K)^{\mathrm{ac}}.
$$

Then over X we get an isomorphism $j' : \Sigma_{K,g}^{(p^{-f_K v_K(\det \delta)})}[\wp_K^m] \xrightarrow{\sim} H[\wp_K^m]$. On the other hand δ gives an isomorphism

$$
\delta : \Sigma_{K,g} \xrightarrow{\ \sim\ } \Sigma_{K,g}^{(p^{-f_K v_K(\det \delta)})}.
$$

Thus over X we get

$$
j' \circ \delta : \Sigma_{K,g}[\wp_K^m] \xrightarrow{\sim} H[\wp_K^m].
$$

This induces a map over S from X to $X_m(H/S)$. Composing this with the inverse of the pullback of $(1 \times \operatorname{Frob}_{v_K}^*)^{v_K(\det \delta)}$ we get the desired automorphism of $X_m(H/S)$. The following diagram

$$
\begin{array}{ccccc}
X_m(H/S) & \longleftarrow & X & \longrightarrow & X_m(H/S) \\
\downarrow & & \downarrow & & \downarrow \\
T \times \operatorname{Spec} k(v_K)^{\mathrm{ac}} & \longleftarrow & T \times \operatorname{Spec} k(v_K)^{\mathrm{ac}} & = & T \times \operatorname{Spec} k(v_K)^{\mathrm{ac}}
\end{array}
$$

(where the leftwards arrow on the bottom row is $(1 \times \operatorname{Frob}_{v_K}^*)^{v_K(\det \delta)}$) illustrates this construction.)

Before going on to the main result of this section let us recall a result from commutative algebra.

Lemma II.1.6 *Let A be an excellent ring and B/A a finite A-algebra. Let I be an ideal in A. Then there is a canonical isomorphism*

$$
(B \otimes_A A_I^\wedge)^{\mathrm{red}} \cong B^{\mathrm{red}} \otimes_A A_I^\wedge.
$$

Proof: As B/A is finite theorem 55 of [Ma] tells us that it suffices to exhibit a canonical isomorphism

$$(B_I^\wedge)^{\mathrm{red}} \cong (B^{\mathrm{red}})_I^\wedge.$$

Theorem 54 (and (23.K)) of [Ma] gives us a surjection

$$B_I^\wedge \twoheadrightarrow (B^{\mathrm{red}})_I^\wedge$$

with nilpotent kernel and hence a map

$$(B^{\mathrm{red}})_I^\wedge \twoheadrightarrow (B_I^\wedge)^{\mathrm{red}}.$$

This map is an isomorphism if $(B^{\mathrm{red}})_I^\wedge$ is reduced, which it is by the excellence of B^{red} (see sections (34.A) and theorem 79 of [Ma]). $\square$

The following proposition is of key importance for us.

Proposition II.1.7 *Let $S/k(v_K)^{\mathrm{ac}}$ be a reduced excellent scheme (e.g. a reduced scheme of finite type) and let H/S be a one-dimensional compatible formal Barsotti-Tate $\mathcal{O}_K$-module of constant height g. Then for each $m \geq 1$, $J^{(m)}(H/S)/S$ is finite etale and Galois with group $(\mathcal{O}_{D_{K,g}}/\wp_K^m \mathcal{O}_{D_{K,g}})^\times$. (N.B. We are not asserting that $J^{(m)}(H/S)$ is connected.)*

Proof: It suffices to show that for any closed point s of S

- the group $(\mathcal{O}_{D_{K,g}}/\wp_K^m \mathcal{O}_{D_{K,g}})^\times$ has a faithful and transitive action on the points of $J^{(m)}(H/S)_s$;

- and if t is any point of $J^{(m)}(H/S)_s$ then $J^{(m)}(H/S)_t^\wedge \xrightarrow{\sim} S_s^\wedge$.

(See for instance theorem 3 of section 5 of chapter 3 of [Mu2].) Equivalently it suffices to check that for all closed points s of S we have

$$J^{(m)}(H/S) \times_S \operatorname{Spec} \mathcal{O}_{S,s}^\wedge \cong (\mathcal{O}_{D_{K,g}}/\wp_K^m \mathcal{O}_{D_{K,g}})^\times_{\operatorname{Spec} \mathcal{O}_{S,s}^\wedge}.$$

We note that by lemma II.1.3

$$H \times_S \operatorname{Spec} \mathcal{O}_{S,s}^\wedge \cong \Sigma_{K,g} \times_{\operatorname{Spec} k(v_K)^{\mathrm{ac}}} \operatorname{Spec} \mathcal{O}_{S,s}^\wedge.$$

Because $\operatorname{Spec} \mathcal{O}_{S,s}^\wedge$ is flat over S, we see that for any closed point s of S we have

$$Y_m(H/S) \times_S \operatorname{Spec} \mathcal{O}_{S,s}^\wedge \cong Y_m(H/\operatorname{Spec} \mathcal{O}_{S,s}^\wedge)$$

and so

$$Y_m(H/S) \times_S \operatorname{Spec} \mathcal{O}_{S,s}^\wedge \cong Y_m(\Sigma_{K,g}/\operatorname{Spec} \mathcal{O}_{S,s}^\wedge)$$
$$\cong \operatorname{Aut}^1(\Sigma_{K,g}[\wp_K^m]) \times \operatorname{Spec} \mathcal{O}_{S,s}^\wedge.$$

In particular $Y_m(H/S) \times_S \operatorname{Spec} \mathcal{O}^\wedge_{S,s}/\operatorname{Spec} \mathcal{O}^\wedge_{S,s}$ is finite and flat, and therefore $Y_m(H/S)$ is finite and flat over S. (Use the fact that $\mathcal{O}^\wedge_{S,s}$ is faithfully flat over $\mathcal{O}_{S,s}$. More precisely it suffices to show that for any closed point s of S the ring $\mathcal{O}_{Y_m(H/S)} \otimes_{\mathcal{O}_S} \mathcal{O}_{S,s}$ is finite and free over $\mathcal{O}_{S,s}$. We have seen that this is true after tensoring with $\mathcal{O}^\wedge_{S,s}$. Thus we can find a morphism of $\mathcal{O}_{S,s}$-modules $\mathcal{O}^a_{S,s} \to \mathcal{O}_{Y_m(H/S)} \otimes_{\mathcal{O}_S} \mathcal{O}_{S,s}$ (for some non-negative integer a) which becomes an isomorphism after tensoring with $\mathcal{O}^\wedge_{S,s}$. Faithful flatness implies that $\mathcal{O}^a_{S,s} \xrightarrow{\sim} \mathcal{O}_{Y_m(H/S)} \otimes_{\mathcal{O}_S} \mathcal{O}_{S,s}$.)

As $Y_m(H/S)$ is finite over S and S is excellent the previous lemma tells us that

$$Y_m(H/S)^{\mathrm{red}} \times \operatorname{Spec} \mathcal{O}^\wedge_{S,s} \cong (Y_m(H/S) \times \operatorname{Spec} \mathcal{O}^\wedge_{S,s})^{\mathrm{red}}.$$

Thus

$$J^{(m)}(H/S) \times_S \operatorname{Spec} \mathcal{O}^\wedge_{S,s} \cong J^{(m)}(\Sigma_{K,g}/\operatorname{Spec} \mathcal{O}^\wedge_{S,s})$$
$$\cong (\mathcal{O}_{D_{K,g}}/\wp^m_K \mathcal{O}_{D_{K,g}})^\times_{\operatorname{Spec} \mathcal{O}^\wedge_{S,s}},$$

which proves the proposition. $\square$

Although we will not need it in this book, it may be of interest to point out the following corollary of the proceeding proposition.

Corollary II.1.8 *Suppose that $S/k(v_K)^{\mathrm{ac}}$ is a reduced, connected scheme of finite type, and suppose that s is a geometric point of S. If H/S is a one-dimensional compatible formal Barsotti-Tate $\mathcal{O}_K$-module of height g then it gives rise to a continuous homomorphism*

$$\rho_H : \pi^{\mathrm{alg}}_1(S,s) \longrightarrow \mathcal{O}^\times_{D_{K,g}}.$$

Thus we get a bijection between isomorphism classes of one-dimensional compatible formal Barsotti-Tate $\mathcal{O}_K$-modules of height g on S and conjugacy classes of continuous homomorphisms

$$\rho : \pi^{\mathrm{alg}}_1(S,s) \longrightarrow \mathcal{O}^\times_{D_{K,g}}.$$

Proof: As we will not use this result elsewhere in this book we will simply sketch the proof.

First we explain the construction of ρ_H from H. Choose a compatible system of geometric points s_m of $J^{(m)}(H/S)$ above s (i.e. if $m' > m$ then $s_{m'}$ maps to s_m under $J^{(m')}(H/S) \to J^{(m)}(H/S)$). If $\sigma \in \pi^{\mathrm{alg}}_1(S,s)$ then

$$\sigma s_m = \rho_{H,m}(\sigma) s_m$$

for some (unique) element $\rho_{H,m}(\sigma) \in (\mathcal{O}_{D_{K,g}}/\wp^m_K \mathcal{O}_{D_{K,g}})^\times$. Moreover for $m' > m$ we have $\rho_{H,m'}(\sigma) \equiv \rho_{H,m}(\sigma) \bmod \wp^m_K$. We set

$$\rho_H = \varprojlim \rho_{H,m} : \pi^{\mathrm{alg}}_1(S,s) \to \mathcal{O}^\times_{D_{K,g}}.$$

It is a continuous homomorphism. The construction appears to depend on the choice of the system $\{s_m\}$, but a different choice simply changes ρ_H by conjugation in $\mathcal{O}_{D_{K,g}}^{\times}$.

Next we explain how one goes from a continuous homomorphism

$$\rho : \pi_1^{\text{alg}}(S, s) \to \mathcal{O}_{D_{K,g}}^{\times}$$

to a one-dimensional compatible formal Barsotti-Tate $\mathcal{O}_K$-module H_ρ/S. The reduction ρ mod $\wp_K^m$ gives rise to a Galois finite etale cover (not necessarily connected) $S_m \to S$ with Galois group $(\mathcal{O}_{D_{K,g}}/\wp_K^m \mathcal{O}_{D_{K,g}})^{\times}$. Consider

$$\Sigma_{K,g}[\wp_K^m] \times S_m \longrightarrow S_m$$

with the diagonal action of $(\mathcal{O}_{D_{K,g}}/\wp_K^m \mathcal{O}_{D_{K,g}})^{\times}$. We may quotient out by the action of this finite group and we obtain a finite flat group scheme H_m/S. We set $H_\rho = \lim_{\to} H_m$.

We leave the reader both to check that H_ρ/S is a one-dimensional compatible formal Barsotti-Tate $\mathcal{O}_K$-module, and that these two constructions are inverse to each other. $\square$

We end this section with some results about lifting extensions of compatible Barsotti-Tate $\mathcal{O}_K$-modules. We are very grateful to Johan de Jong for explaining to us how to prove corollary II.1.10 below.

Lemma II.1.9 *Let A be a noetherian $\mathcal{O}_K$-algebra and J an ideal of A which contains some power of p and which satisfies $J^2 = (0)$. Suppose that over $\operatorname{Spec} A/J$ we have an exact sequence of compatible Barsotti-Tate $\mathcal{O}_K$-modules*

$$(0) \longrightarrow H^0 \longrightarrow H \longrightarrow H^{\text{et}} \longrightarrow (0)$$

with H^0 formal and H^{et} ind-etale. Suppose moreover that $\widetilde{H}^0$ is a lift of H^0 to a compatible Barsotti-Tate $\mathcal{O}_K$-module over $\operatorname{Spec} A$. Then there is an exact sequence

$$(0) \longrightarrow \widetilde{H}^0 \longrightarrow \widetilde{H} \longrightarrow \widetilde{H}^{\text{et}} \longrightarrow (0)$$

of Barsotti-Tate $\mathcal{O}_K$-modules over $\operatorname{Spec} A$, which reduces modulo J to the above exact sequence.

Proof: As $\operatorname{Lie} \widetilde{H}^0 \overset{\sim}{\to} \operatorname{Lie} \widetilde{H}$ compatibility of $\widetilde{H}$ is automatic.

For the construction of $\widetilde{H}$ we use Grothendieck-Messing Dieudonne theory (see [Me] as completed by [I1]). This associates to

$$(0) \longrightarrow H^0 \longrightarrow H \longrightarrow H^{\text{et}} \longrightarrow (0)$$

an exact sequence of crystals in finite locally free modules

$$(0) \longrightarrow D(H^0) \longrightarrow D(H) \longrightarrow D(H^{\mathrm{et}}) \longrightarrow (0)$$

on $\operatorname{Spec} A/J$, with an action of $\mathcal{O}_K$. (For exactness use [BBM] combined with the compatibility of the theories in [Me] and [BBM], which is proved in [BM].) Moreover we have locally free, $\mathcal{O}_K$-invariant submodules $V(H^0) \subset D(H^0)_{A/J}$ (resp. $V(H) \subset D(H)_{A/J}$, resp. $V(H^{\mathrm{et}}) = D(H^{\mathrm{et}})_{A/J}$) with locally free quotients. To $\widetilde{H}^0$ we may associate a locally free $\mathcal{O}_K$-invariant submodule $V(\widetilde{H}^0) \subset D(H^0)_A$ with a locally free quotient and with $V(\widetilde{H}^0)/JV(\widetilde{H}^0) = V(H^0)$. We will look for a locally free $\mathcal{O}_K$-invariant submodule $V \subset D(H)_A$ with locally free quotient on which the two actions of $\mathcal{O}_K$ coincide, such that

- $V(\widetilde{H}^0) \to V$,

- $V \twoheadrightarrow D(H^{\mathrm{et}})_A$,

- and $V/JV = V(H)$.

Assuming we can find such a V we would get a complex of Barsotti-Tate $\mathcal{O}_K$-modules

$$(0) \longrightarrow \widetilde{H}^0 \longrightarrow \widetilde{H} \longrightarrow \widetilde{H}^{\mathrm{et}} \longrightarrow (0)$$

lifting

$$(0) \longrightarrow H^0 \longrightarrow H \longrightarrow H^{\mathrm{et}} \longrightarrow (0).$$

It follows from lemma 4.10 chapter II of [Me] that the lifted sequence is in fact exact.

It remains to construct such a V. This is equivalent to constructing an $A \otimes_{\mathbb{Z}_p} \mathcal{O}_K$-splitting for the exact sequence

$$(0) \longrightarrow D(H^0)_A/V(\widetilde{H}^0) \longrightarrow D(H)_A/V(\widetilde{H}^0) \longrightarrow D(H^{\mathrm{et}})_A \longrightarrow (0)$$

above the splitting of

$$(0) \longrightarrow D(H^0)_{A/J}/V(H^0) \longrightarrow D(H)_{A/J}/V(H^0) \longrightarrow D(H^{\mathrm{et}})_{A/J} \longrightarrow (0)$$

provided by $V(H)/V(H^0)$. As locally on A, $D(H^{\mathrm{et}})_A$ is free over $A \otimes_{\mathbb{Z}_p} \mathcal{O}_K$ we can find such a splitting Zariski locally on $\operatorname{Spec} A$. It is not unique but determined up to an element of

$$\operatorname{Hom}_{A/J \otimes_{\mathbb{Z}_p} \mathcal{O}_K}(D(H^{\mathrm{et}})_{A/J}, J(D(H^0)/V(\widetilde{H}^0))).$$

Thus the obstruction to the existence of a global splitting lies in

$$H^2(\operatorname{Spec} A/J, \operatorname{Hom}_{A/J \otimes_{\mathbb{Z}_p} \mathcal{O}_K}(D(H^{\mathrm{et}})_{A/J}, J(D(H^0)/V(\widetilde{H}^0)))).$$

Because $\operatorname{Spec} A/J$ is affine this group vanishes and so we can find such a splitting globally. $\square$

As an immediate consequence we obtain the following corollary.

Corollary II.1.10 *Let A be a noetherian $\mathcal{O}_K$-algebra complete with respect to the topology defined by an ideal I. Suppose that I contains a power of p. Suppose also that over $\operatorname{Spec} A/I$ we have an exact sequence of compatible Barsotti-Tate $\mathcal{O}_K$-modules*

$$(0) \longrightarrow H^0 \longrightarrow H \longrightarrow H^{\mathrm{et}} \longrightarrow (0)$$

with H^0 formal and H^{et} ind-etale. Suppose moreover that $\widetilde{H}^0$ is a lift of H^0 to a compatible Barsotti-Tate $\mathcal{O}_K$-module over $\operatorname{Spf} A$. Then there is an exact sequence

$$(0) \longrightarrow \widetilde{H}^0 \longrightarrow \widetilde{H} \longrightarrow \widetilde{H}^{\mathrm{et}} \longrightarrow (0)$$

of compatible Barsotti-Tate $\mathcal{O}_K$-modules over $\operatorname{Spf} A$, which reduces modulo I to the previous exact sequence.

II.2 Drinfeld level structures

Suppose that H/S is a Barsotti-Tate $\mathcal{O}_K$-module of constant height h over a scheme S. By a Drinfeld $\wp_K^m$-structure on H/S we shall mean a morphism of $\mathcal{O}_K$-modules

$$\alpha : (\wp_K^{-m}/\mathcal{O}_K)^h \longrightarrow H[\wp_K^m](S)$$

such that the set of $\alpha(x)$ for $x \in (\wp_K^{-m}/\mathcal{O}_K)^h$ forms a full set of sections of $H[\wp_K^m]$ in the sense of [KM] section 1.8. We will collect together here some of the basic properties of Drinfeld level structures.

Lemma II.2.1 *In this lemma S will denote an $\mathcal{O}_K$-scheme and H/S will be a Barsotti-Tate $\mathcal{O}_K$-module of constant height h.*

1. *Suppose that*

$$\alpha : (\wp_K^{-m}/\mathcal{O}_K)^h \longrightarrow H[\wp_K^m](S)$$

   is a Drinfeld $\wp_K^m$-structure and that T/S is any scheme. Then the composite

$$\alpha_T : (\wp_K^{-m}/\mathcal{O}_K)^h \xrightarrow{\alpha} H[\wp_K^m](S) \longrightarrow H[\wp_K^m](T)$$

   is a Drinfeld $\wp_K^m$-structure for $H \times_S T$.

2. *Suppose that $S/\mathbb{F}_p$ is reduced. If H/S is one-dimensional and formal then, for any t, H contains a unique finite flat subgroup scheme of order p^t, namely the kernel of F^t.*

3. *Suppose that $S/\mathbb{F}_p$ is reduced. If H/S is one-dimensional and formal then there is a unique Drinfeld $\wp_K^m$-structure on H/S, namely the trivial homomorphism*

$$\alpha^{\mathrm{triv}} : (\wp_K^{-m}/\mathcal{O}_K)^h \;\longrightarrow\; H(S)$$
$$x \;\longmapsto\; 0$$

for all $x \in (\wp_K^{-m}/\mathcal{O}_K)^h$. (We will refer to this as the trivial Drinfeld $\wp_K^m$-structure.)

4. *Suppose that S is connected and that over S there is an exact sequence of Barsotti-Tate $\mathcal{O}_K$-modules*

$$(0) \longrightarrow H^0 \longrightarrow H \longrightarrow H^{\mathrm{et}} \longrightarrow (0),$$

with H^0 formal and H^{et} ind-etale. Then

$$\alpha : (\wp_K^{-m}/\mathcal{O}_K)^h \to H[\wp_K^m](S)$$

is a Drinfeld $\wp_K^m$-structure if and only if there is a direct summand $\mathcal{O}_K$-submodule $M \subset (\wp_K^{-m}/\mathcal{O}_K)^h$ such that

- *$\alpha|_M : M \to H^0[\wp_K^m](S)$ is a Drinfeld $\wp_K^m$-structure,*
- *α induces an isomorphism*

$$\alpha : ((\wp_K^{-m}/\mathcal{O}_K)^h/M)_S \overset{\sim}{\to} H^{\mathrm{et}}[\wp_K^m].$$

5. *Suppose that S is reduced, connected and that $p = 0$ on S. Suppose also that there is an exact sequence of Barsotti-Tate $\mathcal{O}_K$-modules*

$$(0) \longrightarrow H^0 \longrightarrow H \longrightarrow H^{\mathrm{et}} \longrightarrow (0),$$

over S with H^0 formal and H^{et} ind-etale. If H/S admits a Drinfeld $\wp_K^m$-level structure (with $m \geq 1$) then there is a unique splitting

$$H[\wp_K^m] \cong H^0[\wp_K^m] \times H^{\mathrm{et}}[\wp_K^m]$$

over S. On the other hand if there is a splitting $H[\wp_K^m] \cong H^0[\wp_K^m] \times H^{\mathrm{et}}[\wp_K^m]/S$ then to give a Drinfeld $\wp_K^m$-structure $\alpha : (\wp_K^{-m}/\mathcal{O}_K)^h \to H[\wp_K^m](S)$ is the same as giving a direct summand $M \subset (\wp_K^{-m}/\mathcal{O}_K)^h$ and an isomorphism

$$((\wp_K^{-m}/\mathcal{O}_K)^h/M)_S \overset{\sim}{\to} H^{\mathrm{et}}[\wp_K^m].$$

6. For any $m \geq 0$ there is a scheme $S(m)$ which is finite over S and a
   Drinfeld $\wp_K^m$-structure

$$\alpha^{\mathrm{univ}} : (\wp_K^{-m}/\mathcal{O}_K)^h \longrightarrow H[\wp_K^m](S(m))$$

   on $H \times S(m)$, which is universal in the sense that if T/S is any
   S-scheme and if

$$\alpha : (\wp_K^{-m}/\mathcal{O}_K)^h \longrightarrow H[\wp_K^m](T)$$

   is any Drinfeld $\wp_K^m$-structure on $H \times T$ then $T \to S$ factors uniquely
   through $S(m)$ in such a way that α^{univ} pulls back to α. Moreover
   $S(m)/S$ has a right action of $GL_h(\mathcal{O}_K/\wp_K^m)$, which can be characterised as follows. If $g \in GL_h(\mathcal{O}_K/\wp_K^m)$ then under the morphism
   $g : S(m) \to S(m)$, α^{univ} pulls back to $\alpha^{\mathrm{univ}} \circ g$.

7. Suppose that $S = \mathrm{Spec}\, R$ with R an Artinian local ring in which p
   is nilpotent that H/S is one-dimensional and formal. Then $H \cong$
   $\mathrm{Spf}\, R[[T]]$. Choose a uniformiser $\varpi_K \in \mathcal{O}_K$ and let $f_{\varpi_K^m}(T) \in R[[T]]$
   be the power series representing multiplication by ϖ_K^m (i.e. $f_{\varpi_K^m}(T) =$
   $(\varpi_K^m)^*(T)$). Suppose that $\alpha : (\wp_K^{-m}/\mathcal{O}_K)^h \to H[\wp_K^m](R)$ is a homomorphism of $\mathcal{O}_K$-modules. Then the following are equivalent

   - α is a Drinfeld $\wp_K^m$-level structure,

   - $\prod_{x \in (\wp_K^{-m}/\mathcal{O}_K)^h}(T - T(\alpha(x))) | f_{\varpi_K^m}(T)$,

   - $f_{\varpi_K^m}(T) = g(T) \prod_{x \in (\wp_K^{-m}/\mathcal{O}_K)^h}(T - T(\alpha(x)))$ for some $g(T) \in$
     $R[[T]]^\times$.

8. Suppose that $S = \mathrm{Spec}\, R$ with R an Artinian local ring with residue
   field k of characteristic p. Suppose also that H/S is one-dimensional
   and formal, and that, for some $m \in \mathbb{Z}_{\geq 1}$, the trivial homomorphism
   $\alpha^{\mathrm{triv}} : (\wp_K^{-m}/\mathcal{O}_K)^h \to H(S)$ is a Drinfeld $\wp_K^m$-level structure. Then
   $p = 0$ in R and there is a Barsotti-Tate $\mathcal{O}_K$-module $H_0/\mathrm{Spec}\, k$ such
   that $H \cong H_0 \times_{\mathrm{Spec}\, k} S$.

Proof:

1. This follows from proposition 1.9.1 of [KM].

2. We may suppose that S is connected. From the discussion on page 26
of [Me] we see that $\ker F^s$ is locally isomorphic to $\mathrm{Spec}\, \mathcal{O}_S[T]/(T^{p^s})$. Thus
$\ker F^s$ is indeed a finite flat group scheme of rank p^s. If $A \subset H$ is any
finite flat subgroup scheme then we will show that $A = \ker F^s$ for some
s. Choose s maximal such that $\ker F^s \subset A$. We must show we actually
have equality. Modding out by $\ker F^s$ we may suppose that $s = 0$. We
must show that $A = (0)$. If for any point s of S, $\ker F|_{A_s} = (0)$ then as A_s

is connected we must have $A_s = (0)$. As S is connected the rank of A is constant and we would then have that $A = (0)$ as desired. Thus suppose that for all points s of S we have that $(\ker F_A)_s \neq (0)$. As $\ker F$ is finite flat of rank p we see that we must have $(\ker F_A)_s = (\ker F)_s$ for all points s of S. Over S we have $\ker F|_A$ is a closed subscheme of $\ker F$ which becomes equal to $\ker F$ when pulled back to any point of S. As S is reduced we must have $\ker F = \ker F|_A \subset A$, a contradiction.

3. To see that there is no more than one Drinfeld $\wp_K^m$-structure it suffices to check that $H[\wp_K^m](S) = \{0\}$. This follows because S is reduced and $H[\wp_K^m]/S$ is radical (see proposition 4.4 of chapter II of [Me]). (If $f : T \to S$ is finite and radical and if S is reduced then there is at most one section to f. To see this one reduces to the case that $S = \operatorname{Spec} A$ and $T = \operatorname{Spec} B$. We are looking for sections to $f^* : A \to B$. Suppose g_1^* and g_2^* are two such sections. As S is reduced we can embed A into a product of fields. Thus if $g_1^* \neq g_2^*$ we can find a field k and a homomorphism $\phi^* : A \to k$ such that $\phi^* \circ g_1^* \neq \phi^* \circ g_2^*$. On the other hand we must have $g_1^* \circ f^* = g_2^* \circ f^*$ and so $\phi^* \circ g_1^* \circ f^* = \phi^* \circ g_2^* \circ f^*$. This contradicts the fact that f is radical. (The finiteness hypothesis is presumably unnecessary, but this additional hypothesis does us no harm.))

It remains to show that α^{triv} is indeed a Drinfeld $\wp_K^m$-structure. We must have that $H[\wp_K^m] = \ker F^{f_K hm}$ and hence $H[\wp_K^m]$ is locally isomorphic to $\mathcal{O}_S[T]/(T^{p^{f_K hm}})$. If $f \in \mathcal{O}_S[T]/(T^{p^{f_K hm}})$ and we write $f = f_0 + f_1 T + \cdots + f_{p^{f_K hm}-1}T^{p^{f_K hm}-1}$, then the norm down to $\mathcal{O}_S$ of f is $f_0^{p^{f_K hm}} = f(0)^{p^{f_K hm}}$. This verifies condition (2) on page 33 of [KM].

4. This follows from proposition 1.11.2 and lemma 1.8.3 of [KM].

5. Suppose that $\alpha : (\wp_K^{-m}/\mathcal{O}_K)^h \to H[\wp_K^m](S)$ is a Drinfeld $\wp_K^m$-structure. Let $M = \ker \alpha$. By parts 3 and 4 we see that the composite

$$\alpha : ((\wp_K^{-m}/\mathcal{O}_K)/M)_S \longrightarrow H[\wp_K^m] \longrightarrow H^{\mathrm{et}}[\wp_K^m]$$

is an isomorphism. A splitting of $H^{\mathrm{et}}[\wp_K^m]$ into $H[\wp_K^m]$ is provided by the image of $((\wp_K^{-m}/\mathcal{O}_K)/M)_S$ in $H[\wp_K^m]$. To see the splitting is unique we argue as follows. To give a splitting is the same as giving a morphism $\gamma : ((\wp_K^{-m}/\mathcal{O}_K)/M)_S \to H[\wp_K^m]$ such that the composite

$$\gamma : ((\wp_K^{-m}/\mathcal{O}_K)/M)_S \longrightarrow H[\wp_K^m] \longrightarrow H^{\mathrm{et}}[\wp_K^m]$$

coincides with the map induced by α. To give γ is the same as giving

$$\gamma : ((\wp_K^{-m}/\mathcal{O}_K)/M) \longrightarrow H[\wp_K^m](S) \xrightarrow{\sim} H^{\mathrm{et}}[\wp_K^m](S).$$

Thus there is only one possible choice for γ. The second assertion of this part now follows from parts 3 and 4.

6. Let $S(m)$ be the closed subscheme of $T = H[\wp_K^m]^{(\wp_K^{-m}/\mathcal{O}_K)^h}$ where the tautological map

$$(\wp_K^{-m}/\mathcal{O}_K)^h \longrightarrow H[\wp_K^m](T)$$

is a Drinfeld $\wp_K^m$-structure (use lemma 1.9.1 of [KM]).

7. That $H \cong \operatorname{Spf} R[[T]]$ follows from page 26 of [Me]. Let T^s be the first power of T whose coefficient in $f_{\varpi_K^m}(T)$ is not in the maximal ideal of R. Consider the map

$$\bigoplus_{i=0}^{s-1} RT^i \longrightarrow R[[T]]/(f_{\varpi_K^m}(T)) \cong R^{p^{f_K mh}}.$$

After tensoring with the residue field of R we get an isomorphism. Thus this map is already an isomorphism and $s = p^{f_K mh}$. We conclude that $T^{p^{f_K mh}}$ is a linear combination of $1, T, \ldots, T^{p^{f_K mh}-1}$ in $R[[T]]/f_{\varpi_K^m}(T)$, and again by reducing modulo the maximal ideal of R we see that $T^{p^{f_K mh}}$ is a linear combination of $1, T, \ldots, T^{p^{f_K mh}-1}$ with coefficients in the maximal ideal of R. Put another way we can find a monic polynomial $h(T)$ of degree $p^{f_K mh}$ over R all whose nonleading coefficients are in the maximal ideal of R, and a power series $g(T) \in R[[T]]$ such that $h(T) = g(T)f_{\varpi_K^m}(T)$. We see at once that the constant term of $g(T)$ is a unit in R and hence that $g(T) \in R[[T]]^\times$. We see at once that the second and third conditions are equivalent. The first and third conditions are equivalent by lemma 1.10.2 of [KM].

8. This follows from proposition 5.3.4 of [KM] by the argument of section 5.3.6 of [KM]. $\square$

Suppose that $\mathcal{X}$ is a locally noetherian formal scheme with ideal of definition $\mathcal{I}$. We will let $\mathcal{X}_n$ denote the scheme with underlying topological space $\mathcal{X}$ and structure sheaf $\mathcal{O}_{\mathcal{X}}/\mathcal{I}^n$. By a Drinfeld $\wp_K^m$-structure on a Barsotti-Tate $\mathcal{O}_K$-module $\mathcal{H}/\mathcal{X}$ we shall mean a compatible system of Drinfeld $\wp_K^m$-structures

$$\alpha_n : (\wp_K^{-m}/\mathcal{O}_K)^h \longrightarrow \mathcal{H}[\wp_K^m](\mathcal{X}_n)$$

for $\mathcal{H} \times_{\mathcal{X}} \mathcal{X}_n$. This is easily checked to be canonically independent of the choice of ideal of definition $\mathcal{I}$.

If A is a noetherian ring complete with respect to the I-adic topology for some ideal I and if $H/\operatorname{Spec} A$ is a Barsotti-Tate $\mathcal{O}_K$-module with Drinfeld $\wp_K^m$-structure α, then we obtain a a natural Drinfeld $\wp_K^m$-structure α on the corresponding Barsotti-Tate $\mathcal{O}_K$-module $\mathcal{H}/\operatorname{Spf} A$ (take the push forward of α on $\mathcal{X}_n$ for each n). This establishes a bijection from Drinfeld $\wp_K^m$-structures on $H/\operatorname{Spec} A$ to Drinfeld $\wp_K^m$-structures on $\mathcal{H}/\operatorname{Spf} A$. (Given a compatible system

$$\alpha_n : (\wp_K^{-m}/\mathcal{O}_K)^h \longrightarrow H[\wp_K^m](A/I^n)$$

using the completeness of A we get in the limit a homomorphism

$$\alpha : (\wp_K^{-m}/\mathcal{O}_K)^h \longrightarrow H[\wp_K^m](A).$$

Using lemma 1.9.1 of [KM] we see that α is in fact a Drinfeld $\wp_K^m$-structure.)

Lemma II.2.2 *Let k be a separably closed field of characteristic p. Consider the functor from local Artinian $\mathcal{O}_K$-algebras A with residue field k to the category of sets, which assigns to A the set of isomorphism classes of triples (H, j, α) where H/A is a compatible Barsotti-Tate $\mathcal{O}_K$-module, where $j : \Sigma_{K,g} \overset{\sim}{\to} H \times_{\mathrm{Spec}\,A} \mathrm{Spec}\,k$ and where $\alpha : (\wp_K^{-m}/\mathcal{O}_K)^g \longrightarrow H[\wp_K^m](\mathrm{Spec}\,A)$ is a Drinfeld $\wp_K^m$-structure. This functor is pro-represented by a regular noetherian complete local ring $R_{K,g,m}$ which is finite and flat over $R_{K,g}$ of degree $\#GL_g(\mathcal{O}_K/\wp_K^m)$. We will denote the universal triple as above $(\widetilde{\Sigma}_{K,g}, \widetilde{j}, \widetilde{\alpha})$, which is defined not only over $\mathrm{Spf}\,R_{K,g,m}$ but over $\mathrm{Spec}\,R_{K,g,m}$.*

Proof: Define $R_{K,g,m}$ by

$$\mathrm{Spec}\,R_{K,g,m} = (\mathrm{Spec}\,R_{K,g})(m).$$

Then $R_{K,g,m}$ is local by part 3 of lemma II.2.1 and finite over $R_{K,g}$ by part 6 of lemma II.2.1. Over $R_{K,g} \otimes_{\mathbb{Z}_p} \mathbb{Q}_p$, $\widetilde{\Sigma}_{K,g}$ becaomes ind-etale and a Drinfeld $\wp_K^m$-level structure is nothing but an isomorphism

$$\alpha : (\wp_K^{-m}/\mathcal{O}_K)^h \overset{\sim}{\to} H[\wp_K^m].$$

Thus $(\mathrm{Spec}\,R_{K,g} \otimes_{\mathbb{Z}_p} \mathbb{Q}_p)(m)$ is finite flat (even etale) over $\mathrm{Spec}\,R_{K,g} \otimes_{\mathbb{Z}_p} \mathbb{Q}_p$ of degree $\#GL_g(\mathcal{O}_K/\wp_K^m)$. In particular $R_{K,g}$ and $R_{K,g,m}$ have the same dimension. If we can show that $R_{K,g,m}$ is regular then it will follow that $R_{K,g,m}/R_{K,g}$ is finite flat of the desired degree (because any finite morphism between regular schemes of the same dimension is flat).

Let $e_1, \ldots, e_g$ be an $\mathcal{O}_K/\wp_K^m$-basis of $(\wp_K^{-m}/\mathcal{O}_K)^g$, let $\mathfrak{m}$ denote the maximal ideal of $R_{K,g,m}$ and let T be a parameter on $\widetilde{\Sigma}_{K,g}$, i.e. $\widetilde{\Sigma}_{K,g} \cong \mathrm{Spf}\,R_{K,g,m}[[T]]$. Then $T(\widetilde{\alpha}(e_i)) \in \mathfrak{m}$ for all i and it follows from part 8 of lemma II.2.1 that in fact

$$\mathfrak{m} = (T(\widetilde{\alpha}(e_1)), \ldots, T(\widetilde{\alpha}(e_g)))$$

(as in the proof of theorem 5.3.2 of [KM]). Thus $R_{K,g,m}$ is regular, as desired. $\square$

So far we have followed the approach of Katz and Mazur [KM] to Drinfeld level structures. Their definitions differ slightly from Drinfeld's original one. The next corollary establishes the equivalence so that we will be able to appeal to results from Drinfeld's paper [Dr].

Corollary II.2.3 *Suppose that R is a complete noetherian local $\mathcal{O}_K$-algebra with residue field k of characteristic p. Let $H/\mathrm{Spf}\,R$ be a one-dimensional compatible formal Barsotti-Tate $\mathcal{O}_K$-module, so that there exists an isomorphism $H \cong \mathrm{Spf}\,R[[T]]$. Choose a uniformiser $\varpi_K \in \mathcal{O}_K$ and let $f_{\varpi_K}(T) \in R[[T]]$ be the power series representing multiplication by ϖ_K (in other words $f_{\varpi_K}(T) = (\varpi_K)^*(T)$). Suppose that $\alpha : (\wp_K^{-m}/\mathcal{O}_K)^h \to H[\wp_K^m](R)$ is a homomorphism of $\mathcal{O}_K$-modules. Then the following are equivalent*

- *α is a Drinfeld $\wp_K^m$-level structure,*

- *$\prod_{x \in (\wp_K^{-1}/\mathcal{O}_K)^h}(T - T(\alpha(x)))|f_{\varpi_K}(T)$,*

- *$f_{\varpi_K}(T) = g(T)\prod_{x \in (\wp_K^{-1}/\mathcal{O}_K)^h}(T - T(\alpha(x)))$ for some unit $g(T) \in R[[T]]^\times$.*

Proof: The equivalence of the second two conditions is easily established as in the proof of part 7 of lemma II.2.1. Let us first show that the first condition implies the second condition. Taking the completed tensor product with the Witt vectors of the separable closure of k we may suppose that k is separably closed. Then we may replace R by $R_{K,h,m}$. As $R_{K,h,m}$ is an integral domain (being regular) and $f_{\varpi_K}(T(\tilde{\alpha}(x))) = 0$ for all $x \in (\wp_K^{-1}/\mathcal{O}_K)^h$, it suffices to show that, if $x \neq y$ are elements of $(\wp_K^{-1}/\mathcal{O}_K)^h$ then $T(\tilde{\alpha}(x)) \neq T(\tilde{\alpha}(y))$, i.e. that $\tilde{\alpha}|_{(\wp_K^{-1}/\mathcal{O}_K)^h}$ is injective. As the generic point of $\mathrm{Spec}\,R_{K,h,m}$ is characteristic zero, $\tilde{\alpha}$ is injective.

Finally let us show that the equivalent conditions 2 and 3 imply condition 1. For that purpose we will (temporaily) call a homomorphism α satisfying conditions 2 and 3 an "original Drinfeld $\wp_K^m$-structure". Again we may reduce to the case that k is separably closed and hence that $H \times_{\mathrm{Spf}\,R} k \cong \Sigma_{K,h}$. Now consider the functor which associates to any local Artinian $\mathcal{O}_K$-algebra A with residue field $k(v)^{\mathrm{ac}}$ the set of isomorphism classes of triples (H, j, α) where H/A is a compatible Barsotti-Tate $\mathcal{O}_K$-module, where $j : \Sigma_{K,h} \overset{\sim}{\to} H \times_{\mathrm{Spf}\,A} \mathrm{Spec}\,k$ and where $\alpha : (\wp_K^{-m}/\mathcal{O}_K)^h \longrightarrow H[\wp_K^m](\mathrm{Spec}\,A)$ is an original Drinfeld $\wp_K^m$-structure. [Dr] (particularly proposition 4.3) tells us that this functor is represented by a regular finite flat local $R_{K,h}$-algebra, which we will temporarily denote $R'_{K,h,m}$. Again we may reduce to the case $R = R'_{K,h,m}$. Again, because $R'_{K,h,m}$ is an integral domain and $f_{\varpi_K^m}(T(\alpha(x))) = 0$ for all $x \in (\wp_K^{-m}/\mathcal{O}_K)^h$ it suffices to show that α is injective. Proposition 4.3 of [Dr] tells us that $R'_{K,h,m}/R'_{K,h,1}$ is finite and flat, so that it suffices to check this injectivity in the case $m = 1$. In this case the injectivity is checked in the last sentence of the proof of the lemma in the proof of proposition 4.3 of [Dr]. $\square$

We next turn to deformations of $H_0 = \Sigma_{K,g} \times (K/\mathcal{O}_K)^h/k(v_K)^{\mathrm{ac}}$. Fix a surjection

$$\delta : (\wp_K^{-m}/\mathcal{O}_K)^{g+h} \twoheadrightarrow \wp_K^{-m}TH_0/TH_0;$$

and let $GL_{g+h}(\mathcal{O}_K/\wp_K^m)_\delta$ denote the group of $x \in GL_{g+h}(\mathcal{O}_K/\wp_K^m)$ such that $\delta \circ x = \delta$. Then there is a short exact sequence

$$(0) \to \mathrm{Hom}\,(\wp_K^{-m}TH_0/TH_0, \ker \delta) \to GL_{g+h}(\mathcal{O}_K/\wp_K^m)_\delta$$
$$\to \mathrm{Aut}\,(\ker \delta) \to (0).$$

Consider the functor which associates to any local Artinian $\mathcal{O}_K$-algebra A with residue field $k(v)^{\mathrm{ac}}$ the set of isomorphism classes of triples (H, j, α) where H/A is a compatible Barsotti-Tate $\mathcal{O}_K$-module, where $j : \Sigma_{K,g} \xrightarrow{\sim} H \times_{\mathrm{Spec}\,A} \mathrm{Spec}\,k(v_K)^{\mathrm{ac}}$ and where $\alpha : (\wp_K^{-m}/\mathcal{O}_K)^{g+h} \longrightarrow H[\wp_K^m](\mathrm{Spec}\,A)$ is a Drinfeld $\wp_K^m$-structure such that the composite

$$(\wp_K^{-m}/\mathcal{O}_K)^{g+h} \xrightarrow{\alpha} H[\wp_K^m](\mathrm{Spec}\,A) \longrightarrow H[\wp_K^m](\mathrm{Spec}\,k(v_K)^{\mathrm{ac}}) \xrightarrow{j^{-1}}$$
$$\longrightarrow H_0[\wp_K^m](\mathrm{Spec}\,k(v_K)^{\mathrm{ac}}) = \wp_K^{-m}TH_0/TH_0$$

equals δ. This functor is pro-representable by $(\tilde{H}, \tilde{j}, \tilde{\alpha}_H)/\mathrm{Spf}\,R_\delta^{\mathrm{univ}}$ (by proposition 4.5 of [Dr]).

To describe how it is pro-represented it is convenient to also fix a homomorphism $\gamma : (\wp_K^{-m}/\mathcal{O}_K)^{g+h} \longrightarrow (\wp_K^{-m}/\mathcal{O}_K)^g$ such that

$$\delta \oplus \gamma : (\wp_K^{-m}/\mathcal{O}_K)^{g+h} \xrightarrow{\sim} (\wp_K^{-m}TH_0/TH_0) \oplus (\wp_K^{-m}/\mathcal{O}_K)^g.$$

This first of all gives rise to a splitting $\mathrm{Aut}\,(\ker \delta) \hookrightarrow GL_{g+h}(\mathcal{O}_K/\wp_K^m)_\delta$. Secondly it gives rise to an isomorphism of $\mathrm{Spf}\,R_\delta^{\mathrm{univ}}$ with

$$\mathrm{Hom}\,(\wp_K^{-m}TH_0, \tilde{\Sigma}_{K,g}) \times_{\mathrm{Spf}\,R_{K,g}} \mathrm{Spf}\,R_{K,g,m}.$$

(By $\mathrm{Hom}\,(\wp_K^{-m}TH_0, \tilde{\Sigma}_{K,g})/\mathrm{Spf}\,R_{K,g}$ we mean the formal scheme such that for any Artinian local $R_{K,g}$ algebra A we have $\mathrm{Hom}\,(\wp_K^{-m}TH_0, \tilde{\Sigma}_{K,g})(A) = \mathrm{Hom}\,(\wp_K^{-m}TH_0, \tilde{\Sigma}_{K,g}(A))$. Non-canonically it is isomorphic to $\tilde{\Sigma}_{K,g}^h$ and hence to $\mathrm{Spf}\,R_{K,g}[[T_1, \ldots, T_h]]$.) Over this ring we have both the pullback of the tautological extension

$$(0) \longrightarrow \tilde{\Sigma}_{K,g} \longrightarrow \tilde{H} \longrightarrow TH_0 \otimes (K/\mathcal{O}_K) \longrightarrow (0),$$

from $\mathrm{Hom}\,(TH_0, \tilde{\Sigma}_{K,g})$, and a second extension

$$(0) \longrightarrow \tilde{\Sigma}_{K,g} \longrightarrow \tilde{H}' \longrightarrow TH_0 \otimes (K/\wp_K^{-m}) \longrightarrow (0).$$

There is a natural isogeny $\widetilde{H} \longrightarrow \widetilde{H}'$ whose kernel projects isomorphically to $\wp_K^{-m} T H_0 / T H_0$, and so we get a splitting

$$i : \wp_K^{-m} T H_0 / T H_0 \hookrightarrow \widetilde{H}.$$

Then

$$\widetilde{\alpha}_H = \widetilde{\alpha} \circ \gamma + i \circ \delta.$$

(See the (rather sketchy) proof of proposition 4.5 in [Dr].)

The deformation space Spf R_δ^{univ} has a right action of $GL_{g+h}(\mathcal{O}_K / \wp_K^m)_\delta$ which may be described as follows. The splitting γ makes the group $GL_{g+h}(\mathcal{O}_K / \wp_K^m)_\delta$ the semidirect product of $\mathrm{Hom}\,(\wp_K^{-m} T H_0 / T H_0, \ker \delta)$ and $\mathrm{Aut}\,(\ker \delta)$. It thus suffices to describe the action of both these groups on Spf R_δ^{univ}. Now $\mathrm{Aut}\,(\ker \delta)$ is isomorphic via γ to $GL_h(\mathcal{O}_K / \wp_K^m)$ and simply acts on the factor Spf $R_{K,g,m}$. On the other hand composition with $\widetilde{\alpha}$ gives a map

$$\mathrm{Hom}\,(\wp_K^{-m} T H_0 / T H_0, \ker \delta) \longrightarrow \mathrm{Hom}\,(\wp_K^{-m} T H_0, \widetilde{\Sigma}_{K,g}(R_{K,g,m}))$$
$$= \mathrm{Hom}\,(\wp_K^{-m} T H_0, \widetilde{\Sigma}_{K,g})(R_{K,g.m}).$$

The action of an element $\phi \in \mathrm{Hom}\,(\wp_K^{-m} T H_0 / T H_0, \ker \delta)$ on

$$\mathrm{Hom}\,(\wp_K^{-m} T H_0, \widetilde{\Sigma}_{K,g}) \times_{\mathrm{Spf}\, R_{K,g}} \mathrm{Spf}\, R_{K,g,m}$$

is simply by translation by the image of ϕ in

$$\mathrm{Hom}\,(\wp_K^{-m} T H_0, \widetilde{\Sigma}_{K,g})(R_{K,g.m}).$$

If we write $\mathrm{Hom}\,(T H_0, \widetilde{\Sigma}_{K,g}) = \mathrm{Spf}\, R$ then $\widetilde{H}$ is defined over Spec R. Moreover we can identify $(\mathrm{Spec}\, R)(m)$ with

$$\coprod_\delta \mathrm{Spec}\, R_\delta^{\mathrm{univ}},$$

where the disjoint union is over surjections

$$\delta : (\wp_K^{-m} / \mathcal{O}_K)^{g+h} \twoheadrightarrow \wp_K^{-m} T H_0 / T H_0.$$

If $x \in GL_{g+h}(\mathcal{O}_K / \wp_K^m)$ then x takes R_δ^{univ} isomorphically to $R_{\delta \circ x}^{\mathrm{univ}}$. If $x \in GL_{g+h}(\mathcal{O}_K / \wp_K^m)_\delta$ then the two actions of x on Spec R_δ^{univ} coincide. In particular we see that $(\mathrm{Spec}\, R)(m)$ is regular and is finite and flat over Spec R of degree $\#GL_{g+h}(\mathcal{O}_K / \wp_K^m)$.

We now record a few more basic facts about Drinfeld level structures which can be proved by reduction to the universal case.

Lemma II.2.4 *In this lemma S will denote an $\mathcal{O}_K$-scheme which we will assume is locally noetherian with a dense set of points with residue field algebraic over $k(v)$. Also H/S will be a one-dimensional compatible Barsotti-Tate $\mathcal{O}_K$-module of constant height h.*

1. *$S(m)/S$ is finite, flat of degree $\#GL_h(\mathcal{O}_K/\wp_K^m)$.*

2. *Suppose that*

$$\alpha : (\wp_K^{-m}/\mathcal{O}_K)^h \longrightarrow H[\wp_K^m](S)$$

is a Drinfeld $\wp_K^m$-structure. Suppose also that $M \subset (\wp_K^{-m}/\mathcal{O}_K)^h$ is an $\mathcal{O}_K$-submodule. Then there is a unique $\mathcal{O}_K$-invariant finite flat subgroup scheme $\alpha(M) \subset H[\wp_K^m]$ such that the set of $\alpha(x)$ for $x \in M$ form a full set of sections for $\alpha(M)/S$. If moreover

$$\delta : (\wp_K^{-m'}/\mathcal{O}_K^h) \hookrightarrow (\wp_K^{-m}/\mathcal{O}_K)^h/M$$

is a map of $\mathcal{O}_K$-modules, then $\alpha \circ \delta$ is a Drinfeld $\wp_K^{m'}$ structure for $H/\alpha(M)$.

The construction of $\alpha(M)$ is compatible with base change in the following sense. If T/S is a locally noetherian S-scheme with a dense set of points with residue field algebraic over $k(v)$ and if

$$\alpha_T : (\wp_K^{-m}/\mathcal{O}_K)^h \xrightarrow{\alpha} H[\wp_K^m](S) \longrightarrow (H \times_S T)[\wp_K^m](T),$$

then the set of $\alpha_T(x)$ for $x \in M$ is a full set of sections for $\alpha(M) \times_S T$.

Proof: The first part is proved by a straightforward reduction to the universal formal case. We will prove the second part only, as the argument in this case is slightly more difficult. The last paragraph of the second part follows from lemma 1.9.1 of [KM]. Thus we concentrate on the proof of the first paragraph of part 2.

By corollary 1.10.3 of [KM] there is a unique closed subscheme $\alpha(M) \subset H[\wp_K^m]$ which is locally free over S and for which the set of $\alpha(x)$ for $x \in M$ form a full set of sections. (By uniqueness we need only check this locally on S. To apply corollary 1.10.3 of [KM] it suffices to find (locally on S) a closed immersion of $H[\wp_K^m]$ into the affine line. Suppose that $S = \operatorname{Spec} A$ and $H[\wp_K^m] = \operatorname{Spec} R$. It suffices to find for each maximal ideal $\mathfrak{m}$ of A an element $f \in A - \mathfrak{m}$ and a surjection $A_f[T] \twoheadrightarrow R_f$. By Nakayama's lemma it suffices to find a surjection $A/\mathfrak{m}[T] \twoheadrightarrow R/\mathfrak{m}$, i.e. we have reduced to the case where A is a field. One can further reduce to the case where A is algebraically closed. (Let $e_1 \dots ,e_n$ be a basis of R over A. A morphism $A[T] \to R$ is given by $T \mapsto \sum t_i e_i$ with $t_i \in A$. Surjectivity is expressed as the non-vanishing of a polynomial in the t_i's expressing the fact that the images of $1, T, \ldots, T^{n-1}$ are linearly independent.) In this case as a scheme we have $H[\wp_K^m] = \coprod H^0[\wp_K^m]$ so we may reduce to the case that H is formal and so $R = \operatorname{Spec} A[[T]]/(T^N)$ for some N. Then $A[T] \to A[[T]] \to R$ is surjective.) From the uniqueness it follows that $\alpha(M)$ is invariant by the action of $\mathcal{O}_K^\times$. Thus it suffices to check that

1. $\alpha(M)$ is a subgroup scheme;

2. if

$$\delta : (\wp_K^{-m'}/\mathcal{O}_K^h) \hookrightarrow (\wp_K^{-m}/\mathcal{O}_K)^h/M$$

is a map of $\mathcal{O}_K$-modules, then $\alpha \circ \delta$ is a Drinfeld $\wp_K^{m'}$ structure for $H/\alpha(M)$.

There is a closed subscheme $S' \subset S$ such that for any scheme T/S, $\alpha(M) \times_S T \subset H \times_S T$ has the two properties above if and only if $T \to S$ factors through S'. ($H[\wp_K^m]$ is the spectrum of a sheaf of locally free Hopf algebras $\mathcal{H}/S$. Let $\mathcal{I}$ denote the subsheaf of ideals defining $\alpha(M)$. The first property above is equivalent to the composite map

$$\mathcal{I} \hookrightarrow \mathcal{H} \longrightarrow \mathcal{H} \otimes \mathcal{H} \longrightarrow \mathcal{H}/\mathcal{I} \otimes \mathcal{H}/\mathcal{I},$$

where the middle map is the comultiplication, being zero. It follows that there exists a closed $S'' \subset S$ universal for the truth of the first property. The existence of $S' \subset S''$ follows from proposition 1.9.1 of [KM].) What we must show is that $S' = S$.

We may at once reduce to the case that $S = \operatorname{Spec} A$ for an Artinian local ring A. (Look at $\mathcal{O}_{S,s}/\mathfrak{m}_s^a$ as s runs over a dense set of points and a runs over positive integers.) Then by tensoring with $W(k(v_K)^{\mathrm{ac}})$ we may assume that the residue field of A is $k(v_K)^{\mathrm{ac}}$. Next we may replace H/A by the universal deformation of $H \times \operatorname{Spec} k(v_K)^{\mathrm{ac}}$ together with its Drinfeld $\wp_K^m$-structure, and so we may suppose that $S = \operatorname{Spec} R$ for a complete noetherian local ring which is flat over $\mathcal{O}_{\widehat{K}^{\mathrm{nr}}}$. Then we may replace S by $S \times \operatorname{Spec} \mathbb{Q}_p$, which is dense in S. But $H/S \times \operatorname{Spec} \mathbb{Q}_p$ is ind-etale and the result is easy. $\square$

The following corollary follows readily.

Corollary II.2.5 *In this lemma $\mathcal{X}$ will denote a locally noetherian $\mathcal{O}_K$-formal scheme. We will assume that $\mathcal{X}^{\mathrm{red}}$ has a dense set of points with residue field algebraic over $k(v)$. Also $\mathcal{H}/\mathcal{X}$ will denote a one-dimensional compatible Barsotti-Tate $\mathcal{O}_K$-module of constant height h.*

1. *There is a formal scheme $\mathcal{X}(m)/\mathcal{X}$ and a Drinfeld $\wp_K^m$-level structure α^{univ} on $\mathcal{H} \times_{\mathcal{X}} \mathcal{X}(m)$ which is universal in the following sense. If*

$\mathcal{Y} \to \mathcal{X}$ *is any morphism of formal schemes and if* δ *is a Drinfeld* $\wp_K^m$-*level structure on* $\mathcal{H} \times_{\mathcal{X}} \mathcal{Y}$ *then there is a unique morphism* $\mathcal{Y} \to \mathcal{X}(m)$ *over* $\mathcal{X}$ *under which* α^{univ} *pulls back to* δ. *Moreover* $\mathcal{X}(m)/\mathcal{X}$ *is finite, flat of degree* $\#GL_h(\mathcal{O}_K/\wp_K^m)$.

2. *Suppose that* α *is a Drinfeld* $\wp_K^m$-*structure on* $\mathcal{H}/\mathcal{X}$. *Suppose also that* $M \subset (\wp_K^{-m}/\mathcal{O}_K)^h$ *is a* $\mathcal{O}_K$-*submodule. Then we can find a compatible Barsotti-Tate* $\mathcal{O}_K$-*module* $\mathcal{H}/\alpha(M)$ *over* $\mathcal{X}$ *and a morphism* $\mathcal{H} \to \mathcal{H}/\alpha(M)$ *over* $\mathcal{X}$ *such that when restricted to any closed subscheme* $X \subset \mathcal{X}$ *the morphism*

$$\mathcal{H}|_X \longrightarrow (\mathcal{H}/\alpha(M))|_X$$

is surjective and the set of $\alpha|_X(x)$ *for* $x \in M$ *form a complete set of sections for the kernel.*

The construction of $\mathcal{H}/\alpha(M)$ *is compatible with base change in the following sense. If* $\mathcal{Y}/\mathcal{X}$ *is a locally noetherian* $\mathcal{X}$-*formal scheme such that* $\mathcal{Y}^{\mathrm{red}}$ *has a dense set of points with residue field algebraic over* $k(v)$ *and if* $\mathcal{H}_{\mathcal{Y}}$ *(resp.* $\alpha_{\mathcal{Y}}$*) denotes the pullback of* $\mathcal{H}$ *(resp.* α*) to* $\mathcal{Y}$, *then* $\mathcal{H}_{\mathcal{Y}}/\alpha_{\mathcal{Y}}(M)$ *is canonically the pullback of* $\mathcal{H}/\alpha(M)$ *to* $\mathcal{Y}$.

We have seen that $R_{K,g}$ has a natural continuous left action of $\mathcal{O}_{D_{K,g}}^{\times}$. The same is true of $R_{K,g,m}$ and so in fact $R_{K,g,m}$ has a continuous left action of

$$GL_g(\mathcal{O}_K) \times \mathcal{O}_{D_{K,g}}^{\times} \twoheadrightarrow GL_g(\mathcal{O}_K/\wp_K^m) \times \mathcal{O}_{D_{K,g}}^{\times}.$$

In fact we can extend this action to a continuous left action of $GL_g(K) \times D_{K,g}^{\times}$ on the direct system of the $R_{K,g,m}$ such that

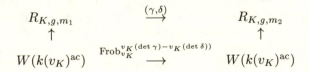

commutes if $m_2 >> m_1$.

To describe this first suppose that $(\gamma, \delta) \in GL_g(K) \times D_{K,g}^{\times}$, that $\gamma^{-1} \in M_g(\mathcal{O}_K)$, that $v_K(\det(\delta)) \leq 0$ and that $\gamma \mathcal{O}_K^g \subset \wp_K^{m_1 - m_2} \mathcal{O}_K^g$. It suffices to define a $W(k(v_K)^{\mathrm{ac}})$-linear map

$$(\gamma, \delta) : R_{K,g,m_1} \longrightarrow R_{K,g,m_2} \widehat{\otimes}_{W(k(v)^{\mathrm{ac}}),\mathrm{Frob}_{v_K}^{v_K(\det \delta) - v_K(\det \gamma)}} W(k(v)^{\mathrm{ac}}).$$

Next, by the universal property of R_{K,g,m_1} it suffices to give a deformation of $(\Sigma_{K,g}, j, \alpha^{\mathrm{triv}})$ to $R_{K,g,m_2} \widehat{\otimes}_{W(k(v)^{\mathrm{ac}}),\mathrm{Frob}_{v_K}^{v_K(\det \delta) - v_K(\det \gamma)}} W(k(v)^{\mathrm{ac}})$. By lemma II.2.4 there is a unique finite flat subgroup scheme $A \subset \widetilde{\Sigma}_{K,g}$ over R_{K,g,m_1} such that the set of $\alpha^{\mathrm{univ}}(x)$ for $x \in (\gamma \mathcal{O}_K^g)/\mathcal{O}_K^g$ are a complete set of sections for A. For our deformation of $\Sigma_{K,g}$ we will take

$$(\widetilde{\Sigma}_{K,g}/A) \times_{\mathrm{Spf}\, W(k(v_K)^{\mathrm{ac}}),(\mathrm{Frob}^*_{v_K})^{v_K \,(\det \delta) - v_K \,(\det \gamma)}} \mathrm{Spf}\, W(k(v_K)^{\mathrm{ac}}).$$

It has a Drinfeld $\wp_K^{m_1}$-structure coming from

$$(\wp_K^{-m_1}/\mathcal{O}_K)^g \overset{\gamma}{\hookrightarrow} \wp_K^{-m_2}\mathcal{O}_K^g/(\gamma \mathcal{O}_K^g) \overset{\alpha^{\mathrm{univ}}}{\longrightarrow} (\widetilde{\Sigma}_{K,g}/A)(R_{K,g,m_2}).$$

(Use lemma II.2.4.) Reducing modulo the maximal ideal of

$$R_{K,g,m_2} \widehat{\otimes}_{W(k(v)^{\mathrm{ac}}),\mathrm{Frob}_{v_K}^{v_K \,(\det \delta) - v_K \,(\det \gamma)}} W(k(v)^{\mathrm{ac}})$$

and using $\widetilde{j}$ we obtain an isomorphism between

$$\Sigma_{K,g}^{(p^{-f_K v_K \,(\det \gamma)})} \times_{\mathrm{Spec}\, k(v_K)^{\mathrm{ac}},(\mathrm{Fr}^*)^{f_K \,(v_K \,(\det \gamma) - v_K \,(\det \delta))}} \mathrm{Spec}\, k(v_K)^{\mathrm{ac}}$$

and $\Sigma_{K,g}^{(p^{-f_K v_K \,(\det \delta)})}$. Finally we identify this with $\Sigma_{K,g}$ via

$$\delta^{-1} : \Sigma_{K,g}^{(p^{-v_K \,(\det \delta)})} \overset{\sim}{\to} \Sigma_{K,g}.$$

We note that if $x \in \mathcal{O}_K$ and $x \neq 0$ then the element (x^{-1}, x^{-1}) acts trivially. Thus we obtain an action of

$$GL_g(K) \times D_{K,g}^\times \twoheadrightarrow (GL_g(K) \times D_{K,g}^\times)/K^\times,$$

where $K^\times$ is embedded diagonally.

Lemma II.2.6 *Let us fix K, g and m. For each positive integer s we can find an integer $N(s)$ such that*

- *$N(s)$ increases monotonically to infinity as $s \to \infty$;*

- *any element of $\varpi_K^{s-m} \mathcal{O}_{D_{K,g}}$ lifts to an endomorphism of*

$$\widetilde{\Sigma}_{K,g} \times (R_{K,g,m}/\mathfrak{m}_{R_{K,g,m}}^{N(s)})$$

- *and $(1 + \varpi_K^s \mathcal{O}_{D_{K,g}})$ acts trivially on $R_{K,g,m}/\mathfrak{m}_{R_{K,g,m}}^{N(s)}$.*

Proof: Note that if we can choose $N(s)$ satisfying the first two conditions, then the third will also be satisfied. (If $\delta \in \mathcal{O}_{D_g}^\times$ with $\delta \equiv 1 \bmod \varpi_K^s$ then $(\delta - 1)/\varpi_K^m$ and $(\delta^{-1} - 1)/\varpi_K^m$ lift to endomorphisms of $\widetilde{\Sigma}_{K,g} \times (R_{K,g,m}/\mathfrak{m}_{R_{K,g,m}}^{N(s)})$. By the uniqueness of such lifts (see part

2 of lemma 1.1.3 of [Kat]) we see that δ lifts to an automorphism of $\widetilde{\Sigma}_{K,g} \times (R_{K,g,m}/\mathfrak{m}_{R_{K,g,m}}^{N(s)})$ which is the identity on ϖ_K^m-torsion.)

Now take $N(s)$ to be the integer part of $\sqrt{(s/e_{K/\mathbb{Q}_p} - m)}$ (or 0 if this is not defined). As $p^{N(s)}$ is zero on $(R_{K,g,m}/\mathfrak{m}_{R_{K,g,m}}^{N(s)})$ the second condition on $N(s)$ follows from part 3 of lemma 1.1.3 of [Kat]. $\square$

The following lemma will be proved in section III.4.

Lemma II.2.7 *We can find an inverse system of proper schemes of finite type $X_m/\mathcal{O}_{\widehat{K}^{\mathrm{nr}}}$ with compatible actions of $GL_g(\mathcal{O}_K)$ and a closed point $x \in X_0$ such that*

1. $\ker(GL_g(\mathcal{O}_K) \to GL_g(\mathcal{O}_K/\wp^m))$ *acts trivially on X_m,*

2. *the generic fibre of X_m/X_0 is finite, etale and Galois with Galois group $GL_g(\mathcal{O}_K/\wp^m)$,*

3. *x has a unique pre-image x_m in X_m,*

4. *and the inverse system of formal schemes $(X_m)^\wedge_{x_m}$, together with its action of $GL_g(\mathcal{O}_K)$ is isomorphic to the inverse system of the Spf $R_{K,g,m}$.*

Suppose that l is a prime not equal to p. We will let $\Psi^i_{K,l,g,m}$ denote $R^i\Psi_\eta(\mathbb{Q}_l^{\mathrm{ac}})_{(\mathrm{Spf}\, R_{K,g,m})_\eta}$, the ith vanishing cycle sheaf with coefficients in $\mathbb{Q}_l^{\mathrm{ac}}$ for Spf $R_{K,g,m}$ in the sense of Berkovich (see section I.5). Thus $\Psi^i_{K,l,g,m}$ is a finite dimensional $\mathbb{Q}_l^{\mathrm{ac}}$-vector space. We set $\Psi^i_{K,l,g} = \lim_{\to m} \Psi^i_{K,l,g,m}$ and we may drop the subscripts K and/or l when no confusion can arise. Recall that $A_{K,g}$ denotes the group of $(\gamma, \delta, \sigma) \in GL_g(K) \times D_{K,g}^\times \times W_K$ such that

$$v_K(\det \gamma) = v_K(\det \delta) + v_K(\sigma).$$

The action of $GL_g(K) \times D_{K,g}^\times$ on the tower of the $R_{K,g,m}$ gives rise to an action of $A_{K,g}$ on the tower $R_{K,g,m} \otimes_{\mathcal{O}_{\widehat{K}^{\mathrm{nr}}}} L$ for any finite extension $L/\widehat{K}^{\mathrm{nr}}$. More precisely $(\gamma, \delta, \sigma) \in A_{K,g}$ acts as $(\gamma, \delta) \otimes \sigma$. In this way we get a left action of $A_{K,g}$ on $\Psi^i_{K,l,g}$.

Lemma II.2.8 *The action of $A_{K,g}$ on $\Psi^i_{K,l,g}$ is admissible/continuous.*

Proof: The action of $GL_g(\mathcal{O}_K)$ is smooth from the definitions. It follows from lemma II.2.6, corollary 4.5 of [Berk3] and lemma II.2.7 that the action of $\mathcal{O}_{D_{K,g}}^\times$ is smooth. Let X_m denote the kernel of $GL_g(\mathcal{O}_K) \to GL_g(\mathcal{O}_K/\wp_K^m)$. It follows from lemmas II.2.7 and I.5.4 that $\Psi^i_{K,l,g,m} \overset{\sim}{\to}$

$(\Psi^i_{K,l,g})^{X_m}$. Finally it follows from lemma II.2.7, the comparison theorem of [Berk3] and lemma I.5.1 that $\Psi^i_{K,l,g,m}$ is finite dimensional and has a continuous action of I_K. $\square$

If ρ is a irreducible admissible representation of $D^\times_{K,g}$ over $\mathbb{Q}^{\mathrm{ac}}_l$ (and hence necessarily finite dimensional) then we set

$$\Psi^i_{K,l,g}(\rho) = \mathrm{Hom}_{\mathcal{O}^\times_{D_{K,g}}}(\rho, \Psi^i_{K,l,g}).$$

This is naturally an admissible $GL_g(K) \times W_K$-module. More precisely

$$((\gamma, \sigma)\phi)(x) = (\gamma, \delta, \sigma)\phi(\rho(\delta)^{-1}x),$$

for any $\delta \in D^\times_{K,g}$ with $v_K(\det \delta) = v_K(\det \gamma) - v_K(\sigma)$. Define a homomorphism $d_g : GL_g(K) \times W_K \to \mathbb{Z}$ by $d_g(\gamma, \sigma) = v_K(\det \gamma) - v_K(\sigma)$. The following lemma is immediate.

Lemma II.2.9 *If* $\psi : \mathbb{Z} \to (\mathbb{Q}^{\mathrm{ac}}_l)^\times$ *then*

$$\Psi^i_{K,l,g}(\rho \otimes (\psi \circ v_K \circ \det)) \cong \Psi^i_{K,l,g}(\rho) \otimes (\psi^{-1} \circ d_g).$$

There is a natural map of $A_{K,g}$-modules

$$\Psi^i_{K,l,g}(\rho) \otimes \rho \longrightarrow \Psi^i_{K,l,g},$$

which sends $f \otimes v$ to $f(v)$. We will denote the image of this map $\Psi^i_{K,l,g}[\rho]$ and we will let $\Psi^i_{K,l,g,m}[\rho]$ denote the preimage of $\Psi^i_{K,l,g}[\rho]$ in $\Psi^i_{K,l,g,m}$. We will call irreducible admissible representations of $D^\times_g$ *inertially equivalent* if they differ by twisting by a character of the form $\psi \circ v_K \circ \det$ for some character $\psi : \mathbb{Z} \to (\mathbb{Q}^{\mathrm{ac}}_l)^\times$. The submodule $\Psi^i_{K,l,g}[\rho]$ only depends on the inertial equivalence class of ρ. Because $\mathcal{O}^\times_{D_{K,g}}$ is compact it is easy to check that

$$\Psi^i_{K,l,g} \cong \bigoplus \Psi^i_{K,l,g}[\rho]$$

where the sum is over one representative of each inertial equivalence class of irreducible admissible representation of $D^\times_{K,g}$.

We will let $e[\rho]$ denote the number of irreducible components of $\rho|_{\mathcal{O}^\times_{D_{K,g}}}$. It also equals the number of characters $\psi : \mathbb{Z} \longrightarrow (\mathbb{Q}^{\mathrm{ac}}_l)^\times$ such that $\rho \cong \rho \otimes (\psi \circ v_K \circ \det)$. Let $\Delta[\rho]$ be a set of $e[\rho]$ elements $\delta \in D^\times_{K,g}$ such that the set of $v_K(\det \delta)$ for $\delta \in \Delta[\rho]$ run over a set of representatives of the congruence classes mod $e[\rho]$. If $\delta \in D^\times_{K,g}$ we will we will let

$$\Psi^i_{K,l,g}[\rho]^\delta$$

denote $\Psi^i_{K,l,g}[\rho]$ but with its $A_{K,g}$ action twisted so that $(\gamma, \epsilon, \sigma)$ acts via $(\gamma, \delta^{-1}\epsilon\delta, \sigma)$. Then there is an isomorphism of $A_{K,g}$-modules

$$\Psi^i_{K,l,g}(\rho) \otimes \rho \xrightarrow{\sim} \bigoplus_{\delta \in \Delta[\rho]} \Psi^i_{K,l,g}[\rho]^\delta,$$

which sends $f \otimes v$ to $(f(\delta^{-1}v))_\delta$.

We also introduce the virtual $GL_g(K) \times W_K$-module

$$[\Psi_{K,l,g}(\rho)] = \sum_{i=0}^{g-1}(-1)^{g-1-i}[\Psi^i_{K,l,g}(\rho)].$$

Lemma II.2.10   *1. If ψ is a character of $K^\times/\mathcal{O}_K^\times$ then*

$$\Psi^0_{K,l,g}(\psi \circ \det)^{GL_g(\mathcal{O}_K)} = \overline{\mathbb{Q}}_l^{ac}(\psi \circ \operatorname{Art}_K^{-1}).$$

2. If $i > 0$ or if ρ does not factor through the map $\det : D_{K,g}^\times \to K^\times/\mathcal{O}_K^\times$ then

$$\Psi^i_{K,l,g}(\rho)^{GL_g(\mathcal{O}_K)} = (0).$$

Proof: This follows easily on combining lemma I.5.5 with the equality

$$(\Psi^i_{K,l,g})^{GL_g(\mathcal{O}_K)} = \Psi^i_{K,l,g,0}$$

(see the proof of lemma II.2.8). $\square$

The following lemma is proved in [Car3] (using Lubin-Tate theory [LT]).

Lemma II.2.11 *If ρ is a character of $K^\times$ then*

$$\Psi_{K,l,1}(\rho) \cong \rho^{-1} \otimes \rho \circ \operatorname{Art}_K^{-1}.$$

Chapter III

Some simple Shimura varieties

In this section we shall introduce some Shimura varieties which will be the main object of study in this book. Our class is a sub-class of the class singled out by Kottwitz in [Ko4]. Our class combines the simple group theoretic properties of Kottwitz's examples with some additional simple geometric properties.

We will use without comment notation established in section I.7.

III.1 Characteristic zero theory

Let us recall a lemma about abelian schemes.

Lemma III.1.1 *Let L be a totally real field, l be a rational prime, S an L-scheme, A/S an abelian scheme and $i : L \to \mathrm{End}^0(A)$ an embedding. Suppose that for all closed points $s \in S$ the Tate module $V_l A$ is a free F_l^+-module. Then locally on S, $\mathrm{Lie}\, A$ is a free $\mathcal{O}_S \otimes_{\mathbb{Q}} L$-module.*

Proof: One may first reduce to the case that S/L is finite type and then using lemma I.6.1 to the case that $S = \mathrm{Spec}\,\mathbb{C}$. Again by lemma I.6.1 we see that $H_1(A(\mathbb{C}), \mathbb{Q})$ is a free L-module and so $\mathrm{Lie}\, A \cong H_1(A(\mathbb{C}), \mathbb{R})$ is a free $\mathbb{R} \otimes_{\mathbb{Q}} L$-module. As L is totally real this implies that $\mathrm{Lie}\, A$ is a free $\mathbb{C} \otimes_{\mathbb{Q}} L$-module. $\square$

Now suppose that S/F is a scheme and that A/S is an abelian scheme of dimension dn^2. Suppose also that $i : B \hookrightarrow \mathrm{End}^0(A)$. Then $\mathrm{Lie}\, A$ is a locally free $\mathcal{O}_S$ module of rank dn^2 with an action of B. We will decompose

$$\mathrm{Lie}\, A = \mathrm{Lie}^+ A \oplus \mathrm{Lie}^- A,$$

where $\mathrm{Lie}^{\pm}A = \mathrm{Lie}\,A \otimes_{\mathcal{O}_S \otimes E} \mathcal{O}_S$ where the implicit map $E \to \mathcal{O}_S$ is the natural map in the $+$ case and the complex conjugate of the natural map in the $-$ case. Both $\mathrm{Lie}^{\pm}A$ are locally free $\mathcal{O}_S$-modules. In this book we will call the pair (A,i) *compatible* if $\mathrm{Lie}^{+}A$ has rank n and the two natural actions of F^{+} (one from the structure map $F^{+} \to \mathcal{O}_S$ and one from the action of B) on $\mathrm{Lie}^{+}A$ coincide. We remind the reader that there are various equivalent formulations of compatibility.

Lemma III.1.2 *Keep the above notation and assumptions. Then the following conditions are equivalent.*

1. *(A,i) is compatible.*

2. *For all $b \in B$ we have*

$$\mathrm{tr}\,(b|_{Lie(A)}) = (c \circ \mathrm{tr}\,_{B/E})(nb) + \mathrm{tr}\,_{B/F}(b) - (c \circ \mathrm{tr}\,_{B/F})(b).$$

3. *For all $b \in F$ we have*

$$\mathrm{tr}\,(b|_{Lie(A)}) = (c \circ \mathrm{tr}\,_{F/E})(n^2 b) + nb - c(nb).$$

4. *For all $b \in F^{+}$ we have*

$$\mathrm{tr}\,(b|_{Lie+(A)}) = nb$$

and

$$\mathrm{tr}\,(b|_{Lie-(A)}) = \mathrm{tr}\,_{F+/\mathbb{Q}}(n^2 b) - nb.$$

5. *For all $b \in F^{+}$ we have*

$$\mathrm{tr}\,(b|_{Lie+(A)}) = nb.$$

Proof: The equivalence of the first and last conditions follows from lemma I.6.1. The equivalence of the last two conditions follows from lemma III.1.1. The equivalence of the third and fourth conditions is elementary by extending the third condition to all $b \in E \otimes_{\mathbb{Q}} F \subset \mathcal{O}_S \otimes_{\mathbb{Q}} F$ and using the decomposition $E \otimes_{\mathbb{Q}} F = F \oplus F$. The equivalence of the second and third conditions follows from lemma I.6.1. $\square$

Let U be a sufficiently small open compact subgroup of $G(\mathbb{A}^{\infty})$. By sufficiently small we shall mean that the projection of U to $G(\mathbb{Q}_x)$ for some prime x contains no element of finite order other than 1. We will define a functor $\mathfrak{X}_U$ from locally noetherian F-schemes to sets. If $S = \coprod S_i$ with S_i connected for each i we set

$$\mathfrak{X}_U(S) = \prod_i \mathfrak{X}_U(S_i).$$

Thus we need only define $\mathfrak{X}_U$ on connected, locally noetherian F-schemes. In fact we will define $\mathfrak{X}_U$ on the category of pairs (S, s) where S is a connected, locally noetherian F-scheme and where s a geometric point of S. We will then remark that $\mathfrak{X}_U(S, s)$ is canonically independent of s. We set $\mathfrak{X}_U(S, s)$ to be the set of equivalence classes of quadruples $(A, \lambda, i, \overline{\eta})$ where

- A is an abelian scheme of dimension dn^2;

- $\lambda : A \to A^\vee$ is a polarisation;

- $i : B \hookrightarrow \mathrm{End}^0(A)$ such that (A, i) is compatible and $\lambda \circ i(b) = i(b^*)^\vee \circ \lambda$ for all $b \in B$;

- $\overline{\eta}$ is a $\pi_1(S, s)$-invariant U-orbit of isomorphisms of $B \otimes_{\mathbb{Q}} \mathbb{A}^\infty$-modules $\eta : V \otimes_{\mathbb{Q}} \mathbb{A}^\infty \to VA_s$ which take the standard pairing $(\ ,\ )$ on V to a $(\mathbb{A}^\infty)^\times$-multiple of the λ-Weil pairing on VA_s (see [Ko3] pages 390-391).

We consider two quadruples $(A, \lambda, i, \overline{\eta})$ and $(A', \lambda', i', \overline{\eta}')$ equivalent if there is an isogeny $\alpha : A \to A'$ which takes λ to a $\mathbb{Q}^\times$ multiple of λ', takes i to i' and takes $\overline{\eta}$ to $\overline{\eta}'$ (see [Ko3] page 390). If s' is a second geometric point of S then there is a canonical bijection between $\mathfrak{X}_U(S, s)$ and $\mathfrak{X}_U(S, s')$ (see [Ko3] page 391).

Because U is sufficiently small this functor is represented by a smooth projective scheme X_U/F (see [Ko3] page 391). If $V \subset U$ there is a natural finite etale map $X_V \twoheadrightarrow X_U$. There is also a natural right action of $G(\mathbb{A}^\infty)$ on the inverse system of the X_U: if $g^{-1}Vg \subset U$ then $g : X_V \to X_U$ by $(A, \lambda, i, \overline{\eta}) \mapsto (A, \lambda, i, \overline{\eta \circ g})$. If U and V are sufficiently small open compact subgroups of $G(\mathbb{A}^\infty)$ and if V is a normal subgroup of U then the finite etale cover $X_V \to X_U$ is Galois with group U/V. Thus if x is a geometric point of X_U we obtain a continuous homomorphism

$$\pi_1^{\mathrm{alg}}(X_U, x) \longrightarrow U.$$

(This map is only determined up to conjugation unless one chooses a compatible choice of liftings of x to all the covers X_V.)

Choose an embedding $\tau : F \hookrightarrow \mathbb{C}$. This allows us to speak of the complex points of X_U, which we will denote $X_{U,\tau}(\mathbb{C})$, a smooth manifold. Let $\mathfrak{H}_\tau$ denote the set of $I \in B^{\mathrm{op}} \otimes_{\mathbb{Q}} \mathbb{R}$ with the following properties.

- $I^2 = -1$.

- $I^{\#_\tau} = -I$.

- The pairing $b_1 \times b_2 \mapsto (b_1, b_2 I)_\tau$ is a positive definite symmetric form on $V \otimes_{\mathbb{Q}} \mathbb{R}$.

- If $\tau' : F \hookrightarrow \mathbb{C}$ coincides with τ on E then

$$\dim_{\mathbb{C}}(V \otimes_{F,\tau'} \mathbb{C})^{I=i}$$

is n if $\tau = \tau'$ and 0 otherwise.

Then $\mathfrak{H}_\tau$ is non-empty. It has transitive action of $G_\tau(\mathbb{R})^+ = \nu^{-1}\mathbb{R}_{>0}^\times \subset G_\tau(\mathbb{R})$. Let U_τ denote the centraliser in $G_\tau(\mathbb{R})^+$ of some $I \in \mathfrak{H}_\tau$, so that

$$\mathfrak{H}_\tau \cong G_\tau(\mathbb{R})^+/U_\tau.$$

In fact U_τ is a maximal connected compact mod centre subgroup of $G_\tau(\mathbb{R})$. We will let $\tilde{U}_\tau$ denote a maximal compact mod centre subgroup of $G_\tau(\mathbb{R})$ containing U_τ. As on page 400 of [Ko3] we can give a more explicit description of the smooth manifold $X_U(\mathbb{C})$. It is the disjoint union of $\# \ker^1(\mathbb{Q}, G_\tau)$ copies of the manifold

$$
\begin{aligned}
G_\tau(\mathbb{Q})^+ \backslash (G(\mathbb{A}^\infty)/U \times \mathfrak{H}_\tau) &\cong G_\tau(\mathbb{Q})^+ \backslash (G(\mathbb{A}^\infty)/U \times G_\tau(\mathbb{R})^+/U_\tau) \\
&\cong G_\tau(\mathbb{Q}) \backslash (G(\mathbb{A}^\infty)/U \times G_\tau(\mathbb{R})/U_\tau).
\end{aligned}
$$

Under this identification the right action of $G(\mathbb{A}^\infty)$ on inverse system of the X_U's corresponds to the action by right translation on the

$$G_\tau(\mathbb{Q}) \backslash (G(\mathbb{A}^\infty)/U \times G_\tau(\mathbb{R})/U_\tau).$$

It will be convenient for the integral theory later if we give some slightly modified definition for X_U. This will be in terms of slightly modified moduli problems $\mathfrak{X}'_U$ and $\mathfrak{X}''_U$, which are equivalent to $\mathfrak{X}_U$. Again we will initially define $\mathfrak{X}'_U$ and $\mathfrak{X}''_U$ on the category of pairs (S, s), where S is a connected, locally noetherian F-scheme and where s is a geometric point of S. We will then extend them to the category of locally noetherian F-schemes exactly as we did for $\mathfrak{X}_U$.

To define $\mathfrak{X}'_U$ let us suppose that $U = U^p \times U_{p,0} \times \prod_{i=1}^r U_{w_i}$, where $U^p \subset G(\mathbb{A}^{\infty,p})$, $U_{p,0} \subset \mathbb{Q}_p^\times$ and for $i \geq 1$, $U_{w_i} \subset (B_{w_i}^{\mathrm{op}})^\times$. Then $\mathfrak{X}'_U$ will take (S, s) to the set of equivalence classes of $(r+5)$-tuples $(A, \lambda, i, \overline{\eta}^p, \overline{\eta}_{p,0}, \overline{\eta}_{w_i})$ where

- A/S is an abelian scheme of dimension dn^2;

- $\lambda : A \longrightarrow A^\vee$ is a polarisation;

- $i : B \hookrightarrow \mathrm{End}^0(A)$ such that (A, i) is compatible and $\lambda \circ i(b) = i(b^*)^\vee \circ \lambda$ for all $b \in B$;

- $\overline{\eta}^p$ is a $\pi_1(S, s)$-invariant U^p-orbit of isomorphisms of $B \otimes_{\mathbb{Q}} \mathbb{A}^{\infty,p}$-modules $\eta^p : V \otimes_{\mathbb{Q}} \mathbb{A}^{\infty,p} \to V^p A_s$ which take the standard pairing $(\ ,\ )$ on $V \otimes_{\mathbb{Q}} \mathbb{A}^{\infty,p}$ to a $(\mathbb{A}^{\infty,p})^\times$ multiple of the λ-Weil pairing on $V^p A_s$;

- $\overline{\eta}_{p,0}$ denotes a $\pi_1(S,s)$-invariant U_p^0-orbit of isomorphisms $\mathbb{Q}_p \xrightarrow{\sim} \mathbb{Q}_p(1)$;

- $\overline{\eta}_{w_1}$ is a $\pi_1(S,s)$-invariant U_{w_1}-orbit of isomorphisms of F_w-modules $\overline{\eta}_{w_1} : \Lambda_{11} \otimes_{\mathbb{Z}_p} \mathbb{Q}_p \xrightarrow{\sim} \varepsilon V_{w_1} A_s$;

- for $i > 1$, $\overline{\eta}_{w_i}$ is a $\pi_1(S,s)$-invariant U_{w_1}-orbit of isomorphisms of B_{w_i}-modules $\overline{\eta}_{w_i} : \Lambda_i \otimes_{\mathbb{Z}_p} \mathbb{Q}_p \xrightarrow{\sim} V_{w_i} A_s$.

We call two $(r+5)$-tuples $(A, \lambda, i, \overline{\eta}^p, \overline{\eta}_{p,0}, \overline{\eta}_{w_i})$ and $(A', \lambda', i', (\overline{\eta}^p)', \overline{\eta}_{p,0}', \overline{\eta}_{w_i}')$ equivalent if there exists an isogeny $\alpha : A \to A'$ and $\gamma \in \mathbb{Q}^\times$ such that α carries λ to $\gamma\lambda'$, i to i', $\overline{\eta}^p$ to $(\overline{\eta}^p)'$, and $\overline{\eta}_{w_i}$ to $\overline{\eta}_{w_i}'$; and such that $\overline{\eta}_{p,0} = \gamma\overline{\eta}_{p,0}'$. The image of this functor is canonically independent of the base point s so we can think of it as a functor on connected locally noetherian F-schemes. We extend it to all locally noetherian schemes over F so that it sends $\coprod S_i$ to the product of the images of the S_i. Then $\mathfrak{X}_U'$ is isomorphic to $\mathfrak{X}_U$ and so it is again represented by X_U/F. The isomorphism between these two moduli problems is given by mapping $(A, \lambda, i, \overline{\eta}^p, \overline{\eta}_{p,0}, \overline{\eta}_{w_i})$ to $(A, \lambda, i, \overline{\eta}')$ where

$$\eta' = \eta^p \times (((\mathrm{Id}_{\mathcal{O}_{F,w}^n} \otimes \eta_{w_1}) \oplus \bigoplus_{i>1} \eta_{w_i}) \oplus ((\mathrm{Id}_{\mathcal{O}_{F,w}^n} \otimes \eta_{w_1}) \oplus \bigoplus_{i>1} \eta_{w_i})^{\vee \eta_{p,0}}).$$

The dual $^{\vee\eta_{p,0}}$ is taken with respect to the canonical pairing on $\Lambda \otimes_{\mathbb{Z}_p} \mathbb{Q}_p$ and $\eta_{p,0}^{-1}$ composed with the λ-Weil pairing on $V_p A$. The action of $G(\mathbb{Q}_p)$ in this picture may be described as follows. An element $(g_0, g_1, \ldots, g_r) \in \mathbb{Q}_p^\times \times \prod_{i=1}^r (B_{w_i}^{\mathrm{op}})^\times$ maps $(A, \lambda, i, \overline{\eta}^p, \overline{\eta}_{p,0}, \overline{\eta}_{w_i})$ to $(A, \lambda, i, \overline{\eta}^p, g_0\overline{\eta}_{p,0}, \overline{\eta}_{w_i} \circ g_i)$.

To define $\mathfrak{X}_U''$ let us suppose that $U = U^p \times \mathbb{Z}_p^\times \times \prod_{i=1}^r U_{w_i}$, where $U^p \subset G(\mathbb{A}^{\infty,p})$ and where, for $i = 1, .., r$ we have

$$U_{w_i} = \ker((\mathcal{O}_{B_{w_i}}^{\mathrm{op}})^\times \longrightarrow (\mathcal{O}_{B_{w_i}}^{\mathrm{op}}/w_i^{m_i})^\times)$$

for some non-negative integers m_i. Then $\mathfrak{X}_U''$ will take (S, s) to the set of equivalence classes of $(r+4)$-tuples $(A, \lambda, i, \overline{\eta}^p, \alpha_i)$ where

- A/S is an abelian scheme of dimension dn^2;

- $\lambda : A \longrightarrow A^\vee$ is a prime to p polarisation;

- $i : B \hookrightarrow \mathrm{End}^0(A)$ such that (A, i) is compatible and $\lambda \circ i(b) = i(b^*)^\vee \circ \lambda$ for all $b \in B$;

- $\overline{\eta}^p$ is a $\pi_1(S,s)$-invariant U^p-orbit of isomorphisms of $B \otimes_{\mathbb{Q}} \mathbb{A}^{\infty,p}$-modules $\eta^p : V \otimes_{\mathbb{Q}} \mathbb{A}^{\infty,p} \to V^p A_s$ which take the standard pairing $(\ ,\ )$ on $V \otimes_{\mathbb{Q}} \mathbb{A}^{\infty,p}$ to a $(\mathbb{A}^{\infty,p})^\times$ multiple of the λ-Weil pairing on $V^p A_s$;

- $\alpha_1 : (w_1^{-m_1} \Lambda_{11}/\Lambda_{11})_S \xrightarrow{\sim} \varepsilon A[w_1^{m_1}]$ is an isomorphism of S-schemes with $\mathcal{O}_F$-actions;

- for $i > 1$, $\alpha_i : (w_i^{-m_i} \Lambda_i/\Lambda_i)_S \xrightarrow{\sim} A[w_i^{m_i}]$ is an isomorphism of S-schemes with $\mathcal{O}_B$-actions.

We call two $(r+4)$-tuples $(A, \lambda, i, \overline{\eta}^p, \alpha_i)$ and $(A', \lambda', i', (\overline{\eta}^p)', \alpha_i')$ equivalent if there exists a prime to p isogeny $\delta : A \to A'$ and $\gamma \in \mathbb{Z}_{(p)}^\times$ such that δ carries λ to $\gamma\lambda'$, i to i', $\overline{\eta}^p$ to $(\overline{\eta}^p)'$, and α_i to α_i'. The image of this functor is canonically independent of the base point s so we can think of it as a functor on connected locally noetherian F-schemes. We extend it to all locally noetherian schemes over F so that it sends $\coprod S_i$ to the product of the images of the S_i. Then $\mathfrak{X}_U''$ is isomorphic to $\mathfrak{X}_U'$ and so it is again represented by X_U/F. (Note that because we are now assuming that $\mathbb{Z}_p^\times \subset U$, we no longer require an analogue of $\overline{\eta}_{p,0}$.)

III.2   Cohomology

We must now define ceratin etale sheaves on the X_U. This is a standard construction, which we now briefly review.

Suppose that X is a scheme,

$$\ldots \longrightarrow X_i \longrightarrow X_{i-1} \longrightarrow \ldots \longrightarrow X_1 \longrightarrow X$$

is a family of Galois etale covers of X, $\Gamma = \lim_{\leftarrow} \mathrm{Gal}\,(X_i/X)$ and ρ is a continuous representation of Γ on a finite dimensional $\mathbb{Q}_l^{\mathrm{ac}}$ vector space W_ρ. (We define $\mathrm{Gal}\,(X_i/X)$ so it acts on the right on X_i.) To this data we can canonically attach a lisse etale $\mathbb{Q}_l^{\mathrm{ac}}$-sheaf $\mathcal{L}_\rho/X$ as follows. By the Baire category theorem ρ is defined over some finite extension L of $\mathbb{Q}_l$ and hence also over $\mathcal{O}_L$. That is we can find a finite free $\mathcal{O}_L$ module $\Lambda \subset W_\rho$ which is Γ invariant and such that $\Lambda \otimes_{\mathcal{O}_L} \mathbb{Q}_l^{\mathrm{ac}} \xrightarrow{\sim} W_\rho$. For each positive integer i we will compatibly define a lisse etale sheaf of $\mathcal{O}_L/l^i$-modules $\mathcal{L}_{\rho,\Lambda,i}/X$ and we set

$$\mathcal{L}_\rho = (\lim_{\leftarrow} \mathcal{L}_{\rho,\Lambda,i}) \otimes_{\mathcal{O}_L} \mathbb{Q}_l^{\mathrm{ac}}.$$

For each i choose a positive integer $j(i)$ such that the action of Γ on Λ/l^i factors through $\mathrm{Gal}\,(X_{j(i)}/X)$. Then, if $U \to X$ is a finite etale cover of a Zariski open subset of X, we define $\mathcal{L}_{\rho,\Lambda,i}$ to be the set of functions

$$f : \pi_0(X_{j(i)} \times_X U) \longrightarrow \Lambda/l^i$$

such that for all $\sigma \in \mathrm{Gal}\,(X_{j(i)}/X)$ and all $C \in \pi_0(X_{j(i)} \times_X U)$ we have

$$f(C\sigma) = \rho(\sigma)^{-1} f(C).$$

The definition of $\mathcal{L}_\rho$ is independent of all choices.

By a morphism $\vec{g} = (g, \{g_{ij}\}, g_*, g^*)$ between two such sets of data:

$$\vec{g} : (X, \{X_i\}, \Gamma, \rho) \longrightarrow (Y, \{Y_i\}, \Delta, \rho'),$$

we mean

- a morphism $g : X \to Y$,

- for any j and for $i >> j$ compatible morphisms $g_{ij} : X_i \to Y_j$ over $g : X \to Y$,

- a continuous homomorphism $g_* : \Gamma \to \Delta$ such that for any $i >> j$ and any $\sigma \in \Gamma$ we have $g_{ij} \circ \sigma = g_*(\sigma) \circ g_{ij}$,

- and a linear map $g^* : W_{\rho'} \to W_\rho$ such that for all $\sigma \in \Gamma$ we have $\rho(\sigma) \circ g^* = g^* \circ \rho'(g_*(\sigma))$.

Such a morphism $\vec{g}$ gives rise to a morphism

$$\vec{g}^* : g^* \mathcal{L}_{\rho'} \longrightarrow \mathcal{L}_\rho$$

as follows. Choose L and $\Lambda' \subset W_{\rho'}$ as above, and set $\Lambda = g^* \Lambda'$. It suffices to define compatible maps

$$\vec{g}^* : g^* \mathcal{L}_{\rho', \Lambda', i'} \longrightarrow \mathcal{L}_{\rho, \Lambda, i}$$

for any $i' \geq i$. For this it suffices to give compatible maps

$$\vec{g}^* : \mathcal{L}_{\rho', \Lambda', i'}(V) \longrightarrow \mathcal{L}_{\rho, \Lambda, i}(U)$$

whenever we have a commutative diagram

$$
\begin{array}{ccc}
U & \xrightarrow{\tilde{g}} & V \\
\downarrow & & \downarrow \\
X & \xrightarrow{g} & Y
\end{array}
$$

with the vertical maps etale. Choose $j(i')$ and $j(i)$ as above with $j(i) >> j(i')$. If an element of $\mathcal{L}_{\rho', \Lambda', i'}(V)$ is represented by $f : \pi_0(Y_{j(i')} \times_Y V) \to \Lambda'/l^{i'}$ then we define $\vec{g}^*(f)$ to be the composite

$$\pi_0(X_{j(i)} \times_X U) \xrightarrow{g_{j(i), j(i')} \times \tilde{g}} \pi_0(Y_{j(i')} \times_Y V) \xrightarrow{f} \Lambda'/l^{i'} \xrightarrow{g^*} \Lambda/l^i.$$

Now suppose we have morphisms

$$\vec{g} : (X, \{X_i\}, \Gamma, \rho) \longrightarrow (Y, \{Y_i\}, \Delta, \rho')$$

and

$$\vec{h} : (Y, \{Y_i\}, \Delta, \rho') \longrightarrow (Z, \{Z_i\}, \Sigma, \rho'').$$

We define the composite $\vec{h} \circ \vec{g}$ to be

$$(h \circ g, \{h_{jk} \circ g_{ij}\}, h_* \circ g_*, g^* \circ h^*).$$

Then it is tedious but straightforward to check that

$$(\vec{h} \circ \vec{g})^* = \vec{g}^* \circ g^*(\vec{h}^*).$$

Choose a nested collection $\mathcal{U}_{l,i}$ of open compact subgroups of $G(\mathbb{Q}_l)$ with trivial intersection. If U is a sufficiently small open compact subgroup of $G(\mathbb{A}^\infty)$ we let U_i denote the set of $u \in U$ such that $u_l \in \mathcal{U}_{l,i}$ and let U_l denote the projection of U to $G(\mathbb{Q}_l)$. Recall that ξ is a finite dimensional representation of G on a $\mathbb{Q}_l^{ac}$-vector space W_ξ. The collection $(X_U, \{X_{U_i}\}, U_l, \xi)$ defines a lisse etale $\mathbb{Q}_l^{ac}$-sheaf $\mathcal{L}_\xi / X_U$. It is canonically independent of the choice of $\mathcal{U}_{l,i}$. If $g \in G(\mathbb{A}^\infty)$ and U, V are sufficiently small open compact subgroups of $G(\mathbb{A}^\infty)$ with $g^{-1}Vg \subset U$ then we have the morphism

$$\vec{g} = (g, \{g\}, c(g_l), \xi(g_l)) : (X_V, \{X_{V_i}\}, V_l, \xi) \longrightarrow (X_U, \{X_{U_i}\}, U_l, \xi)$$

where $c(g_l)$ denotes the conjugation $v \mapsto g_l^{-1}vg_l$. Thus we get a morphism of sheaves on X_V

$$\vec{g}^* : g^*\mathcal{L}_\xi \longrightarrow \mathcal{L}_\xi,$$

which we will simply denote by g. We see that

$$(\vec{gh})^* = \vec{g}^* \circ g^*(\vec{h}^*).$$

We will set

$$H^i(X, \mathcal{L}_\xi) = \varinjlim_U H^i_{et}(X_U \times_F F^{ac}, \mathcal{L}_\xi).$$

If $g \in G(\mathbb{A}^\infty)$ and U, V are sufficiently small open compact subgroups of $G(\mathbb{A}^\infty)$ with $g^{-1}Vg \subset U$ then we get a morphism

$$g : H^i_{et}(X_U \times_F F^{ac}, \mathcal{L}_\xi) \longrightarrow H^i_{et}(X_V \times_F F^{ac}, \mathcal{L}_\xi).$$

If $V \subset U$ then we see that

$$H^i_{et}(X_U \times_F F^{ac}, \mathcal{L}_\xi) \cong H^i_{et}(X_V \times_F F^{ac}, \mathcal{L}_\xi)^{U/V}.$$

Thus $H^i(X, \mathcal{L}_\xi)$ becomes an admissible $G(\mathbb{A}^\infty)$-module, in fact an admissible/continuous $G(\mathbb{A}^\infty) \times \mathrm{Gal}\,(F^{\mathrm{ac}}/F)$-module. We will let $[H(X, \mathcal{L}_\xi)]$ denote the virtual $G(\mathbb{A}^\infty) \times \mathrm{Gal}\,(F^{\mathrm{ac}}/F)$-module

$$\sum_i (-1)^{n-1-i} [H^i(X, \mathcal{L}_\xi)]$$

(see section I.2). We will also decompose

$$[H(X, \mathcal{L}_\xi)] = \sum_\pi [\pi][R_\xi(\pi)]$$

as π runs over irreducible representations of $G(\mathbb{A}^\infty)$ and $[R_\xi(\pi)]$ is in the Grothendieck group of continuous $\mathrm{Gal}\,(F^{\mathrm{ac}}/F)$-modules.

To discuss the cohomology of these sheaves in more detail we will need a parametrisation of irreducible representations ξ of G over $\mathbb{Q}_l^{\mathrm{ac}}$. First recall that irreducible representations of GL_n are classified by n-tuples $\vec{a} = (a_1, \ldots, a_n) \in \mathbb{Z}^n$ with $a_1 \leq \cdots \leq a_n$. We will let $X(\mathbb{G}_m^n)^+$ denote the set of such n-tuples. The n-tuple $\vec{a}$ corresponds to the irreducible representation with extremal weight sending a diagonal matrix with entries $(t_1, \ldots, t_n)$ to $t_1^{a_1} \ldots t_n^{a_n}$. Let Sd denote the standard representation of GL_n, i.e. the one parametrised by $(0, 0, \ldots, 1) \in X(\mathbb{G}_m^n)^+$. If $\vec{a} \in X(\mathbb{G}_m^n)^+$ then we may find non-negative integers $t_{\vec{a}}$ and $m_{\vec{a}}$ and an idempotent $\varepsilon_{\vec{a}} \in \mathbb{Q}[S_{m_{\vec{a}}}]$ such that the representation parametrised by $\vec{a}$ can be given explicitly as

$$(\wedge^n \mathrm{Sd}^\vee)^{\otimes t_{\vec{a}}} \otimes \varepsilon_{\vec{a}} \mathrm{Sd}^{\otimes m_{\vec{a}}}.$$

Now fix $\sigma_0 : E \hookrightarrow \mathbb{Q}_l^{\mathrm{ac}}$, so that we get a decomposition

$$G \times \mathbb{Q}_l^{\mathrm{ac}} \cong \mathbb{G}_m \times \prod_\sigma GL_n$$

where σ runs over embeddings $F \hookrightarrow \mathbb{Q}_l^{\mathrm{ac}}$ which extend σ_0 (see section I.7). Thus, given the choice σ_0, we get a bijection between irreducible representations of G over $\mathbb{Q}_l^{\mathrm{ac}}$ and $(d+1)$-tuples $(a_0, \vec{a}_\sigma)$, where $a_0 \in \mathbb{Z}$, where $\vec{a}_\sigma \in X(\mathbb{G}_m^n)^+$ and where σ runs over embeddings $F \hookrightarrow \mathbb{Q}_l^{\mathrm{ac}}$ extending σ_0. We will write $(\xi)_{\sigma_0} = (a_0(\xi, \sigma_0), \vec{a}(\xi, \sigma_0)_\sigma)$ for the $(d+1)$-tuple corresponding to ξ. Conversely we will write $\xi_{(a_0, \vec{a}_\sigma), \sigma_0}$ for the irreducible representation corresponding to $(a_0, \vec{a}_\sigma)$. We will call $V^\vee \otimes \mathbb{Q}_l^{\mathrm{ac}}$ the standard representation of $G \times \mathbb{Q}_l^{\mathrm{ac}}$. Then $V^\vee \otimes \mathbb{Q}_l^{\mathrm{ac}}$ decomposes as the sum of n copies of each of the following irreducibles:

- $a_0 = 0$, $\vec{a}_\sigma = \vec{0}$ for all but one σ and for that exceptional σ we have $\vec{a}_\sigma = (-1, 0, \ldots, 0)$;

- $a_0 = -1$, $\vec{a}_\sigma = \vec{0}$ for all but one σ and for that exceptional σ we have $\vec{a}_\sigma = (0, \ldots, 0, 1)$.

Moreover the action of $B \otimes \mathbb{Q}_l^{ac}$ can be used to decompose $V^\vee \otimes \mathbb{Q}_l^{ac}$ into irreducibles. Thus, if ξ is irreducible, we may find non-negative integers t_ξ and m_ξ and an idempotent ε_ξ in the $\mathbb{Q}_l^{ac}$-algebra generated by $(B \otimes \mathbb{Q}_l^{ac})^{\otimes m_\xi}$ and the symmetric group S_{m_ξ} such that

$$W_\xi \cong \nu^{t_\xi} \otimes \varepsilon_\xi (V^\vee \otimes \mathbb{Q}_l^{ac})^{\otimes m_\xi}.$$

Note that

$$2t_\xi - m_\xi = 2a_0(\xi, \sigma_0) + \sum_{\sigma,i} a(\xi, \sigma_0)_{\sigma,i},$$

which we will in future denote $-w(\xi)$.

Continue to assume that ξ is irreducible. Let $\pi : \mathcal{A} \to X_U$ denote the universal abelian variety and $\pi^m : \mathcal{A}^m \to X_U$ its m-fold self-product over X_U. Set

$$\mathcal{L} = R^1 \pi_* \mathbb{Q}_l^{ac}.$$

Let $\varepsilon(m)$ denote the idempotent

$$\prod_{x=1}^m \prod_{y \neq 1} ([2]_x - 2^y)/(2 - 2^y),$$

where $[2]_x$ denotes the endomorphism generated by multiplication by 2 on the xth factor in $\mathcal{A}^m$, and where y runs from 0 to $2dn^2$, but excluding 1. Then

$$\mathcal{L}^{\otimes m} \cong \varepsilon(m) R^m (\pi^m)_* \mathbb{Q}_l^{ac},$$

while for $j \neq m$ we have

$$\varepsilon(m) R^j (\pi^m)_* \mathbb{Q}_l^{ac} = (0).$$

We see that we get a natural isomorphism

$$\mathcal{L}_\xi \cong (\varepsilon_\xi \mathcal{L}^{\otimes m_\xi})(t_\xi) \cong \varepsilon_\xi \varepsilon(m_\xi) R^{m_\xi} (\pi^{m_\xi})_* \mathbb{Q}_l^{ac}(t_\xi).$$

We also have a spectral sequence

$$H^i(X_U \times F^{ac}, R^j(\pi^m)_* \mathbb{Q}_l^{ac}(t)) \Longrightarrow H^{i+j}(\mathcal{A}^m \times F^{ac}, \mathbb{Q}_l^{ac}(t)),$$

and hence we get isomorphisms

$$H^i(X_U \times F^{ac}, \mathcal{L}_\xi) \xrightarrow{\sim} \varepsilon_\xi \varepsilon(m) H^{i+m_\xi}(\mathcal{A}^{m_\xi} \times F^{ac}, \mathbb{Q}_l^{ac}(t_\xi)).$$

In particular $H^i(X_U \times F^{\mathrm{ac}}, \mathcal{L}_\xi)$ is pure of weight $i + w(\xi)$, in the sense that for almost all primes x of F and for all eigenvalues α of Frob_x on $H^i(X_U \times F^{\mathrm{ac}}, \mathcal{L}_\xi)$, α is algebraic and for all embeddings of α into $\mathbb{C}$ we have $|\alpha|^2 = (\#k(x))^{i+w(\xi)}$. Moreover, if x is any prime of F not dividing l, if $\sigma \in W_{F_x}$ and if α is an eigenvalue of σ on $H^i(X_U \times F^{\mathrm{ac}}, \mathcal{L}_\xi)$ then $\alpha \in \mathbb{Q}^{\mathrm{ac}}$ and for any embedding $\mathbb{Q}^{\mathrm{ac}} \hookrightarrow \mathbb{C}$ we have

$$|\alpha|^2 \in (\#k(x))^{\mathbb{Z}}$$

(see lemma I.5.7).

Next we wish to discuss the actions of the decomposition groups at primes above l on $H^i(X_U \times F^{\mathrm{ac}}, \mathcal{L}_\xi)$. We will start by placing ourselves in the following general setting. Let R be a finite dimensional $\mathbb{Q}_l^{\mathrm{ac}}$-vector space with a continuous linear action of $\mathrm{Gal}(F^{\mathrm{ac}}/F)$. For each prime λ of F above l, the space R is also a representation of the decomposition group G_λ and so we can speak of R being crystalline, semi-stable or potentially semi-stable at λ (see [Fo]). If R is potentially semi-stable at all primes $\lambda|l$, then we shall simply say that R is potentially semistable. In that case the space

$$D_{\mathrm{DR},\lambda}(R) = (R \otimes_{\mathbb{Q}_l} B_{\mathrm{DR}})^{G_{F_\lambda}}$$

(see [Fo]) is a filtered free $\mathbb{Q}_l^{\mathrm{ac}} \otimes_{\mathbb{Q}_l} F_\lambda \cong (\mathbb{Q}_l^{\mathrm{ac}})^{\mathrm{Hom}\,(F_\lambda, \mathbb{Q}_l^{\mathrm{ac}})}$-module. If $\sigma : F \hookrightarrow \mathbb{Q}_l^{\mathrm{ac}}$ gives rise to a prime $\lambda|l$ then we will set

$$D_{\mathrm{DR},\sigma}(R) = D_{\mathrm{DR},\lambda}(R) \otimes_{\mathbb{Q}_l^{\mathrm{ac}} \otimes_{\mathbb{Q}_l} F_\lambda, 1 \otimes \sigma} \mathbb{Q}_l^{\mathrm{ac}}.$$

We will call the integers j such that

$$\mathrm{gr}^j D_{\mathrm{DR},\sigma}(R) \neq (0)$$

the Hodge-Tate numbers of R with respect to σ (and we will say that j occurs with multiplicity $\dim \mathrm{gr}^j D_{\mathrm{DR},\sigma}(R)$).

It follows from the above analysis and the conjecture C_{pst} (now a theorem due to many people) that $H^i(X_U \times F^{\mathrm{ac}}, \mathcal{L}_\xi)$ is potentially semi-stable (see for instance [Bert]). Moreover it follows from a theorem of Faltings (see [Fa2]) that we have functorial isomorphisms

$$
\begin{aligned}
&\mathrm{gr}^j D_{\mathrm{DR},\sigma}(H^i(X_U \times F^{\mathrm{ac}}, \mathcal{L}_\xi))\\
&\cong\ \mathrm{gr}^j D_{\mathrm{DR}}(\varepsilon_\xi \varepsilon(m_\xi) H^{i+m_\xi}(\mathcal{A}^{m_\xi} \times F_\lambda^{\mathrm{ac}}, \mathbb{Q}_l^{\mathrm{ac}}(t_\xi))) \otimes_{\mathbb{Q}_l^{\mathrm{ac}} \otimes_{\mathbb{Q}_l} F_\lambda, 1 \otimes \sigma} \mathbb{Q}_l^{\mathrm{ac}}\\
&\cong\ \mathrm{gr}^{j+t_\xi} \varepsilon_\xi \varepsilon(m_\xi) D_{\mathrm{DR}}(H^{i+m_\xi}(\mathcal{A}^{m_\xi} \times F_\lambda^{\mathrm{ac}}, \mathbb{Q}_l^{\mathrm{ac}})) \otimes_{\mathbb{Q}_l^{\mathrm{ac}} \otimes_{\mathbb{Q}_l} F_\lambda, 1 \otimes \sigma} \mathbb{Q}_l^{\mathrm{ac}}\\
&\cong\ \mathrm{gr}^{j+t_\xi} \varepsilon_\xi \varepsilon(m_\xi)\\
&\qquad (H_{\mathrm{DR}}^{i+m_\xi}(\mathcal{A}^{m_\xi}/F) \otimes_F (F_\lambda \otimes_{\mathbb{Q}_l} \mathbb{Q}_l^{\mathrm{ac}})) \otimes_{\mathbb{Q}_l^{\mathrm{ac}} \otimes_{\mathbb{Q}_l} F_\lambda, 1 \otimes \sigma} \mathbb{Q}_l^{\mathrm{ac}}\\
&\cong\ \mathrm{gr}^{j+t_\xi} \varepsilon_\xi \varepsilon(m_\xi) H_{\mathrm{DR}}^{i+m_\xi}(\mathcal{A}^{m_\xi}/F) \otimes_{F,\sigma} \mathbb{Q}_l^{\mathrm{ac}}\\
&\cong\ \mathrm{gr}^{j+t_\xi} \imath(\varepsilon_\xi) \varepsilon(m_\xi) H_{\mathrm{DR}}^{i+m_\xi}(\mathcal{A}^{m_\xi}/F) \otimes_{F,\imath\sigma} \mathbb{C}.
\end{aligned}
$$

More explicit computation of the groups

$$H^i(X_U \times F^{\mathrm{ac}}, \mathcal{L}_\xi)$$

and

$$\mathrm{gr}^j D_{\mathrm{DR},\sigma}(H^i(X_U \times F^{\mathrm{ac}}, \mathcal{L}_\xi))$$

depends on comparison with the analytic theory.

Choose $\tau : F \hookrightarrow \mathbb{C}$. Using our isomorphism $\imath : \overline{\mathbb{Q}}_l^{\mathrm{ac}} \xrightarrow{\sim} \mathbb{C}$ the $\overline{\mathbb{Q}}_l^{\mathrm{ac}}$-sheaf $\mathcal{L}_\xi$ corresponds to a locally constant sheaf $\mathcal{L}_{\imath\xi}^{\mathrm{top}}/X_{U,\tau}(\mathbb{C})$. More precisely if ξ' is a representation of G_τ over $\mathbb{C}$ then we define a locally constant sheaf $\mathcal{L}_{\xi'}^{\mathrm{top}}/X_{U,\tau}(\mathbb{C})$ by

$$\mathcal{L}_{\xi'}^{\mathrm{top}} = G_\tau(\mathbb{Q}) \backslash (G(\mathbb{A}^\infty)/U \times G_\tau(\mathbb{R})/U_\tau \times W_{\xi'}),$$

i.e. if π denotes the projection from $G(\mathbb{A}^\infty)/U \times G_\tau(\mathbb{R})/U_\tau$ to $X_{U,\tau}$ and if $\mathcal{U} \subset X_{U,\tau}(\mathbb{C})$ is an open set then $\mathcal{L}_{\xi'}^{\mathrm{top}}(\mathcal{U})$ is the set of locally constant functions

$$f : \pi^{-1}\mathcal{U} \longrightarrow W_{\xi'}$$

such that

$$f(\gamma x) = \gamma f(x)$$

for all $\gamma \in G_\tau(\mathbb{Q})$ and all $x \in \pi^{-1}\mathcal{U}$. The isomorphism $\imath$ induces a functorial isomorphism

$$\imath : H^i(X_U \times F^{\mathrm{ac}}, \mathcal{L}_\xi) \xrightarrow{\sim} H^i(X_{U,\tau}(\mathbb{C}), \mathcal{L}_{\imath(\xi)}^{\mathrm{top}}).$$

In fact the sheaves $\mathcal{L}_{\xi'}^{\mathrm{top}}$ have a natural structure of a variation of mixed $\mathbb{C}$-Hodge structure. Let $Q \subset GL_n$ denote the parabolic subgroup consisting of matrices with last row of the form $(0, \ldots, 0, *)$ and let $Q_\tau \subset G_\tau \times \mathbb{C}$ denote the parabolic subgroup

$$\mathbb{G}_m \times Q \times \prod_{\tau'} GL_n \subset \mathbb{G}_m \times GL_n \times \prod_{\tau'} GL_n$$

where τ' runs over embeddings $F \hookrightarrow \mathbb{C}$ such that $\tau'|_E = \tau|_E$ but $\tau' \neq \tau$. Thus $(G_\tau \times \mathbb{C})/Q_\tau$ is isomorphic to $(\mathbb{P}^{n-1})^\vee$, the Grassmanian of hyperplanes in the affine n-space $V \otimes_{F,\tau} \mathbb{C}$. There is a natural embedding $\mathcal{H}_\tau \hookrightarrow G_\tau(\mathbb{C})/Q_\tau(\mathbb{C})$ as an open subset, and this gives rise to the complex structure on $\mathcal{H}_\tau$. It sends I to the $I = -i$ subspace in $V \otimes_{F,\tau} \mathbb{C}$.

Define a morphism wt : $\mathbb{G}_m \to G_\tau$ to be the composite of $t \mapsto t^{-1}$ with the inclusion $\mathbb{G}_m \hookrightarrow Z(G_\tau)$ coming from $\mathbb{Q} \hookrightarrow B^{\mathrm{op}}$. Also define a morphism hdg : $\mathbb{G}_m \to Q_\tau$ which sends

$$t \longmapsto \left(t^{-1}, \begin{pmatrix} 1_{n-1} & 0 \\ 0 & t^{-1} \end{pmatrix}, 1, \ldots, 1 \right) \in \mathbb{G}_m \times Q \times \prod_{\tau'} GL_n.$$

If ξ' is a representation of G_τ define a filtration $W_m(W_{\xi'})$ on $W_{\xi'}$ by setting $W_m(W_{\xi'})$ to be the sum over $m' \leq m$ of the $t \mapsto t^{m'}$ weight spaces for $\xi' \circ \mathrm{wt}$. This induces a filtration $W_m \mathcal{L}_{\xi'}^{\mathrm{top}}$ of $\mathcal{L}_{\xi'}^{\mathrm{top}}$. If μ is a representation of Q_τ define a filtration $F^m W_\mu$ on W_μ by setting $F^m W_\mu$ to be the sum over $m' \geq m$ of the $t \mapsto t^{m'}$ weight spaces for $\mu \circ \mathrm{hdg}$.

If μ is a representation of Q_τ then we obtain a coherent sheaf $\mathcal{E}_\mu$ over $G_\tau(\mathbb{C})/Q_\tau(\mathbb{C})$ as the sheaf of holomorphic sections of the vector bundle

$$(G_\tau(\mathbb{C}) \times W_\mu)/Q_\tau(\mathbb{C}) \longrightarrow G_\tau(\mathbb{C})/Q_\tau(\mathbb{C}),$$

where $q : (g, x) \mapsto (gq, \mu(q)^{-1}x)$. If $\gamma \in G_\tau(\mathbb{Q})$ then $\gamma^* \mathcal{E}_\mu$ is naturally isomorphic to $\mathcal{E}_\mu$ (as $(g, x) \mapsto (\gamma g, x)$ gives a map of vector bundles above $g \mapsto \gamma g$). Thus $\mathcal{E}_\mu$ descends to $X_{U,\tau}(\mathbb{C})$. The filtration $F^m W_\mu$ induces a filtration $F^m \mathcal{E}_\mu$ of $\mathcal{E}_\mu$. If ξ' is a representation of G_τ over $\mathbb{C}$ we get an isomorphism

$$\mathcal{E}_{\xi'} \cong \mathcal{L}_{\xi'}^{\mathrm{top}} \otimes_\mathbb{C} \mathcal{O}_{X_{U,\tau}(\mathbb{C})}$$

coming from the map of vector bundles over $G_\tau(\mathbb{C})/Q_\tau(\mathbb{C})$:

$$\begin{array}{ccc} G_\tau(\mathbb{C})/Q_\tau(\mathbb{C}) \times W_{\xi'} & \longrightarrow & (G_\tau(\mathbb{C}) \times W_{\xi'})/Q_\tau(\mathbb{C}) \\ (g, x) & \longmapsto & (g, \xi'(g)^{-1}x). \end{array}$$

Because $W_{\xi'} \otimes_{\mathbb{C},c} \mathbb{C} \cong W_{(\xi')^c}$ as $G_\tau(\mathbb{Q})$-modules we also see that

$$\mathcal{L}_{\xi'} \otimes_{\mathbb{C},c} \mathbb{C} \cong \mathcal{L}_{(\xi')^c}$$

and that

$$\mathcal{E}_{(\xi')^c} \cong \mathcal{L}_{\xi'}^{\mathrm{top}} \otimes_{\mathbb{C},c} \mathcal{O}_{X_{U,\tau}(\mathbb{C})}.$$

We make $\mathcal{L}_{\xi'}^{\mathrm{top}}$ a variation of mixed $\mathbb{C}$-Hodge structures by taking $W_m \mathcal{L}_{\xi'}^{\mathrm{top}}$ to be the (canonically split) weight filtration, $F^m \mathcal{E}_{\xi'}$ to be the Hodge filtration and defining $\overline{F}^m \mathcal{L}_{\xi'}^{\mathrm{top}} \otimes_{\mathbb{C},c} \mathcal{O}_{X_{U,\tau}(\mathbb{C})}$ to be the subsheaf corresponding to $F^m \mathcal{E}_{(\xi')^c}$. In particular $H^i(X_{U,\tau}(\mathbb{C}), \mathcal{L}_{\xi'}^{\mathrm{top}})$ has a canonical mixed $\mathbb{C}$-Hodge structure.

Now suppose again that ξ is an irreducible representation of G over $\mathbb{Q}_l^{\mathrm{ac}}$. We see that

$$\mathcal{L}_{\imath\xi}^{\mathrm{top}} \cong \imath(\varepsilon_\xi)\varepsilon(m_\xi)R^{m_\xi}(\pi^{m_\xi})_*\mathbb{C}(t_\xi)$$

as locally constant sheaves which are variations of $\mathbb{C}$-Hodge structures. In particular we get isomorphisms (depending on the choice of $\imath$)

$$
\begin{aligned}
&\operatorname{gr}^j D_{\mathrm{DR},\sigma}(H^i(X_U \times F^{\mathrm{ac}}, \mathcal{L}_\xi)) \\
\cong\ &\operatorname{gr}^{j+t_\xi}\imath(\varepsilon_\xi)\varepsilon(m_\xi) H_{\mathrm{DR}}^{i+m_\xi}(\mathcal{A}^{m_\xi}/F) \otimes_{F,\imath\sigma} \mathbb{C} \\
\cong\ &\operatorname{gr}_F^j \imath(\varepsilon_\xi)\varepsilon(m_\xi) H^{i+m_\xi}(\mathcal{A}_{\imath\sigma}^{m_\xi}(\mathbb{C}), \mathbb{C}(t_\xi)) \\
\cong\ &\operatorname{gr}_F^j H^i(X_{U,\imath\sigma}(\mathbb{C}), \mathcal{L}_{\imath\xi}^{\mathrm{top}}).
\end{aligned}
$$

Moreover we see that the Hodge-de Rham spectral sequence

$$
H^i(X_{U,\imath\sigma}(\mathbb{C}), \operatorname{gr}_F^j(\mathcal{L}_{\imath\xi}^{\mathrm{top}} \otimes \Omega^\bullet_{X_{U,\imath\sigma}(\mathbb{C})})) \Rightarrow H^{i+j}(X_{U,\imath\sigma}(\mathbb{C}), \mathcal{L}_{\imath\xi}^{\mathrm{top}})
$$

degenerates at E_1 (being a direct summand of the Hodge spectral sequence for $H^{i+j}(\mathcal{A}_{\imath\sigma}^{m_\xi}(\mathbb{C}), \mathbb{C})$). Note that $\operatorname{gr}_F^j(\mathcal{L}_{\imath\xi}^{\mathrm{top}} \otimes \Omega^\bullet_{X_{U,\imath\sigma}(\mathbb{C})})$ is the complex with

$$
(\operatorname{gr}_F^{j-p}(\mathcal{L}_{\imath\xi}^{\mathrm{top}}) \otimes \mathcal{O}_{X_{U,\imath\sigma}(\mathbb{C})}) \otimes \Omega^p_{X_{U,\imath\sigma}(\mathbb{C})}
$$

in degree p. If we identify $\Omega^p_{X_{U,\imath\sigma}(\mathbb{C})}$ with

$$
\mathcal{E}_{\wedge^p(\operatorname{Lie}(G_\tau \times \mathbb{C})/\operatorname{Lie} Q_\tau)^\vee}
$$

then this is just the grading from the filtration F^m defined above on

$$
\mathcal{L}_{\imath\xi}^{\mathrm{top}} \otimes_{\mathbb{C}} \mathcal{E}_{\wedge^\bullet(\operatorname{Lie}(G_\tau \times \mathbb{C})/\operatorname{Lie} Q_\tau)^\vee}.
$$

In [Fa1] Faltings defines a subcomplex

$$
\mathcal{K}_{\imath\xi}^\bullet \subset \mathcal{L}_{\imath\xi}^{\mathrm{top}} \otimes \Omega^\bullet_{X_{U,\imath\sigma}(\mathbb{C})},
$$

which has the same cohomology and which is a direct summand as a filtered complex. Thus we get a degenerating spectral sequence

$$
H^i(X_{U,\imath\sigma}(\mathbb{C}), \operatorname{gr}_F^j \mathcal{K}_{\imath\xi}^\bullet) \Rightarrow H^i(X_{U,\imath\sigma}(\mathbb{C}), \mathcal{L}_{\imath\xi}^{\mathrm{top}})
$$

and hence isomorphisms

$$
\operatorname{gr}^j D_{\mathrm{DR},\sigma}(H^i(X_U \times F^{\mathrm{ac}}, \mathcal{L}_\xi)) \cong H^i(X_{U,\imath\sigma}(\mathbb{C}), \operatorname{gr}_F^j \mathcal{K}_{\imath\xi}^\bullet).
$$

In fact [Fa1] defines $\mathcal{K}_{\imath\xi}^p$ as $\mathcal{E}_{\mu^p(\imath\xi)}$ with the above defined filtration F^m for a certain representation $\mu^p(\imath\xi)$ of $Q_{\imath\sigma}$ modulo its unipotent radical $N_{\imath\sigma}$. Irreducible representations of

$$
Q_{\imath\sigma}/N_{\imath\sigma} \cong \mathbb{G}_m \times (GL_{n-1} \times GL_1) \times \prod_{\tau'} GL_n
$$

(where τ' runs over embeddings $F \hookrightarrow \mathbb{C}$ with $\tau'|_E = (\imath\sigma)|_E$ but $\tau' \neq \imath\sigma$) are parametrised by tuples

$$
(b_0, (\vec{b}_{\imath\sigma}, b_1), \vec{b}_{\tau'})
$$

where

- $b_0, b_1 \in \mathbb{Z}$;

- $\vec{b}_{\iota\sigma} \in X(\mathbb{G}_m^{n-1})^+$; and

- $\vec{b}_{\tau'} \in X(\mathbb{G}_m^n)^+$.

Let $\rho = ((1-n)/2, (3-n)/2, \ldots, (n-1)/2)$. Also for $p = 0, \ldots, n-1$ set

$$w_p(a_1, \ldots, a_n) = (a_1, \ldots, a_{n-p-1}, a_{n-p+1}, \ldots, a_n, a_{n-p}).$$

In our case the explicit formulae in [Fa1] (see in particular page 73) tell us that $\mu^p(\iota\xi)$ is irreducible and parametrised by

$$(a_0(\xi, \sigma|_E), w_p(\vec{a}(\xi, \sigma|_E)_\sigma + \rho) - \rho, \vec{a}(\xi, \sigma|_E)_{\iota^{-1}\tau'}).$$

Note that if $\vec{a}(\xi, \sigma|_E)_\sigma = (a_1, \ldots, a_n)$ then

$$w_p(\vec{a}(\xi, \sigma|_E)_\sigma + \rho) - \rho =$$
$$(a_1, \ldots, a_{n-p-1}, a_{n-p+1} + 1, \ldots, a_n + 1, a_{n-p} - p).$$

In particular $\mathcal{K}_{\iota\xi}^p$ contributes only to the

$$p - \vec{a}(\xi, \sigma|_E)_{\sigma, n-p} - a_0(\xi, \sigma|_E)$$

graded piece. As $\vec{a}(\xi, \sigma|_E)_\sigma \in X(\mathbb{G}_m^n)^+$ we see that $p - \vec{a}(\xi, \sigma|_E)_{\sigma, n-p} - a_0(\xi, \sigma|_E)$ strictly increases with p. In particular if $j = p - \vec{a}(\xi, \sigma|_E)_{\sigma, n-p} - a_0(\xi, \sigma|_E)$ for a necessarily unique $p \in \{0, \ldots, n-1\}$ then

$$\mathrm{gr}^j D_{\mathrm{DR},\sigma}(H^i(X_U \times F^{\mathrm{ac}}, \mathcal{L}_\xi)) \cong H^{i-p}(X_{U,\iota\sigma}(\mathbb{C}), \mathcal{E}_{\mu^p(\iota\xi)})$$

while for all other j

$$\mathrm{gr}^j D_{\mathrm{DR},\sigma}(H^i(X_U \times F^{\mathrm{ac}}, \mathcal{L}_\xi)) = (0).$$

Matsushima's formula gives us an isomorphism

$$\varinjlim_U H^i(X_{U,\tau}(\mathbb{C}), \mathcal{L}_{\iota(\xi)}^{\mathrm{top}})$$
$$\cong \bigoplus_\pi \pi^\infty \otimes H^i(\mathrm{Lie}\, G_\tau(\mathbb{R}), U_\tau, \pi_\infty \otimes \iota(\xi))^{\#\ker^1(\mathbb{Q}, G_\tau)},$$

where π runs over irreducible constituents of the space of automorphic forms for $G_\tau(\mathbb{Q}) \backslash G_\tau(\mathbb{A})$, each taken with its multiplicity in the space of automorphic forms. (See [Ko4] page 655.) In particular $H^i(X, \mathcal{L}_\xi)$ is a semi-simple $G(\mathbb{A}^\infty)$-module and we may decompose

$$H^i(X, \mathcal{L}_\xi) = \bigoplus_\pi \pi \otimes R_\xi^i(\pi),$$

where π runs over irreducible representations of $G(\mathbb{A}^\infty)$ and where $R^i_\xi(\pi)$ is a continuous finite dimensional representation of $\mathrm{Gal}\,(F^{\mathrm{ac}}/F)$. We have

$$[R_\xi(\pi)] = \sum_i (-1)^{n-1-i}[R^i_\xi(\pi)].$$

Moreover $R^i_\xi(\pi)$ has dimension

$$\#\ker^1(\mathbb{Q}, G_\tau)\sum_{\pi_\infty} m_\tau(\imath(\pi)\otimes\pi_\infty)\dim H^i(\mathrm{Lie}\,G_\tau(\mathbb{R}), U_\tau, \pi_\infty\otimes\imath(\xi))$$

where π_∞ runs over irreducible representations of $G_\tau(\mathbb{R})$ and where $m_\tau(\pi)$ is the multiplicity of π in the space of automorphic forms on $G_\tau(\mathbb{Q})\backslash G_\tau(\mathbb{A})$.

Similarly (see [Har1]) $\lim_{\to U} H^{i-p}(X_{U,\imath\sigma}(\mathbb{C}), \mathcal{E}_{\mu^p(\imath\xi)})$ is isomorphic to

$$\bigoplus_\pi \pi^\infty\otimes H^{i-p}(\mathrm{Lie}\,Q_{\imath\sigma}, U_{\imath\sigma}, \pi_\infty\otimes\mu^p(\imath\xi))^{\ker^1(\mathbb{Q}, G_{\imath\sigma})},$$

where π runs over irreducible constituents of the space of automorphic forms on $G_{\imath\sigma}(\mathbb{Q})\backslash G_{\imath\sigma}(\mathbb{A})$, each taken with its multiplicity in the space of automorphic forms. Thus for any irreducible representation π of $G(\mathbb{A}^\infty)$ we have

$$\dim\mathrm{gr}^j D_{\mathrm{DR},\sigma}(R^i_\xi(\pi)) = \#\ker^1(\mathbb{Q}, G_{\imath\sigma})$$
$$\sum_{\pi_\infty} m_{\imath\sigma}(\imath(\pi)\otimes\pi_\infty)\dim H^{i-p}(\mathrm{Lie}\,Q_{\imath\sigma}, U_{\imath\sigma}, \pi_\infty\otimes\mu^p(\imath\xi))$$

if $j = p - \vec{a}(\xi, \sigma|_E)_{\sigma,n-p} - a_0(\xi, \sigma|_E)$ for a necessarily unique $p\in\{0,\ldots,n-1\}$, and

$$\dim\mathrm{gr}^j D_{\mathrm{DR},\sigma}(R^i_\xi(\pi)) = 0$$

otherwise. Here again π_∞ runs over irreducible representations of $G_{\imath\sigma}(\mathbb{R})$ and $m_{\imath\sigma}(\pi)$ denotes the multiplicity of π is the space of automorphic forms on $G_{\imath\sigma}(\mathbb{Q})\backslash G_{\imath\sigma}(\mathbb{A})$.

Summarising the above discussion we get the following proposition.

Proposition III.2.1 *Let ξ be an irreducible representation of G over $\mathbb{Q}^{\mathrm{ac}}_l$. Recall that we have fixed an isomorphism $\imath : \mathbb{Q}^{\mathrm{ac}}_l \overset{\sim}{\to} \mathbb{C}$.*

1. We have a decomposition

$$H^i(X, \mathcal{L}_\xi) = \bigoplus_\pi \pi\otimes R^i_\xi(\pi),$$

where π runs over irreducible representations $G(\mathbb{A}^\infty)$ over $\mathbb{Q}^{\mathrm{ac}}_l$ and where $R^i_\xi(\pi)$ is a finite dimensional continuous representation of the Galois group $\mathrm{Gal}\,(F^{\mathrm{ac}}/F)$.

2. *For any* $\tau : F \hookrightarrow \mathbb{C}$ *we have*

$$\dim R_\xi^i(\pi) = \# \ker^1(\mathbb{Q}, G_\tau) \sum_{\pi_\infty} m_\tau(\imath(\pi) \otimes \pi_\infty)$$
$$\dim H^i(\operatorname{Lie} G_\tau(\mathbb{R}), U_\tau, \pi_\infty \otimes \imath(\xi))$$

where π_∞ *runs over irreducible representations of* $G_\tau(\mathbb{R})$. *Also* $m_\tau(\pi)$
denotes the multiplicity of π *in the space of automorphic forms on*
$G_\tau(\mathbb{A})$.

3. *If* x *is any prime of* F *not dividing* l, *if* $\sigma \in W_{F_x}$ *and if* α *is an*
eigenvalue of $R_\xi^i(\pi)(\sigma)$ *then* $\alpha \in \mathbb{Q}^{\mathrm{ac}}$ *and for any embedding* $\mathbb{Q}^{\mathrm{ac}} \hookrightarrow \mathbb{C}$
we have $|\alpha|^2 \in (\#k(x))^{\mathbb{Z}}$.

4. $R_\xi^i(\pi)$ *is pure of weight* $i+w(\xi)$, *in the sense that for almost all primes*
x *of* F, *for all eigenvalues* α *of* $R_\xi^i(\pi)(\mathrm{Frob}_x)$ *and for all embeddings*
of α *into* $\mathbb{C}$ *we have* $|\alpha|^2 = (\#k(x))^{i+w(\xi)}$.

5. $R_\xi^i(\pi)$ *is potentially semi-stable.*

6. *For any embedding* $\sigma : F \hookrightarrow \mathbb{Q}_l^{\mathrm{ac}}$ *we have*

$$\dim \operatorname{gr}^j D_{\mathrm{DR},\sigma}(R_\xi^i(\pi)) = \# \ker^1(\mathbb{Q}, G_{\imath\sigma}) \sum_{\pi_\infty} m_{\imath\sigma}(\imath(\pi) \otimes \pi_\infty)$$
$$\dim H^{i-p}(\operatorname{Lie} Q_{\imath\sigma}, U_{\imath\sigma}, \pi_\infty \otimes \mu^p(\imath\xi))$$

if j *is of the form* $p - \vec{a}(\xi, \sigma|_E)_{\sigma,n-p} - a_0(\xi, \sigma|_E)$ *for a necessarily*
unique $p \in \{0, \ldots, n-1\}$, *and*

$$\dim \operatorname{gr}^j D_{\mathrm{DR},\sigma}(R_\xi^i(\pi)) = 0$$

otherwise. Here again π_∞ *runs over irreducible representations of*
$G_{\imath\sigma}(\mathbb{R})$ *and* $m_{\imath\sigma}(\pi)$ *denotes the multiplicity of* π *is the space of au-*
tomorphic forms on $G_{\imath\sigma}(\mathbb{Q}) \backslash G_{\imath\sigma}(\mathbb{A})$.

For the rest of this book we will assume that ξ (our chosen representation of G over $\mathbb{Q}_l^{ac}$) is *irreducible*.

III.3 The trace formula

In this section let us give a group theoretic expresssion for the trace of an
element $\varphi \in C_c^\infty(G(\mathbb{A}^\infty))$ on $H(X, \mathcal{L}_\xi)$. For simplicity of references we will
derive this as a simple specialisation of the main result of [Arth], which
relies on Arthur's trace formula. However in our case this expression could
be derived much more simply (because X_U is proper), either by combining
Matsushima's formula with the Selberg trace formula (which is basically
what Arthur does, except in our case there are no boundary contributions
which provide the key difficulty for his work), or by using the Lefschetz
trace formula (cf [KS] or [Bew]).

Proposition III.3.1 *Suppose that $\varphi \in C_c^\infty(G(\mathbb{A}^\infty))$ and that $\tau : F \hookrightarrow \mathbb{C}$. Then*

$$\operatorname{tr}(\varphi|H(X,\mathcal{L}_\xi)) = (-1)^n n \kappa_B \sum_\gamma (-1)^{n/[F(\gamma):F]}$$
$$[F(\gamma):F]^{-1}\operatorname{vol}(Z_{G_\tau}(\gamma)(\mathbb{R})_0^1)^{-1}O_\gamma^{G(\mathbb{A}^\infty)}(\varphi)\operatorname{tr}\xi(\gamma),$$

unless $F^+ = \mathbb{Q}$ and $n = 2$ in which case we drop the factors $[F(\gamma):F]^{-1}$. Here

- *γ runs over a set of representatives of $G_\tau(\mathbb{A})$-conjugacy classes in $G_\tau(\mathbb{Q})$ which are elliptic in $G_\tau(\mathbb{R})$;*

- *$\kappa_B = 2$ if $[B:\mathbb{Q}]/2$ is odd and $\kappa_B = 1$ otherwise;*

- *$Z_{G_\tau}(\gamma)(\mathbb{R})_0$ denotes the compact mod centre inner form of $Z_{G_\tau}(\gamma)(\mathbb{R})$ and $Z_{G_\tau}(\gamma)(\mathbb{R})_0^1$ the kernel of*

$$|\nu| : Z_{G_\tau}(\gamma)_0 \longrightarrow \mathbb{R}_{>0}^\times;$$

- *and we use measures on $Z_{G_\tau}(\gamma)(\mathbb{R})_0^1$ and $Z_{G_\tau}(\gamma)(\mathbb{A}^\infty)$ compatible with*

  - *Tamagawa measure on $Z_{G_\tau}(\gamma)(\mathbb{A})$,*
  - *the decomposition*

    $$Z_{G_\tau}(\gamma)(\mathbb{A}) = Z_{G_\tau}(\gamma)(\mathbb{A}^\infty) \times Z_{G_\tau}(\gamma)(\mathbb{R}),$$

  - *association of measures on $Z_{G_\tau}(\gamma)(\mathbb{R})_0$ and $Z_{G_\tau}(\gamma)(\mathbb{R})$,*
  - *and the measure dt/t on $\mathbb{R}_{>0}^\times$.*

Proof: By theorem 6.1 of [Arth] and by remark 3 following that theorem we see that

$$\operatorname{tr}(\varphi|H(X,\mathcal{L}_\xi)) = (-1)^n \delta_B \#\ker^1(\mathbb{Q}, G_\tau)$$
$$\sum_\gamma (-1)^{n/[F(\gamma):F]}(n/[F(\gamma):F])\operatorname{vol}(Z_{G_\tau}(\gamma)(\mathbb{Q})\mathbb{R}_{>0}^\times\backslash Z_{G_\tau}(\gamma)(\mathbb{A}))$$
$$\operatorname{vol}(\mathbb{R}_{>0}^\times\backslash Z_{G_\tau}(\gamma)(\mathbb{R})_0)^{-1}O_\gamma^{G(\mathbb{A}^\infty)}(\varphi)\operatorname{tr}\xi(\gamma),$$

where

- the sum runs over a set of representatives of $G_\tau(\mathbb{Q})$-conjugacy classes in $G_\tau(\mathbb{Q})$;

- $\delta_B = 2$ if $F^+ = \mathbb{Q}$ and $n = 2$ and $\delta_B = 1$ otherwise;

- and we use associated measures on $Z_{G_\tau}(\gamma)(\mathbb{R})$ and $Z_{G_\tau}(\gamma)(\mathbb{R})_0$.

(Note that in the notation of [Arth] we have

- $\delta_B = [K_{\mathbb{R}} : K_{\mathbb{R}}^0]$,

- $q(M_\gamma) = n/[F(\gamma) : F] - 1$,

- $|\mathcal{D}(M_\gamma, B)| = n/[F(\gamma) : F]$, unless $n = 2$ and $F^+ = \mathbb{Q}$ in which case $|\mathcal{D}(M_\gamma, B)| = 1$,

- $|i^{G_\tau}(\gamma)| = 1$,

- $\Phi_{G_\tau}(\gamma, \xi) = \operatorname{tr} \xi(\gamma)$,

- and $H_{G_\tau}(\gamma) = O_\gamma^{G(\mathbb{A}^\infty)}(\varphi)$.)

Note that
$$\operatorname{vol}(Z_{G_\tau}(\gamma)(\mathbb{Q})\mathbb{R}_{>0}^\times \backslash Z_{G_\tau}(\gamma)(\mathbb{A}))\operatorname{vol}(\mathbb{R}_{>0}^\times \backslash Z_{G_\tau}(\gamma)(\mathbb{R})_0)^{-1} =$$
$$\operatorname{vol}(Z_{G_\tau}(\gamma)(\mathbb{Q})\backslash Z_{G_\tau}(\gamma)(\mathbb{A})^1)\operatorname{vol}(Z_{G_\tau}(\gamma)(\mathbb{R})_0^1)^{-1},$$

where $Z_{G_\tau}(\gamma)(\mathbb{A})^1$ denotes the elements $g \in Z_{G_\tau}(\gamma)(\mathbb{A})$ with $|\nu(g)| = 1$. By the main theorem of [Ko5] we see that if we use Tamagawa measure then

$$\operatorname{vol}(Z_{G_\tau}(\gamma)(\mathbb{Q})\backslash Z_{G_\tau}(\gamma)(\mathbb{A})^1) = \kappa_B/\# \ker^1(\mathbb{Q}, Z_{G_\tau}(\gamma)).$$

(Using the fact that $[B : \mathbb{Q}]/2$ and $[Z_B(\gamma) : \mathbb{Q}]/2$ have the same parity, a direct calculation shows that $\#A(Z_{G_\tau}(\gamma)) = \kappa_B$.) Moreover the $G_\tau(\mathbb{A})$-conjugacy class of γ in $G_\tau(\mathbb{Q})$ contains

$$\# \ker(\ker^1(\mathbb{Q}, Z_{G_\tau}(\gamma)) \longrightarrow \ker^1(\mathbb{Q}, G_\tau))$$

$G_\tau(\mathbb{Q})$-conjugacy classes. Let Z (resp. Z') denote the centre of G_τ (resp. $Z_{G_\tau}(\gamma)$). Then we have homomorphisms

$$\ker^1(\mathbb{Q}, Z) \longrightarrow \ker^1(\mathbb{Q}, Z') \longrightarrow \ker^1(\mathbb{Q}, Z_{G_\tau}(\gamma)) \longrightarrow \ker^1(\mathbb{Q}, G_\tau),$$

where as on pages 393 and 394 of [Ko3] we see that

$$\ker^1(\mathbb{Q}, Z') \xrightarrow{\sim} \ker^1(\mathbb{Q}, Z_{G_\tau}(\gamma))$$

and

$$\ker^1(\mathbb{Q}, Z) \xrightarrow{\sim} \ker^1(\mathbb{Q}, G_\tau).$$

Thus

$$\ker^1(\mathbb{Q}, Z_{G_\tau}(\gamma)) \twoheadrightarrow \ker^1(\mathbb{Q}, G_\tau)$$

and

$$\# \ker(\ker^1(\mathbb{Q}, Z_{G_\tau}(\gamma)) \longrightarrow \ker^1(\mathbb{Q}, G_\tau))$$
$$= \# \ker^1(\mathbb{Q}, Z_{G_\tau}(\gamma))/\# \ker^1(\mathbb{Q}, G_\tau).$$

Hence we may rewrite Arthur's formula to give the formula in the proposition. □

III.4 Integral models

In this section we will describe and begin the analysis of integral models for the X_U over $\mathcal{O}_{F,w}$.

Suppose that $S/\mathrm{Spec}\,\mathcal{O}_{F,w}$ is a scheme, that A/S is an abelian scheme and that $i : \mathcal{O}_B \hookrightarrow \mathrm{End}\,(A) \otimes_{\mathbb{Z}} \mathbb{Z}_{(p)}$. We will call (A, i) *compatible* if $\mathrm{Lie}\,A \otimes_{\mathbb{Z}_p \otimes \mathcal{O}_E} \mathcal{O}_{E,u}$ is locally free of rank n and the two actions of $\mathcal{O}_F$ (one from the structure map $\mathcal{O}_{F,w} \to \mathcal{O}_S$ and one from the $\mathcal{O}_B$-action) coincide. If (A, i) is compatible then for $i > 1$ the Barsotti-Tate $\mathcal{O}_{F,w_i}$-module $A[w_i^\infty]$ is ind-etale. We will write

$$\mathfrak{G}_A = \varepsilon A[w^\infty],$$

a Barsotti-Tate $\mathcal{O}_{F,w}$-module. If p is locally nilpotent on S then we see that (A, i) is compatible if and only if

- $\mathfrak{G}_A$ is a compatible, one dimensional Barsotti-Tate $\mathcal{O}_{F,w}$-module, and

- $A[w_i^\infty]$ is ind-etale for $i > 1$.

Now suppose that $U^p \subset G(\mathbb{A}^{\infty,p})$ is a sufficiently small open compact subgroup and that $m = (m_1, \ldots, m_r) \in \mathbb{Z}_{\geq 0}^r$. Then we will let $U^p(m) \subset G(\mathbb{A}^\infty)$ denote the product

$$U^p \times \mathbb{Z}_p^\times \times \prod_{i=1}^r \ker((\mathcal{O}_{B_{w_i}}^{\mathrm{op}})^\times \longrightarrow (\mathcal{O}_{B_{w_i}}^{\mathrm{op}}/w_i^{m_i})^\times).$$

For simplicity we will restrict attention to open compact subgroups of this form. Given U^p and m as above we shall consider the functor $\mathfrak{X}_{U^p,m}$ from the category of pairs (S, s), where S is a connected, locally noetherian $\mathcal{O}_{F,w}$-scheme and s is a geometric point of S, to sets, which sends (S, s) to the set of equivalence classes of $(r + 4)$-tuples $(A, \lambda, i, \overline{\eta}^p, \alpha_i)$ where

- A/S is an abelian scheme of dimension dn^2;

- $\lambda : A \longrightarrow A^\vee$ is a prime to p polarisation;

- $i : \mathcal{O}_B \hookrightarrow \mathrm{End}\,(A) \otimes_{\mathbb{Z}} \mathbb{Z}_{(p)}$ such that (A, i) is compatible and $\lambda \circ i(b) = i(b^*)^\vee \circ \lambda$ for all $b \in \mathcal{O}_B$;

- $\overline{\eta}^p$ is a $\pi_1(S, s)$-invariant U^p-orbit of isomorphisms of $B \otimes_{\mathbb{Q}} \mathbb{A}^{\infty,p}$-modules $\eta^p : V \otimes_{\mathbb{Q}} \mathbb{A}^{\infty,p} \to V^p A_s$ which take the standard pairing $(\ ,\ )$ on $V \otimes_{\mathbb{Q}} \mathbb{A}^{\infty,p}$ to a $(\mathbb{A}^{\infty,p})^\times$ multiple of the λ-Weil pairing on $V^p A_s$;

- $\alpha_1 : w^{-m_1}\Lambda_{11}/\Lambda_{11} \to \mathfrak{G}_A[w^{m_1}]$ is a Drinfeld w^{m_1}-structure;

- for $i > 1$, $\alpha_i : (w_i^{-m_i} \Lambda_i / \Lambda_i)_S \overset{\sim}{\to} A[w_i^{m_i}]$ is an isomorphism of S-schemes with $\mathcal{O}_B$-actions.

Two $(r+4)$-tuples $(A, \lambda, i, \overline{\eta}^p, \alpha_i)$ and $(A', \lambda', i', (\overline{\eta}^p)', \alpha_i')$ are equivalent if there exists a prime to p isogeny $\delta : A \to A'$ and $\gamma \in \mathbb{Z}_{(p)}^{\times}$ such that δ carries λ to $\gamma\lambda'$, i to i', $\overline{\eta}^p$ to $(\overline{\eta}^p)'$, and α_i to α_i'. Again $\mathfrak{X}_{U^p, m}(S, s)$ is canonically independent of s so we obtain a functor from connected, locally noetherian $\mathcal{O}_{F,w}$-schemes to sets. We extend it to all locally noetherian $\mathcal{O}_{F,w}$-schemes by setting $\mathfrak{X}_{U^p, m}(\coprod S_i) = \prod \mathfrak{X}_{U^p, m}(S_i)$. On locally noetherian F_w-schemes we have natural isomorphisms $\mathfrak{X}_{U^p, m} \cong \mathfrak{X}_{U^p(m)}'' \cong \mathfrak{X}_{U^p(m)}$.

If $m_1 = 0$ then it is known that this functor is represented by a projective scheme $X_{U^p, m}/\mathcal{O}_{F,w}$. (Representability and quasi-projectivity follow as on page 391 of [Ko3] or as in section 5.3 of [Car1]. Properness follows from the valuative criterion as in section 5.5 of [Car1], the point being that if A is an abelian variety of dimension dn^2 with an action of an order in B over the field of fractions of a DVR and if A has semistable reduction then A has good reduction (otherwise the toric part of the reduction has too small a dimension to have an action of an order in B). The level structure then extends uniquely to the Neron model $\widetilde{A}$ of A, because $\widetilde{A}[\mathfrak{n}]$ is etale over the DVR for $\mathfrak{n}$ supported on $w_2, \ldots, w_r$ and the primes not dividing p (use the fact that $\mathrm{Lie}\, A[w_i]^{\infty} = (0)$ for $i > 1$).) Hence by II.2.1, this functor is represented for all m by a projective scheme $X_{U^p, m}/\mathcal{O}_{F,w}$. We have a canonical isomorphism

$$X_{U^p, m} \times_{\mathrm{Spec}\, \mathcal{O}_{F,w}} \mathrm{Spec}\, F_w \cong X_{U^p(m)} \times_{\mathrm{Spec}\, F} \mathrm{Spec}\, F_w.$$

The inverse system of the $X_{U^p, m}/\mathcal{O}_{F,w}$ again has an action of $G(\mathbb{A}^{\infty})$. The action of $g \in G(\mathbb{A}^{\infty, p})$ just sends $(A, \lambda, i, \overline{\eta}^p, \alpha_i)$ to $(A, \lambda, i, \overline{\eta}^p \circ g, \alpha_i)$. The action of $(g_0, g_1, \ldots, g_r) \in G(\mathbb{Q}_p)$ is slightly trickier to describe. To do so let us suppose that for each $i \geq 1$ we have the following integrality conditions

- $g_i^{-1} \in \mathcal{O}_{B, w_i}^{\mathrm{op}}$,

- $g_0^{-1} g_i \in \mathcal{O}_{B, w_i}^{\mathrm{op}}$,

- $w_i^{m_i - m_i'} g_i \in \mathcal{O}_{B, w_i}^{\mathrm{op}}$.

Under these assumptions we will define a morphism

$$(g_i) : X_{U^p, m} \longrightarrow X_{U^p, m'}.$$

It will send $(A, \lambda, i, \overline{\eta}^p, \alpha_i)$ to $(A/(C \oplus C^{\perp}), p^{\mathrm{val}\, p(g_0)}\lambda, i, \overline{\eta}^p, \alpha_i \circ g_i)$, where

- $C_1 \subset \varepsilon A[w_1^{m_1}]$ is the unique closed subscheme for which the set of $\alpha_1(x)$ with $x \in g_1 \Lambda_{11}/\Lambda_{11}$ is a complete set of sections;

- for $i > 1$, $C_i = \alpha_i(g_i\Lambda_i/\Lambda_i)$;

- $C = (\mathcal{O}^n_{F,w} \otimes_{\mathcal{O}_{F,w}} C_1) \oplus \bigoplus^r_{i=2} C_i \subset A[u^{-\operatorname{val}p(g_0)}]$;

- $C^\perp$ is the annihilator of $C \subset A[u^{-\operatorname{val}p(g_0)}]$ inside $A[(u^c)^{-\operatorname{val}p(g_0)}]$ under the λ-Weil pairing;

- $p^{\operatorname{val}p(g_0)}\lambda$ is the polarisation $A/(C \oplus C^\perp) \to (A/(C \oplus C^\perp))^\vee$ which makes the following diagram commute

$$
\begin{array}{ccc}
A & \xrightarrow{p^{-\operatorname{val}p(g_0)}\lambda} & A^\vee \\
\downarrow & & \uparrow \\
A/(C \oplus C^\perp) & \longrightarrow & (A/(C \oplus C^\perp))^\vee;
\end{array}
$$

- $\alpha_1 \circ g_1 : w_1^{-m'_1}\Lambda_{11}/\Lambda_{11} \to (\varepsilon A[w_1^\infty]/C_1)(S)$ is the homomorphism making the following diagram commute

$$
\begin{array}{ccc}
w_1^{-m'_1}\Lambda_{11}/\Lambda_{11} & \longrightarrow & \varepsilon A[w_1^\infty]/C_1(S) \\
\downarrow & & \uparrow \\
w_1^{-m'_1}g_1\Lambda_{11}/g_1\Lambda_{11} & \longrightarrow & (\varepsilon A[w_1^\infty]/C_1)[w_1^{m'_1}](S) \\
\downarrow & & \downarrow \\
w_1^{-m_1}\Lambda_{11}/g_1\Lambda_{11} & \longrightarrow & (\varepsilon A[w_1^{m_1}]/C_1)(S) \\
\uparrow & & \uparrow \\
w_1^{-m_1}\Lambda_{11}/\Lambda_{11} & \xrightarrow{\alpha_1} & \varepsilon A[w_1^{m_1}](S);
\end{array}
$$

- for $i > 1$, $\alpha_i \circ g_i : w_i^{-m'_i}\Lambda_i/\Lambda_i \to A[w_i^\infty]/C_i$ is the homomorphism making the following diagram commute

$$
\begin{array}{ccc}
w_i^{-m'_i}\Lambda_i/\Lambda_i & \longrightarrow & A[w_i^\infty]/C_i \\
\downarrow & & \uparrow \\
w_i^{-m'_i}g_i\Lambda_i/g_i\Lambda_i & \xrightarrow{\sim} & (A[w_i^\infty]/C_i)[w_i^{m'_i}] \\
\downarrow & & \downarrow \\
w_i^{-m_i}\Lambda_i/g_i\Lambda_i & \xrightarrow{\sim} & A[w_i^{m_i}]/C_i \\
\uparrow & & \uparrow \\
w_i^{-m_i}\Lambda_i/\Lambda_i & \xrightarrow{\alpha_i} & A[w_i^{m_i}].
\end{array}
$$

It is tedious but straightforward to check that this does define an action. We see that $(p^{-2}, p^{-1}, \ldots, p^{-1})$ acts in the same way as $p \in G(\mathbb{A}^{\infty,p})$ and so acts invertibly on the inverse system. Thus this definition can be extended to the whole of $G(\mathbb{Q}_p)$. We also see that on the generic fibre (i.e. over F_w) this definition (when it makes sense) agrees with the action previously defined. (A less tedious argument is to first note that this definition coincides with the previously defined action on the generic fibre and then use the fact that the generic fibre is Zariski dense in $X_{U^p,m}$ to check the first two assertions. That the generic fibre is indeed dense follows at once from lemma III.4.1 below.)

We next establish some important pieces of notation. We will let $\mathcal{A}/X_{U^p,m}$ denote the universal abelian variety. We write simply $\mathcal{G}/X_{U^p,m}$ for $\mathfrak{G}_{\mathcal{A}}$. If s is a closed geometric point of $X_{U^p,m}$ we will let $h(s)$ denote the height of $\mathcal{G}_s^{\text{et}}$. We will let $\overline{X}_{U^p,m}$ denote the reduction $X_{U^p,m} \times_{\operatorname{Spec}\mathcal{O}_{F,w}}$ $\operatorname{Spec} k(w)$. We will let $\overline{X}_{U^p,m}^{[h]}$ denote the reduced closed subscheme of $\overline{X}_{U^p,m}$ which is the closure of the set of closed geometric points s with $h(s) \leq h$. We will also let

$$\overline{X}_{U^p,m}^{(h)} = \overline{X}_{U^p,m}^{[h]} - \overline{X}_{U^p,m}^{[h-1]}.$$

The action of $G(\mathbb{A}^\infty)$ on the inverse system of the $X_{U^p,m}$ takes the inverse system of locally closed subschemes $\overline{X}_{U^p,m}^{(h)}$ to itself (because they are defined in an invariant manner).

Lemma III.4.1 *Throughout this lemma we suppose that U^p is sufficiently small. Let m, m' and m'' be r-tuples of non-negative integers with $m_1 = 0$ and $m_i' = m_i$ for $i > 1$. Let s be a closed point of $\overline{X}_{U^p,m} \times_{\operatorname{Spec} k(w)} k(w)^{\text{ac}}$ and fix an isomorphism $\mathcal{G}_s^0 \xrightarrow{\sim} \Sigma_{F_w,n-h(s)}$.*

1. *The formal completion of $X_{U^p,m} \times_{\operatorname{Spec}\mathcal{O}_{F,w}} \operatorname{Spec}\mathcal{O}_{\widehat{F}_w^{\text{nr}}}$ at s is isomorphic to the universal formal deformation space for the Barsotti-Tate $\mathcal{O}_{F,w}$-module $\mathcal{G}_s$. Thus we get an identification*

$$(X_{U^p,m} \times_{\operatorname{Spec}\mathcal{O}_{F,w}} \operatorname{Spec}\mathcal{O}_{\widehat{F}_w^{\text{nr}}})_s^\wedge \cong \operatorname{Hom}(T\mathcal{G}_s, \widetilde{\Sigma}_{F_w,n-h(s)});$$

while $(\overline{X}_{U^p,m}^{(h(s))} \times_{\operatorname{Spec} k(w)} \operatorname{Spec} k(w)^{\text{ac}})_s^\wedge$ is identified to the closed formal subscheme

$$\operatorname{Hom}(T\mathcal{G}_s, \Sigma_{F_w,n-h(s)}) \subset \operatorname{Hom}(T\mathcal{G}_s, \widetilde{\Sigma}_{F_w,n-h(s)}).$$

2. *$X_{U^p,m}/\operatorname{Spec}\mathcal{O}_{F,w}$ is smooth. Moreover each $\overline{X}_{U^p,m}^{(h)}/\operatorname{Spec} k(w)$ is either empty or smooth of dimension h.*

3. *The closed points of $\overline{X}_{U^p,m'} \times_{\operatorname{Spec} k(w)} k(w)^{\text{ac}}$ above s are in natural bijection with the surjective homomorphisms*

$$\delta : w^{-m_1'}\Lambda_{11}/\Lambda_{11} \twoheadrightarrow \mathcal{G}_s^{\text{et}}[w^{m_1'}](k(s)).$$

We will write s_δ for the point corresponding to δ. Then we can identify the formal completion of $X_{U^p,m'} \times_{\operatorname{Spec}\mathcal{O}_{F,w}} \operatorname{Spec}\mathcal{O}_{\widehat{F}_w^{\text{nr}}}$ at s_δ with

$$\operatorname{Hom}(w^{-m_1'}T\mathcal{G}_s, \widetilde{\Sigma}_{F_w,n-h(s)}) \times_{\operatorname{Spf} R_{F_w,n-h(s)}} \operatorname{Spf} R_{F_w,n-h(s),m_1'},$$

such that the morphism

$$(X_{U^p,m'} \times_{\operatorname{Spec} \mathcal{O}_{F,w}} \operatorname{Spec} \mathcal{O}_{\widehat{F}_w^{nr}})^{\wedge}_{s_\delta} \longrightarrow (X_{U^p,m} \times_{\operatorname{Spec} \mathcal{O}_{F,w}} \operatorname{Spec} \mathcal{O}_{\widehat{F}_w^{nr}})^{\wedge}_s$$

corresponds to the natural morphism

$$\operatorname{Hom}(w^{-m'_1}T\mathcal{G}_s, \widetilde{\Sigma}_{F_w,n-h(s)}) \times_{\operatorname{Spf} R_{F_w,n-h(s)}} \operatorname{Spf} R_{F_w,n-h(s),m'_1}$$
$$\downarrow$$
$$\operatorname{Hom}(T\mathcal{G}_s, \widetilde{\Sigma}_{F_w,n-h(s)}).$$

Moreover the formal completion of $\overline{X}_{U^p,m'}^{(h(s))}$ *at* s_δ *corresponds to the closed formal subscheme* $\operatorname{Hom}(w^{-m'_1}T\mathcal{G}_s, \Sigma_{F_w,n-h(s)})$ *inside*

$$\operatorname{Hom}(w^{-m'_1}T\mathcal{G}_s, \widetilde{\Sigma}_{F_w,n-h(s)}) \times_{\operatorname{Spf} R_{F_w,n-h(s)}} \operatorname{Spf} R_{F_w,n-h(s),m'_1}.$$

4. $X_{U^p,m'}/\mathcal{O}_{F,w}$ *is regular and flat.*

5. $\overline{X}_{U^p,m'}^{(h)}/k(w)$ *is smooth and the morphism* $\overline{X}_{U^p,m'}^{(h)} \to \overline{X}_{U^p,m}^{(h)}$ *is finite and flat of degree* $\#GL_n(\mathcal{O}_{F,w}/w^{m'_1})/\#GL_{n-h}(\mathcal{O}_{F,w}/w^{m'_1})$.

6. *Suppose that* $(U^p)'' \subset U^p$ *and that for all* i *we have* $m''_i \geq m'_i$. *Then the natural morphism*

$$X_{(U^p)'',m''} \longrightarrow X_{U^p,m'}$$

is finite and flat of degree

$$[U^p : (U^p)''] \prod_{i=1}^r \#GL_n(\mathcal{O}_{F,w_i}/w_i^{m''_i}\mathcal{O}_{F,w_i})/$$
$$(\prod_{i=1}^r \#GL_n(\mathcal{O}_{F,w_i}/w_i^{m'_i}\mathcal{O}_{F,w_i})).$$

If $m''_1 = m'_1$ *then this morphism is in fact etale.*

Proof: First of all it is standard that $(X_{U^p,m} \times_{\operatorname{Spec} \mathcal{O}_{F,w}} \operatorname{Spec} \mathcal{O}_{\widehat{F}_w^{nr}})^{\wedge}_s$ is the formal deformation space for $(r+2)$-tuples deforming $(\mathcal{A}_s, \lambda_s, i_s, \alpha_{i,s})$ (where $i > 1$). By the Serre-Tate theorem this is the same as deformations of the $(r+2)$-tuple $(\mathcal{A}_s[p^\infty], \lambda_s, i_s, \alpha_{i,s})$. As $\lambda : \mathcal{A}_s[u^\infty] \overset{\sim}{\to} \mathcal{A}_s[(u^c)^\infty]$ we see that this is the same as deformations of the $(r+1)$-tuple $(\mathcal{A}_s[u^\infty], i_s, \alpha_{i,s})$. As $\mathcal{A}_s[w_i^\infty]$ is ind-etale for $i > 1$ it has a unique deformation over any Artinian local ring with residue field $k(s)$ as does $\alpha_{i,s}$. Thus we need only consider deformations of the pair $(\mathcal{A}_s[w^\infty], i_s)$. As $\mathcal{O}_{B,w} \cong M_n(\mathcal{O}_{F,w})$ this is the same as deformations of $\mathcal{G}_s = \varepsilon \mathcal{A}_s[w^\infty]$ with its $\mathcal{O}_{F,w}$-action. This proves the first assertion of the lemma.

The rest of the first part of the lemma follows from the discussion before lemma II.1.3 and from corollary II.1.4. The second part of the lemma follows from the first.

The first assertion of the third part of the lemma follows from lemma II.2.1. The second assertion follows from the discussion proceeding lemma II.2.4. The scheme $\overline{X}_{U^p,m'}^{(h(s))}$ can be constructed as the reduced subscheme of the fibre product of $\overline{X}_{U^p,m}^{(h(s))}$ and $X_{U^p,m'}$ over $X_{U^p,m}$. Thus $(\overline{X}_{U^p,m'}^{(h(s))})^\wedge_{s_\delta}$ is the reduced formal subscheme of the fibre product over $\mathrm{Hom}\,(T\mathcal{G}_s, \widetilde{\Sigma}_{F_w,n-h(s)})$ of $\mathrm{Hom}\,(T\mathcal{G}_s, \Sigma_{F_w,n-h(s)})$ and

$$\mathrm{Hom}\,(w^{-m_1'}T\mathcal{G}_s, \widetilde{\Sigma}_{F_w,n-h(s)}) \times_{\mathrm{Spf}\,R_{F_w,n-h(s)}} \mathrm{Spf}\,R_{F_w,n-h(s),m_1'}.$$

(Here we make use of lemma II.1.6.) Hence $(\overline{X}_{U^p,m'}^{(h(s))})^\wedge_{s_\delta}$ is the reduced formal subscheme of

$$\mathrm{Hom}\,(w^{-m_1'}T\mathcal{G}_s, \Sigma_{F_w,n-h(s)}) \times_{\mathrm{Spf}\,R_{F_w,n-h(s)}} \mathrm{Spf}\,R_{F_w,n-h(s),m_1'},$$

i.e. $\mathrm{Hom}\,(w^{-m_1'}T\mathcal{G}_s, \Sigma_{F_w,n-h(s)})$.

The fourth part now follows on applying proposition 4.3 of [Dr] because both these properties can be detected on formal completions at closed points. (If A is a noetherian local ring with maximal ideal $\mathfrak{m}$ then $\dim A_{\mathfrak{m}}^\wedge = \dim A$, $\mathfrak{m}/\mathfrak{m}^2 \overset{\sim}{\to} \mathfrak{m}^\wedge/(\mathfrak{m}^\wedge)^2$ and $A_{\mathfrak{m}}^\wedge/A$ is faithfully flat.) As for the fifth part, finiteness follows from lemma II.2.1. Smoothness and flatness follow from the computation of the formal completions. The degree can also be computed on formal completions: suppose that s is a closed point of $\overline{X}_{U^p,m}^{(h)} \times \mathrm{Spec}\,k(w)^{\mathrm{ac}}$. The number of closed points of $\overline{X}_{U^p,m'}^{(h)} \times \mathrm{Spec}\,k(w)^{\mathrm{ac}}$ above s is the number of surjective homomorphisms from $(\mathcal{O}_{F,w}/w^{m_1'})^n$ to $(\mathcal{O}_{F,w}/w^{m_1'})^h$. If s_δ is one of these points the degree of $(\overline{X}_{U^p,m'}^{(h)} \times \mathrm{Spec}\,k(w)^{\mathrm{ac}})^\wedge_{s_\delta}$ over $(\overline{X}_{U^p,m}^{(h)} \times \mathrm{Spec}\,k(w)^{\mathrm{ac}})^\wedge_s$ is the rank of $\Sigma_{F_w,n-h}^h[w^{m_1'}]$. Thus the degree of $\overline{X}_{U^p,m'}^{(h)}$ over $\overline{X}_{U^p,m}^{(h)}$ is

$$(\#\mathcal{O}_{F,w}/w^{m_1'})^{h(n-h)}(\#GL_n(\mathcal{O}_{F,w}/w^{m_1'})/\#GL_n(\mathcal{O}_{F,w}/w^{m_1'})_\delta) =$$
$$= \#GL_n(\mathcal{O}_{F,w}/w^{m_1'})/\#GL_{n-h}(\mathcal{O}_{F,w}/w^{m_1'}).$$

We can divide the proof of the sixth part into two cases: the case where $m_1'' = m_1'$ and the case where $U^p = (U^p)''$ and $m_i'' = m_i'$ for $i > 1$. In the second of these two cases it is standard that the morphism is etale of the stated degree. In the first case it follows from lemma II.2.4. $\square$

(We remark that one can use the results of Drinfeld's paper [Dr] to show that in fact if $m_1 = 0$ then $\overline{X}_{U^p,m}^{[h]}$ is smooth. We will not give details here as we will not need this result. It seems to us an interesting question whether this remains true for $m_1 > 0$.)

The universal abelian variety $\mathcal{A}$ extends over $X_{U^p,m}$. If $m = 0$ then it is smooth over $\mathcal{O}_{F,w}$. We have seen in section III.2 that $R_\xi^i(\pi)$ is a Tate twist of a direct summand of the cohomology of $\mathcal{A}^m$ for suitable m. Thus we deduce the following lemma.

Lemma III.4.2 *Suppose that π is an irreducible representation of $G(\mathbb{A}^\infty)$ with $\pi_{p,0}$ and π_w unramified. If $l = p$ then $R^i_\xi(\pi)$ is crystalline at w. If $l \neq p$ then $R^i_\xi(\pi)$ is unramified at w, and if α is an eigenvalue of $R^i_\xi(\pi)(\mathrm{Frob}_w)$ then α is algebraic and for each embedding of α in $\mathbb{C}$ we have $|\alpha|^2 = (\#k(w))^{w(\xi)}$.*

Now and in the rest of the book we assume that $p \neq l$. In this case, the lisse $\overline{\mathbb{Q}}_l^{ac}$ sheaf $\mathcal{L}_\xi$ can be defined over the whole of $X_{U^p,m}$ in exactly the same manner it was defined over the generic fibre $X_{U^p(m)}$. If $g \in G(\mathbb{A}^\infty)$ maps $X_{U^p,m}$ to $X_{(U^p)',m'}$ then again $\xi(g_l)$ induces a morphism of sheaves

$$g : g^* \mathcal{L}_\xi \longrightarrow \mathcal{L}_\xi$$

over $X_{U^p,m}$.

The next lemma will be proved in section V.4 below. (It follows from corollary V.4.5.)

Lemma III.4.3 *The scheme $\overline{X}^{(0)}_{U^p,m}$ is non-empty.*

As a first application of this lemma we have the following corollary.

Corollary III.4.4 *The scheme $\overline{X}^{(h)}_{U^p,m}$ for $h = 0, \ldots, n-1$ is smooth of pure dimension h.*

As a second application we now provide the postponed proof of lemma II.2.7.

Proof of lemma II.2.7: Choose a totally real field F^+ with a place w above p such that $F^+_w \cong K$ and choose an imaginary quadratic field E in which p splits. We may then choose $u, B, *, (\, , \,)$ and Λ_i as in section I.7 and such that $\dim_F B = g^2$. Also choose a sufficiently small open compact subgroup $U^p \subset G(\mathbb{A}^{\infty,p})$. Let

$$X_m = X_{U^p,(m,0,\ldots,0)} \times_{\mathrm{Spec}\, \mathcal{O}_{F,w}} \mathrm{Spec}\, \mathcal{O}_{\widehat{F}^{nr}_w}$$

and let x be any closed point of

$$\overline{X}^{(g)}_{U^p,(0,0,\ldots,0)} \times_{\mathrm{Spec}\, k(w)} \mathrm{Spec}\, k(w)^{ac} \subset X_0.$$

(The existence of x follows from the last lemma.) That x and the collection of the X_m have the asserted properties follows from lemma III.4.1. $\square$

Now let Φ^i denote the vanishing cycles for $\overline{X}_{U^p,m} \subset X_{U^p,m}$. Then we have a spectral sequence

$$H^i(\overline{X}_{U^p,m} \times k(w)^{ac}, \Phi^j \otimes \mathcal{L}_\xi) \Rightarrow H^{i+j}(X_{U^p,m} \times F^{ac}_w, \mathcal{L}_\xi).$$

(See lemma I.5.2.) If $(g, \sigma) \in G(\mathbb{A}^\infty) \times W_{F_w}$ then we have a natural map

$$(g, \sigma) : (g \times \mathrm{Frob}_w^{w(\sigma)})^* \Phi^j = (g \circ (\mathrm{Fr}^*)^{f_1 w(\sigma)} \times 1)^* \Phi^j \otimes \mathcal{L}_\xi \longrightarrow \Phi^j \otimes \mathcal{L}_\xi.$$

Thus we get a smooth/continuous action of $G(\mathbb{A}^\infty) \times W_{F_w}$ on

$$\varinjlim_{U^p, m} H^i(\overline{X}_{U^p, m} \times k(w)^{\mathrm{ac}}, \Phi^j \otimes \mathcal{L}_\xi)$$

which is compatible with the action on $H^{i+j}(X, \mathcal{L}_\xi)$ and the above spectral sequence.

For any $0 < h \leq n - 1$ we get a long exact sequence (see for example [FK] I.8.7 (3))

$$\dots \longrightarrow H_c^i(\overline{X}_{U^p, m}^{(h)} \times k(w)^{\mathrm{ac}}, \Phi^j \otimes \mathcal{L}_\xi) \longrightarrow H^i(\overline{X}_{U^p, m}^{[h]} \times k(w)^{\mathrm{ac}}, \Phi^j \otimes \mathcal{L}_\xi)$$
$$\longrightarrow H^i(\overline{X}_{U^p, m}^{[h-1]} \times k(w)^{\mathrm{ac}}, \Phi^j \otimes \mathcal{L}_\xi) \longrightarrow \dots$$

Combining these two observations we obtain the following lemma.

Lemma III.4.5 *Suppose that for each* $0 \leq h \leq n - 1$, $0 \leq i \leq 2h$ *and* $0 \leq j \leq n - 1$ *the* $G(\mathbb{A}^\infty) \times W_{F_w}$*-module* $\varinjlim_{U^p, m} H_c^i(\overline{X}_{U^p, m}^{(h)} \times k(w)^{\mathrm{ac}}, \Phi^j \otimes \mathcal{L}_\xi)$ *is admissible/continuous. Then the same is true for each* $\varinjlim_{U^p, m} H^i(\overline{X}_{U^p, m} \times k(w)^{\mathrm{ac}}, \Phi^j \otimes \mathcal{L}_\xi)$ *and we have an equality of virtual* $G(\mathbb{A}^\infty) \times W_{F_w}$*-modules*

$$\sum_i (-1)^i [H^i(X, \mathcal{L}_\xi)^{\mathbb{Z}_p^\times}]$$
$$= \sum_{i,j,h} (-1)^{i+j} [\varinjlim_{U^p, m} H_c^i(\overline{X}_{U^p, m}^{(h)} \times k(w)^{\mathrm{ac}}, \Phi^j \otimes \mathcal{L}_\xi)].$$

We wish to further analyse the structure of $\overline{X}_{U^p, m}$. For s a closed point of $\overline{X}_{U^p, m}^{(h)}$ let M_s denote the kernel of the composite

$$\alpha_1 : w^{-m_1} \Lambda_{11} / \Lambda_{11} \longrightarrow \mathcal{G}_s[w_1^{m_1}](k(s)) \longrightarrow \mathcal{G}_s^{\mathrm{et}}[w_1^{m_1}](k(s)).$$

Then M_s is a direct summand of $w^{-m_1} \Lambda_{11} / \Lambda_{11}$ which is free over $\mathcal{O}_{F,w}/w^{m_1}$ of rank $(n - h)$. The function $s \mapsto M_s$ is locally constant on $\overline{X}_{U^p, m}^{(h)}$. (Suppose that H is a finite abelian group, that S is a connected scheme and that $\mathcal{F}$ is a lisse etale sheaf on S. If $\alpha : H_S \to \mathcal{F}$ is a morphism of etale sheaves then there is a subgroup $H' \subset H$ such that $\ker \alpha = H_S$ (cf page 49 of [KM]).) Thus we have a decomposition

$$\overline{X}_{U^p, m}^{(h)} = \coprod_M \overline{X}_{U^p, m, M},$$

where M runs over free $\mathcal{O}_{F,w}/w^{m_1}$-submodules of $w^{-m_1}\Lambda_{11}/\Lambda_{11}$ of rank $n - h$ and where $M_s = M$ for s a closed point of $\overline{X}_{U^p,m,M}$. If $g \in (\mathcal{O}_{B,w}^{\mathrm{op}})^\times$ then g gives an isomorphism

$$g : \overline{X}_{U^p,m,M} \xrightarrow{\sim} \overline{X}_{U^p,m,g^{-1}M}.$$

The following lemma follows at once (using lemma III.4.1).

Lemma III.4.6 *Suppose that $m_1 = 0$ and $m'_i = m_i$ for $i > 1$. Then $\overline{X}_{U^p,m',M}^{(h)}$ is smooth of dimension h and $\overline{X}_{U^p,m',M}^{(h)}/\overline{X}_{U^p,m}^{(h)}$ is finite flat of degree*

$$\#P_M(\mathcal{O}_{F,w}/w^{m'_1})/\#GL_{n-h}(\mathcal{O}_{F,w}/w^{m'_1}) =$$
$$= \#(\mathcal{O}_{F,w}/w^{m'_1})^{h(n-h)}\#GL_h(\mathcal{O}_{F,w}/w^{m'_1}).$$

Now suppose that $M \subset \Lambda_{11}$ is a $\mathcal{O}_{F,w}$-submodule which is both a direct summand and free of rank $n - h$. We will let $P_M \subset \mathrm{Aut}\,(\Lambda_{11})$ denote the maximal parabolic subgroup which stabilises M. Then we will set

$$G_M(\mathbb{A}^\infty) = G(\mathbb{A}^{\infty,p}) \times \mathbb{Q}_p^\times \times P_M(F_w) \times \prod_{r=2}^r (B_{w_i}^{\mathrm{op}})^\times.$$

We will also set

$$\overline{X}_{U^p,m,M} = \overline{X}_{U^p,m,w^{-m_1}M/M}.$$

For fixed M the inverse system of the $\overline{X}_{U^p,m,M}$ inherits an action of $G_M(\mathbb{A}^\infty)$. Then

$$H_c^i(\overline{X}_M, \Phi^j \otimes \mathcal{L}_\xi) = \varinjlim_{U^p,m} H_c^i(\overline{X}_{U^p,m,M} \times \mathrm{Spec}\, k(w)^{\mathrm{ac}}, \Phi^j \otimes \mathcal{L}_\xi)$$

is a smooth/continuous $G_M(\mathbb{A}^\infty) \times W_{F_w}$-module.

We will next describe a natural map

$$\mathrm{Ind}_{P_M(\mathcal{O}_{F,w}/w^{m_1})}^{(\mathcal{O}_{B,w}^{\mathrm{op}}/w^{m_1})^\times} H_c^i(\overline{X}_{U^p,m,M} \times \mathrm{Spec}\, k(w)^{\mathrm{ac}}, \Phi^j \otimes \mathcal{L}_\xi)$$
$$\downarrow$$
$$H_c^i(\overline{X}_{U^p,m}^{(h)} \times \mathrm{Spec}\, k(w)^{\mathrm{ac}}, \Phi^j \otimes \mathcal{L}_\xi).$$

A typical element of $\mathrm{Ind}_{P_M(\mathcal{O}_{F,w}/w^{m_1})}^{(\mathcal{O}_{B,w}^{\mathrm{op}}/w^{m_1})^\times} H_c^i(\overline{X}_{U^p,m,M} \times \mathrm{Spec}\, k(w)^{\mathrm{ac}}, \Phi^j \otimes \mathcal{L}_\xi)$ is represented by a function

$$f : (\mathcal{O}_{B,w}^{\mathrm{op}}/w^{m_1})^\times \longrightarrow H_c^i(\overline{X}_{U^p,m,M} \times \mathrm{Spec}\, k(w)^{\mathrm{ac}}, \Phi^j \otimes \mathcal{L}_\xi)$$

such that

$$f(\gamma g) = \gamma f(g)$$

for all $\gamma \in P_M(\mathcal{O}_{F,w}/w^{m_1})$ and all $g \in (\mathcal{O}^{\mathrm{op}}_{B,w}/w^{m_1})^{\times}$. We map f to

$$(\#P_M(\mathcal{O}_{F,w}/w^{m_1})/\#(\mathcal{O}^{\mathrm{op}}_{B,w}/w^{m_1})^{\times})^{-1} \sum_g g^{-1}f(g),$$

where g runs over $P_M(\mathcal{O}_{F,w}/w^{m_1})\backslash(\mathcal{O}^{\mathrm{op}}_{B,w}/w^{m_1})^{\times}$. This map is easily checked to be a well defined isomorphism of $(\mathcal{O}^{\mathrm{op}}_{B,w})^{\times}$-modules.

Similarly we can define a natural map

$$\mathrm{Ind}^{(B^{\mathrm{op}}_w)^{\times}}_{P_M(F_w)} H^i_c(\overline{X}_M, \Phi^j \otimes \mathcal{L}_\xi)$$

$$\longrightarrow \varinjlim_{U^p,m} H^i_c(\overline{X}^{(h)}_{U^p,m} \times \mathrm{Spec}\, k(w)^{\mathrm{ac}}, \Phi^j \otimes \mathcal{L}_\xi)$$

as follows. A typical element of $\mathrm{Ind}^{(B^{\mathrm{op}}_w)^{\times}}_{P_M(F_w)} H^i_c(\overline{X}_M, \Phi^j \otimes \mathcal{L}_\xi)$ is represented by a locally constant function

$$f : (B^{\mathrm{op}}_w)^{\times} \longrightarrow H^i_c(\overline{X}_M, \Phi^j \otimes \mathcal{L}_\xi)$$

such that

$$f(\gamma g) = \gamma f(g)$$

for all $\gamma \in P_M(F_w)$ and all $g \in (B^{\mathrm{op}}_w)^{\times}$. We map f to

$$\int_{P_M(F_w)\backslash(B^{\mathrm{op}}_w)^{\times}} g^{-1}f(g)dg,$$

where dg is the quotient of a Haar measure on $(B^{\mathrm{op}}_w)^{\times}$ by a right Haar measure on $P_M(F_w)$ normalised so that $P_M(F_w)\backslash(B^{\mathrm{op}}_w)^{\times}$ has volume 1, and where the integral makes sense as it is in fact a finite sum (as f is locally constant and $P_M(F_W)\backslash(B^{\mathrm{op}}_w)^{\times}$ is compact). This is a morphism of $G(\mathbb{A}^{\infty}) \times W_{F_w}$-modules.

Lemma III.4.7 *Suppose that $H^i_c(\overline{X}_M, \Phi^j \otimes \mathcal{L}_\xi)$ is an admissible $P_M(\mathbb{A}^{\infty})$-module. Then the above map gives an isomorphism of $G(\mathbb{A}^{\infty}) \times W_{F_w}$-modules*

$$\mathrm{Ind}^{(B^{\mathrm{op}}_w)^{\times}}_{P_M(F_w)} H^i_c(\overline{X}_M, \Phi^j \otimes \mathcal{L}_\xi)$$

$$\longrightarrow \varinjlim_{U^p,m} H^i_c(\overline{X}^{(h)}_{U^p,m} \times \mathrm{Spec}\, k(w)^{\mathrm{ac}}, \Phi^j \otimes \mathcal{L}_\xi).$$

In particular $H^i_c(\overline{X}^{(h)}_{U^p,m} \times \mathrm{Spec}\, k(w)^{\mathrm{ac}}, \Phi^j \otimes \mathcal{L}_\xi)$ is an admissible $G(\mathbb{A}^{\infty}) \times W_{F_w}$-module.

Proof: Recall the Iwasawa decomposition $(B_w^{\mathrm{op}})^\times = P_M(F_w)(\mathcal{O}_{B,w}^{\mathrm{op}})^\times$. It follows that

$$P_M(F_w)\backslash (B_w^{\mathrm{op}})^\times \cong P_M(\mathcal{O}_{F,w})\backslash (\mathcal{O}_{B,w}^{\mathrm{op}})^\times$$

surjects to

$$P_M(\mathcal{O}_{F,w}/w^{m_1})\backslash (\mathcal{O}_{B,w}^{\mathrm{op}}/w^{m_1})^\times.$$

This gives rise to maps

$$\mathrm{Ind}_{P_M(\mathcal{O}_{F,w}/w^{m_1})}^{(\mathcal{O}_{B,w}^{\mathrm{op}}/w^{m_1})^\times} H_c^i(\overline{X}_{U^p,m,M} \times \mathrm{Spec}\, k(w)^{\mathrm{ac}}, \Phi^j \otimes \mathcal{L}_\xi)$$
$$\downarrow$$
$$\mathrm{Ind}_{P_M(F_w)}^{(B_w^{\mathrm{op}})^\times} H_c^i(\overline{X}_M, \Phi^j \otimes \mathcal{L}_\xi)$$

which are compatible with the maps

$$H_c^i(\overline{X}_{U^p,m}^{(h)} \times \mathrm{Spec}\, k(w)^{\mathrm{ac}}, \Phi^j \otimes \mathcal{L}_\xi)$$
$$\downarrow$$
$$\lim{}_{\to U^p,m}\, H_c^i(\overline{X}_{U^p,m}^{(h)} \times \mathrm{Spec}\, k(w)^{\mathrm{ac}}, \Phi^j \otimes \mathcal{L}_\xi).$$

As each of the maps

$$\mathrm{Ind}_{P_M(\mathcal{O}_{F,w}/w^{m_1})}^{(\mathcal{O}_{B,w}^{\mathrm{op}}/w^{m_1})^\times} H_c^i(\overline{X}_{U^p,m,M} \times \mathrm{Spec}\, k(w)^{\mathrm{ac}}, \Phi^j \otimes \mathcal{L}_\xi)$$
$$\downarrow$$
$$H_c^i(\overline{X}_{U^p,m}^{(h)} \times \mathrm{Spec}\, k(w)^{\mathrm{ac}}, \Phi^j \otimes \mathcal{L}_\xi).$$

is an isomorphism, the lemma will follow on passing to the limit as long as we can check that the map

$$\lim{}_{\to U^p,m}\, \mathrm{Ind}_{P_M(\mathcal{O}_{F,w}/w^{m_1})}^{(\mathcal{O}_{B,w}^{\mathrm{op}}/w^{m_1})^\times} H_c^i(\overline{X}_{U^p,m,M} \times \mathrm{Spec}\, k(w)^{\mathrm{ac}}, \Phi^j \otimes \mathcal{L}_\xi)$$
$$\downarrow$$
$$\mathrm{Ind}_{P_M(F_w)}^{(B_w^{\mathrm{op}})^\times} H_c^i(\overline{X}_M, \Phi^j \otimes \mathcal{L}_\xi)$$

is an isomorphism. Injectivity is straightforward. As for surjectivity any $f \in \mathrm{Ind}_{P_M(F_w)}^{(B_w^{\mathrm{op}})^\times} H_c^i(\overline{X}_M, \Phi^j \otimes \mathcal{L}_\xi)$ being locally constant factors through one of the finite quotients $P_M(\mathcal{O}_{F,w}/w^{m_1})\backslash (\mathcal{O}_{B,w}^{\mathrm{op}}/w^{m_1})^\times$. Then f will be in the image of $\mathrm{Ind}_{P_M(\mathcal{O}_{F,w}/w^{m_1'})}^{(\mathcal{O}_{B,w}^{\mathrm{op}}/w^{m_1'})^\times} H_c^i(\overline{X}_{U^p,m',M} \times \mathrm{Spec}\, k(w)^{\mathrm{ac}}, \Phi^j \otimes \mathcal{L}_\xi)$ for some U^p and m'. $\square$

Putting together the analysis of this section we obtain the following proposition.

Proposition III.4.8 *For $h = 0, \ldots, n-1$ choose a direct summand $M_h \subset \Lambda_{11}$ of rank $n - h$. Suppose that for each $0 \le h \le n-1$, $0 \le j \le n-1$ and $0 \le i \le 2h$ the $G_{M_h}(\mathbb{A}^\infty)$-module $H_c^i(\overline{X}_{M_h}, \Phi^j \otimes \mathcal{L}_\xi)$ is admissible. Then we have an equality of virtual $G(\mathbb{A}^\infty) \times W_{F_w}$-modules*

$$[H(X, \mathcal{L}_\xi)^{\mathbb{Z}_p^\times}] = \sum_{h,j,i} (-1)^{n-1+i+j} \operatorname{Ind}_{P_{M_h}(F_w)}^{(B_w^{\mathrm{op}})^\times} [H_c^i(\overline{X}_{M_h}, \Phi^j \otimes \mathcal{L}_\xi)].$$

We will let $(X_{U^p,m}^{(h)})^\wedge$ (resp. $X_{U^p,m,M}^\wedge$) denote the formal completion of $X_{U^p,m}$ along the locally closed subscheme $\overline{X}_{U^p,m}^{(h)}$ (resp. $\overline{X}_{U^p,m,M}$). The comparison theorem of [Berk3] implies that $\Phi^i|_{\overline{X}_{U^p,m}^{(h)}}$ (resp. $\Phi^i|_{\overline{X}_{U^p,m,M}}$) coincides with the formal vanishing cycles for $\overline{X}_{U^p,m}^{(h)} \subset (X_{U^p,m}^{(h)})^\wedge$ (resp. $\overline{X}_{U^p,m,M} \subset X_{U^p,m,M}^\wedge$) (see section I.5). In terms of $\mathcal{G}/\overline{X}_{U^p,m}^{(h)}$ (resp. $\mathcal{G}/\overline{X}_{U^p,m,M}$) the formal completion is completely characterised by the following useful universal property.

Lemma III.4.9 *Suppose that $\mathcal{X}$ is a locally noetherian formal scheme over $\mathcal{O}_{F,w}$ and assume $p = 0$ on $\mathcal{X}^{\mathrm{red}}$. Suppose also that $\mathcal{H}/\mathcal{X}$ is a Barsotti-Tate $\mathcal{O}_{F,w}$-module and that γ is a Drinfeld w^{m_1}-structure on $\mathcal{H}/\mathcal{X}$. Moreover suppose that we are given a morphism $f : \mathcal{X}^{\mathrm{red}} \to \overline{X}_{U^p,m}^{(h)}$ (resp. $\overline{X}_{U^p,m,M}$) under which $\mathcal{G}$ with its canonical Drinfeld w^{m_1}-structure pulls back to $\mathcal{H}|_{\mathcal{X}^{\mathrm{red}}}$ with the Drinfeld w^{m_1}-structure $\gamma|_{\mathcal{X}^{\mathrm{red}}}$. Then there is a unique extension of f to a morphism $\tilde{f} : \mathcal{X} \to X_{U^p,m,M}^\wedge$ under which $\mathcal{G}$ with its canonical Drinfeld w^{m_1}-structure pulls back to $\mathcal{H}$ and γ respectively.*

Proof: Let $(A, \lambda, i, \overline{\eta}^p, \alpha_i)$ be the pullback to $\mathcal{X}^{\mathrm{red}}$ of the universal object over $X_{U^p,m}$. Exactly as in the first paragraph of the proof of lemma III.4.1 we see that deformations of $(A, \lambda, i, \overline{\eta}^p, \alpha_i)$ to $\mathcal{X}$ are in natural bijection with the deformations of $(f^*\mathcal{G}, f^*\alpha_1)$. Thus we have a unique deformation over $\mathcal{X}$ of $(A, \lambda, i, \overline{\eta}^p, \alpha_i)$ which gives rise to $(\mathcal{H}, \gamma)$. Call it $(A', \lambda', i', (\overline{\eta}^p)', \alpha_i')$. Thus we have a unique morphism $\tilde{f} : \mathcal{X} \to X_{U^p,m}$ such that the universal $(r + 4)$-tuple pulls back to $(A', \lambda', i', (\overline{\eta}^p)', \alpha_i')$. This morphism restricts on $\mathcal{X}^{\mathrm{red}}$ to f and so must factor through $X_{U^p,m,M}^\wedge$. We see that $\tilde{f}$ is also the unique such morphism extending f under which $(\mathcal{G}, \alpha_1)$ pulls back to $(\mathcal{H}, \gamma)$. $\square$

Chapter IV

Igusa varieties

In our setting there seem to be two natural analogues of the familiar Igusa curves in the theory of elliptic modular curves. We will call these Igusa varieties of the first and second kind. When we refer to these Igusa varieties we will refer only to the analogue of the ordinary locus on the usual Igusa curves. We have not looked at the question of whether our Igusa varieties admit natural smooth compactifications, although we feel this is a natural and interesting question. In the case of elliptic modular curves, the Weil pairing on the p-divisible group of an elliptic curve allows one to identify these two kinds of Igusa variety.

IV.1   Igusa varieties of the first kind

In this section we introduce the more naive notion of Igusa variety of the first kind in the context of the Shimura varieties we are studying. To this end fix an integer h in the range $0 \le h \le n-1$. Also if $m = (m_1, \ldots, m_r) \in \mathbb{Z}_{\ge 0}^r$ then let $\overline{m}$ denote $(0, m_2, \ldots, m_r)$.

By an *Igusa variety of the first kind*

$$I_{U^p,m}^{(h)}/\overline{X}_{U^p,\overline{m}}^{(h)}$$

we shall mean the moduli space for isomorphisms

$$\alpha_1^{\mathrm{et}} : (w^{-m_1}\mathcal{O}_{F,w}/\mathcal{O}_{F,w})_{\overline{X}_{U^p,\overline{m}}^{(h)}}^h \xrightarrow{\sim} \mathcal{G}^{\mathrm{et}}[w^{m_1}].$$

Thus $I_{U^p,m}^{(h)}/\overline{X}_{U^p,\overline{m}}^{(h)}$ is Galois (and in particular finite etale, but not necessarily connected) with Galois group $GL_h(\mathcal{O}_{F,w}/w^{m_1})$. The morphism $I_{U^p,m}^{(h)} \to \overline{X}_{U^p,\overline{m}}^{(h)}$ factors naturally through $I_{U^p,m'}^{(h)}$ if $m_1' < m_1$ and $m_i' =$

m_i for $i > 1$. The inverse system of the $I_{U^p,m}^{(h)}$ has a natural action of $G(\mathbb{A}^{\infty,p}) \times GL_h(\mathcal{O}_{F,w}) \times \prod_{i=2}^r (\mathcal{O}_{B,w_i}^{op})^\times$.

Let $(\mathbb{Z} \times GL_h(F_w))^+$ denote the sub-semigroup of elements $(c, g) \in \mathbb{Z} \times GL_h(F_w)$ such that ϖ_w to the integral part of $-c/(n - h)$ times g is integral. Then the inverse system of the $I_{U^p,m}^{(h)}$ has an action of $G(\mathbb{A}^{\infty,p}) \times \mathbb{Q}_p^\times \times (\mathbb{Z} \times GL_h(F_w))^+ \times \prod_{i=2}^r (B_{w_i}^{op})^\times$ extending that of $G(\mathbb{A}^{\infty,p}) \times GL_h(\mathcal{O}_{F,w}) \times \prod_{i=2}^r (\mathcal{O}_{B,w_i}^{op})^\times$. We leave the action of $G(\mathbb{A}^{\infty,p})$ to the reader. First suppose that $(g_0, c, g_1^{et}, g_i) \in \mathbb{Q}_p^\times \times (\mathbb{Z} \times GL_h(F_w))^+ \times \prod_{i=2}^r (B_{w_i}^{op})^\times$ also satisfies

- for $i > 1$ we have $g_i^{-1} \in \mathcal{O}_{B,w_i}^{op}$ and $g_0^{-1} g_i \in \mathcal{O}_{B,w_i}^{op}$,

- $(g_1^{et})^{-1} \in M_h(\mathcal{O}_{F,w})$ and $g_0^{-1} g_1^{et} \in M_h(\mathcal{O}_{F,w})$,

- $(n - h)w(g_0) \le c \le 0$,

- for $i > 1$ we have $w_i^{m_i - m_i'} g_i \in \mathcal{O}_{B,w_i}^{op}$,

- $w_1^{m_1 - m_1'} g_1^{et} \in M_h(\mathcal{O}_{F,w})$.

Under these assumptions we will define a morphism

$$(g_0, c, g_1^{et}, g_i) : I_{U^p,m}^{(h)} \longrightarrow I_{U^p,m'}^{(h)}.$$

It will send $(A, \lambda, i, \overline{\eta}^p, \alpha_1^{et}, \alpha_i)$ to $(A/(C \oplus C^\perp), p^{\operatorname{val} p(g_0)} \lambda, i, \overline{\eta}^p, \alpha_1^{et} \circ g_1^{et}, \alpha_i \circ g_i)$, where we have set

- $C_1 \subset \varepsilon A[w_1^{m_1}]$ is the unique closed subscheme for which there is an exact sequence

$$(0) \longrightarrow \ker F^{-f_1 c} \longrightarrow C_1 \longrightarrow \alpha_1^{et}(F_w^h / \mathcal{O}_{F,w}^h [(g_1^{et})^{-1}]) \longrightarrow (0),$$

  (this makes sense as if d denotes the integral part of $-c/(n - h)$ then $\ker F^{-f_1 c} \supset G_A^0[w^d]$ and $G_A^{et}[w^d] \supset \alpha(F_w^h / \mathcal{O}_{F,w}^h [(g_1^{et})^{-1}])$ (we are using the fact that $(c, g_1) \in (\mathbb{Z} \times GL_h(F_w))^+$));

- for $i > 1$, $C_i = \alpha_i(g_i \Lambda_i / \Lambda_i)$;

- $C = (\mathcal{O}_{F,w}^n \otimes_{\mathcal{O}_{F,w}} C_1) \oplus \bigoplus_{i=2}^r C_i \subset A[u^{-\operatorname{val} p(g_0)}]$;

- $C^\perp$ is the annihilator of $C \subset A[u^{-\operatorname{val} p(g_0)}]$ inside $A[(u^c)^{-\operatorname{val} p(g_0)}]$ under the λ-Weil pairing;

- $p^{\operatorname{val} p(g_0)} \lambda$ is the polarisation $A/(C \oplus C^\perp) \to (A/(C \oplus C^\perp))^\vee$ which makes the following diagram commute

$$
\begin{array}{ccc}
A & \xrightarrow{\; p^{-\operatorname{val} p(g_0)} \lambda \;} & A^\vee \\
\downarrow & & \uparrow \\
A/(C \oplus C^\perp) & \longrightarrow & (A/(C \oplus C^\perp))^\vee;
\end{array}
$$

- $\alpha_1^{\text{et}} \circ g_1^{\text{et}} : (w_1^{-m_1'}\mathcal{O}_{F,w}/\mathcal{O}_{F,w})_{I_{U^p,m}^{(h)}}^h \to (\varepsilon A[w_1^\infty]/C_1)^{\text{et}}$ is the homomorphism making the following diagram commute

$$
\begin{array}{ccc}
(w_1^{-m_1'}\mathcal{O}_{F,w}/\mathcal{O}_{F,w})^h & \longrightarrow & (\varepsilon A[w_1^\infty]/C_1)^{\text{et}} \\
\downarrow & & \uparrow \\
w_1^{-m_1'}g_1^{\text{et}}\mathcal{O}_{F,w}^h/g_1^{\text{et}}\mathcal{O}_{F,w}^h & \longrightarrow & (\varepsilon A[w_1^\infty]/C_1)^{\text{et}}[w_1^{m_1'}] \\
\downarrow & & \downarrow \\
w_1^{-m_1}\mathcal{O}_{F,w}^h/g_1^{\text{et}}\mathcal{O}_{F,w}^h & \longrightarrow & (\varepsilon A[w_1^{m_1}]/C_1)^{\text{et}} \\
\downarrow & & \downarrow \\
w_1^{-m_1}\mathcal{O}_{F,w}^h/g_1^{\text{et}}\mathcal{O}_{F,w}^h & \longrightarrow & \varepsilon A[w_1^{m_1}]^{\text{et}}/\alpha_1^{\text{et}}((F_w/\mathcal{O}_{F,w})^h[(g_1^{\text{et}})^{-1}]) \\
\uparrow & & \uparrow \\
(w_1^{-m_1}\mathcal{O}_{F,w}/\mathcal{O}_{F,w})^h & \xrightarrow{\alpha_1^{\text{et}}} & \varepsilon A[w_1^{m_1}]^{\text{et}};
\end{array}
$$

- for $i > 1$, $\alpha_i \circ g_i : w_i^{-m_i'}\Lambda_i/\Lambda_i \to A[w_i^\infty]/C_i$ is the homomorphism making the following diagram commute

$$
\begin{array}{ccc}
w_i^{-m_i'}\Lambda_i/\Lambda_i & \longrightarrow & A[w_i^\infty]/C_i \\
\downarrow & & \uparrow \\
w_i^{-m_i'}g_i\Lambda_i/g_i\Lambda_i & \xrightarrow{\sim} & (A[w_i^\infty]/C_i)[w_i^{m_i'}] \\
\downarrow & & \downarrow \\
w_i^{-m_i}\Lambda_i/g_i\Lambda_i & \xrightarrow{\sim} & A[w_i^{m_i}]/C_i \\
\uparrow & & \uparrow \\
w_i^{-m_i}\Lambda_i/\Lambda_i & \xrightarrow{\alpha_i} & A[w_i^{m_i}].
\end{array}
$$

It is tedious but straightforward to check that this does define an action. We see that $(p^{-2}, p^{-1}, \ldots, p^{-1})$ acts in the same way as $p \in G(\mathbb{A}^{\infty,p})$ and so acts invertibly on the inverse system. Thus this definition can be extended to the whole of $\mathbb{Q}_p^\times \times (\mathbb{Z} \times GL_h(F_w))^+ \times \prod_{i=2}^r (B_{w_i}^{\text{op}})^\times$. (A less tedious argument is to use the compatibility described below with the action of $G(\mathbb{A}^{\infty,p}) \times \mathbb{Q}_p^\times \times P_M(F_w) \times \prod_{i=2}^r (B_{w_i}^{\text{op}})^\times$ on the inverse system of the $\overline{X}_{U^p,m,M}$.)

We will denote by $\widetilde{\text{Fr}^*}^{f_1}$ the element

$$(p^{-f_1}, -1, 1) \in \mathbb{Q}_p^\times \times (\mathbb{Z} \times GL_h(F_w))^+.$$

We see that

1. $\mathbb{Q}_p^\times \times \mathbb{Z} \times GL_h(F_w)$ is generated by $\mathbb{Q}_p^\times \times (\mathbb{Z} \times GL_h(F_w))^+$ and $\widetilde{\text{Fr}^*}^{-f_1}$;

2. $\widetilde{\text{Fr}^*}^{f_1} : I_{U^p,m}^{(h)} \to I_{U^p,m}^{(h)}$ is just $(\text{Fr}^*)^{f_1}$. (Note that according to the definitions above $\widetilde{\text{Fr}^*}^{f_1}$ does take $I_{U^p,m}^{(h)}$ to itself.)

Now fix $j : \Lambda_{11} \twoheadrightarrow \mathcal{O}^h_{F,w}$ with kernel M. This induces a homomorphism

$$
\begin{array}{rcl}
j_* : P_M(F_w) & \twoheadrightarrow & \mathbb{Z} \times GL_h(F_w) \\
g & \longmapsto & (w \circ \det(g|_M), j \circ g \circ j^{-1}).
\end{array}
$$

We will define a morphism $j^* : I^{(h)}_{U^p,m} \to \overline{X}_{U^p,m,M}$ such that

$$
\begin{array}{ccc}
I^{(h)}_{U^p,m} & \xrightarrow{\ j^*\ } & \overline{X}_{U^p,m,M} \\
\downarrow & & \downarrow \\
\overline{X}^{(h)}_{U^p,\overline{m}} & \xrightarrow{(\mathrm{Fr}^*)^{f_1(n-h)m_1}} & \overline{X}^{(h)}_{U^p,\overline{m}}
\end{array}
$$

commutes. More precisely j^* is the map which takes $(A, \lambda, i, \overline{\eta}^p, \alpha^{\mathrm{et}}_1, \alpha_i)$ to $(A^{(p^{m_1 f_1(n-h)})}, \lambda^{(p^{m_1 f_1(n-h)})}, i, \overline{\eta}^p, \alpha_1, F^{m_1 f_1(n-h)} \circ \alpha_i)$ where

$$
\alpha_1(x) = F^{m_1 f_1(n-h)} \circ \alpha^{\mathrm{et}}_1 \circ j(x).
$$

(To see that α_1 is well defined and that it is a Drinfeld level structure use lemma II.2.1.) Because $\overline{X}_{U^p,m,M}/\overline{X}^{(h)}_{U^p,\overline{m}}$ is finite flat of degree

$$
\#(\mathcal{O}_{F,w}/w^{m'_1})^{h(n-h)} \# GL_h(\mathcal{O}_{F,w}/w^{m'_1})
$$

(see lemma III.4.6), because $\overline{X}_{U^p,m,M}$ is smooth and hence normal (see lemma III.4.6), and because the composite

$$
I^{(h)}_{U^p,m} \longrightarrow \overline{X}^{(h)}_{U^p,\overline{m}} \xrightarrow{(\mathrm{Fr}^*)^{f_1(n-h)m_1}} \overline{X}^{(h)}_{U^p,\overline{m}}
$$

is also finite flat of the same degree we see that j^* is an isomorphism.

Suppose that $g \in G(\mathbb{A}^{\infty,p}) \times \mathbb{Q}^\times_p \times P_M(F_w) \times \prod^r_{i=2}(B^{\mathrm{op}}_{w_i})^\times$ and that $j_*(g) \in G(\mathbb{A}^{\infty,p}) \times \mathbb{Q}^\times_p \times (\mathbb{Z} \times GL_h(F_w))^+ \times \prod^r_{i=2}(B^{\mathrm{op}}_{w_i})^\times$. Suppose also that $(U^p)' \supset g^{-1}U^p g$. If for each i we have $m_i >> m'_i$ then

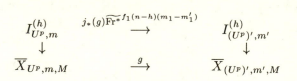

$$
\begin{array}{ccc}
I^{(h)}_{U^p,m} & \xrightarrow{j_*(g)\widetilde{\mathrm{Fr}^*}^{f_1(n-h)(m_1-m'_1)}} & I^{(h)}_{(U^p)',m'} \\
\downarrow & & \downarrow \\
\overline{X}_{U^p,m,M} & \xrightarrow{\ g\ } & \overline{X}_{(U^p)',m',M}
\end{array}
$$

commutes.

We now look at natural formal extensions of these Igusa varieties. In particular we will let $(I^{(h)}_{U^p,m})^\wedge/(X^{(h)}_{U^p,\overline{m}})^\wedge$ denote the unique etale covering with reduced subschemes $I^{(h)}_{U^p,m}/\overline{X}^{(h)}_{U^p,\overline{m}}$ (see [Berk2]). If $t \in \mathbb{Z}_{\geq 0}$ then we will let $(I^{(h)}_{U^p,m})^\wedge(t)/(I^{(h)}_{U^p,m})^\wedge$ denote the moduli space for Drinfeld w^t-structures on $\mathcal{G}^0/(I^{(h)}_{U^p,m})^\wedge$.

Lemma IV.1.1   1. *The natural morphism* $(I^{(h)}_{U^p,m})^\wedge(t) \longrightarrow (I^{(h)}_{U^p,m})^\wedge$ *is finite and flat of degree* $\#GL_{n-h}(\mathcal{O}_{F,w}/w^t)$.

2. $(I^{(h)}_{U^p,m})^\wedge(t)/(I^{(h)}_{U^p,\overline{m}})^\wedge(t)$ *is the unique etale cover with reduced subschemes* $I^{(h)}_{U^p,m}/\overline{X}^{(h)}_{U^p,\overline{m}}$.

3. $(I^{(h)}_{U^p,m})^\wedge(t)$ *has the following universal property. Suppose that $\mathcal{X}$ is a locally noetherian $\mathcal{O}_{F,w}$-formal scheme and assume $p = 0$ on $\mathcal{X}^{\mathrm{red}}$. Suppose also that $\mathcal{H}/\mathcal{X}$ is a Barsotti-Tate $\mathcal{O}_{F,w}$-module and that we are given a morphism $f : \mathcal{X}^{\mathrm{red}} \to I^{(h)}_{U^p,m}$ under which $\mathcal{G}$ pulls back to $\mathcal{H}|_{\mathcal{X}^{\mathrm{red}}}$. Then we have an exact sequence*

$$(0) \to \mathcal{H}^0 \to \mathcal{H} \to \mathcal{H}^{\mathrm{et}} \to (0)$$

over $\mathcal{X}$, with $\mathcal{H}^0$ formal and $\mathcal{H}^{\mathrm{et}}$ ind-etale. Suppose finally that γ is a Drinfeld w^t-structure on $\mathcal{H}^0/\mathcal{X}$. Then there is a unique extension of f to a morphism $\widetilde{f} : \mathcal{X} \to (I^{(h)}_{U^p,m})^\wedge$ under which $\mathcal{G}$ pulls back to $\mathcal{H}$ and the canonical Drinfeld w^t-structure on $\mathcal{G}^0/(I^{(h)}_{U^p,m})^\wedge(t)$ pulls back to γ.

Proof: The first part follows from corollary II.2.5. The second part follows because

$$(I^{(h)}_{U^p,m})^\wedge(t) \overset{\sim}{\longrightarrow} (I^{(h)}_{U^p,m})^\wedge \times_{(I^{(h)}_{U^p,\overline{m}})^\wedge} (I^{(h)}_{U^p,\overline{m}})^\wedge(t).$$

From the definition of $(I^{(h)}_{U^p,m})^\wedge(t)$ the third part reduces to the special case $t = 0$. In this case, by lemma III.4.9, we obtain a unique morphism $\mathcal{X} \to (I^{(h)}_{U^p,\overline{m}})^\wedge$ under which $\mathcal{G}$ pulls back to $\mathcal{H}$. The third part of the lemma now follows from lemma I.5.8. $\square$

We will let $(GL_{n-h}(F_w) \times GL_h(F_w))^+$ denote the set of pairs (g^0, g^{et}) in $GL_{n-h}(F_w) \times GL_h(F_w)$ for which there exists a scalar $a \in F_w^\times$ such that both $ag^{\mathrm{et}} \in M_h(\mathcal{O}_{F,w})$ and $(ag^0)^{-1} \in M_{n-h}(\mathcal{O}_{F,w})$. This is a subsemigroup of $GL_{n-h}(F_w) \times GL_h(F_w)$. There is a natural homomorphism from

$$G(\mathbb{A}^{\infty,p}) \times \mathbb{Q}_p^\times \times GL_{n-h}(F_w) \times GL_h(F_w) \times \prod_{i=2}^{r}(B_{w_i}^{\mathrm{op}})^\times$$

to

$$G(\mathbb{A}^{\infty,p}) \times \mathbb{Q}_p^\times \times \mathbb{Z} \times GL_h(F_w) \times \prod_{i=2}^{r}(B_{w_i}^{\mathrm{op}})^\times$$

under which

$$(g^p, g_{p,0}, g_w^0, g_w^{\text{et}}, g_{w_i}) \longmapsto (g^p, g_{p,0}, w(\det g_w^0), g_w^{\text{et}}, g_{w_i}).$$

We will denote this map $g \mapsto [g]$. Under this homomorphism

$$G(\mathbb{A}^{\infty,p}) \times \mathbb{Q}_p^\times \times (GL_{n-h}(F_w) \times GL_h(F_w))^+ \times \prod_{i=2}^r (B_{w_i}^{\text{op}})^\times$$

is taken to

$$G(\mathbb{A}^{\infty,p}) \times \mathbb{Q}_p^\times \times (\mathbb{Z} \times GL_h(F_w))^+ \times \prod_{i=2}^r (B_{w_i}^{\text{op}})^\times.$$

If ϖ is a uniformiser in $\mathcal{O}_{F,w}$ then we will let

$$\widetilde{\text{Fr}_\varpi^*}^{f_1(n-h)} = (1, p^{-f_1(n-h)}, \varpi^{-1}, 1, 1)$$

in

$$G(\mathbb{A}^{\infty,p}) \times \mathbb{Q}_p^\times \times (GL_{n-h}(F_w) \times GL_h(F_w))^+ \times \prod_{i=2}^r (B_{w_i}^{\text{op}})^\times.$$

Then

$$[\widetilde{\text{Fr}_\varpi^*}^{f_1(n-h)}] = \widetilde{\text{Fr}^*}^{f_1(n-h)},$$

and $G(\mathbb{A}^{\infty,p}) \times \mathbb{Q}_p^\times \times GL_{n-h}(F_w) \times GL_h(F_w) \times \prod_{i=2}^r (B_{w_i}^{\text{op}})^\times$ is generated as a semi-group by $G(\mathbb{A}^{\infty,p}) \times \mathbb{Q}_p^\times \times (GL_{n-h}(F_w) \times GL_h(F_w))^+ \times \prod_{i=2}^r (B_{w_i}^{\text{op}})^\times$ and $\widetilde{\text{Fr}_\varpi^*}^{-f_1(n-h)}$.

The inverse system of the $(I_{U^p,m}^{(h)})^\wedge(t)$ has a natural action of $G(\mathbb{A}^{\infty,p}) \times \mathbb{Q}_p^\times \times (GL_{n-h}(F_w) \times GL_h(F_w))^+ \times \prod_{i=2}^r (B_{w_i}^{\text{op}})^\times$, which is compatible via $[\ ]$ with the action of $G(\mathbb{A}^{\infty,p}) \times \mathbb{Q}_p^\times \times (\mathbb{Z} \times GL_h(F_w))^+ \times \prod_{i=2}^r (B_{w_i}^{\text{op}})^\times$ on the inverse system of the $I_{U^p,m}^{(h)}$. We will leave the action of $G(\mathbb{A}^{\infty,p})$ to the reader and describe the action of $\mathbb{Q}_p^\times \times (GL_{n-h}(F_w) \times GL_h(F_w))^+ \times \prod_{i=2}^r (B_{w_i}^{\text{op}})^\times$. To this end suppose that

$$(g_0, g_1^0, g_1^{\text{et}}, g_i) \in \mathbb{Q}_p^\times \times (GL_{n-h}(F_w) \times GL_h(F_w))^+ \times \prod_{i=2}^r (B_{w_i}^{\text{op}})^\times,$$

and that

- for $i > 1$ we have $g_i^{-1} \in \mathcal{O}_{B,w_i}^{\text{op}}$ and $g_0^{-1} g_i \in \mathcal{O}_{B,w_i}^{\text{op}}$,

- $(g_1^{\mathrm{et}})^{-1} \in M_h(\mathcal{O}_{F,w})$ and $g_0^{-1} g_1^{\mathrm{et}} \in M_h(\mathcal{O}_{F,w})$,

- $g_0^{-1} g_1^0 \in M_{n-h}(\mathcal{O}_{F,w})$,

- for $i > 1$ we have $w_i^{m_i - m_i'} g_i \in \mathcal{O}_{B,w_i}^{\mathrm{op}}$,

- $w_1^{m_1 - m_1'} g_1^{\mathrm{et}} \in M_h(\mathcal{O}_{F,w})$,

Choose $a \in \mathbb{Z}_{\geq 0}$ maximally such that $w^a g_1^{\mathrm{et}} \in M_h(\mathcal{O}_{F,w})$ and $(w^a g_1^0)^{-1} \in M_{n-h}(\mathcal{O}_{F,w})$. Finally also suppose that

- $w^{t-t'+a} g_0 \in M_{n-h}(\mathcal{O}_{F,w})$.

We will define a morphism

$$(g_0, g_1^0, g_1^{\mathrm{et}}, g_i) : (I_{U^p,m}^{(h)})^{\wedge}(t) \longrightarrow (I_{U^p,m'}^{(h)})^{\wedge}(t'),$$

which extends

$$[(g_0, g_1^0, g_1^{\mathrm{et}}, g_i)] : I_{U^p,m}^{(h)} \longrightarrow I_{U^p,m'}^{(h)}.$$

Let $\overline{C}_1$ be the unique closed subscheme of $\mathcal{G}[w^{m_1}]/I_{U^p,m}^{(h)}$ for which there is a short exact sequence

$$(0) \longrightarrow \ker F^{-f_1 w(\det g_1^0)} \longrightarrow \overline{C}_1 \longrightarrow \alpha_1^{\mathrm{et}}(F_w^h / \mathcal{O}_{F,w}^h [(g_1^{\mathrm{et}})^{-1}]) \longrightarrow (0).$$

To define the desired extension $(g_0, g_1^0, g_1^{\mathrm{et}}, g_i)$ of $[(g_0, g_1^0, g_1^{\mathrm{et}}, g_i)]$ it suffices (by the universal property of $(I_{U^p,m'}^{(h)})^{\wedge}(t')$) to specify a lifting $\mathcal{G}'$ of $\mathcal{G}/\overline{C}_1$ from $I_{U^p,m}^{(h)}$ to $(I_{U^p,m}^{(h)})^{\wedge}(t)$ together with a Drinfeld $w^{t'}$-level structure on $(\mathcal{G}')^0$.

We now explain the construction of $\mathcal{G}'$ and the Drinfeld $w^{t'}$-structure on $(\mathcal{G}')^0$. To do so fix a uniformiser ϖ of $\mathcal{O}_{F,w}$. Note that we have an embedding

$$(F_w/\mathcal{O}_{F,w})^h [(g_1^{\mathrm{et}})^{-1}] \xrightarrow{\alpha_1^{\mathrm{et}}} \mathcal{G}^{\mathrm{et}}[w^a] \longrightarrow \mathcal{G}/\mathcal{G}^0[w^a].$$

(By an embedding we mean a compatible system of embeddings over each closed subscheme of $(I_{U^p,m}^{(h)})^{\wedge}(t)$.) We also have a Drinfeld w^t-structure $\varpi^{-a}\alpha_1^0$ on $(\mathcal{G}/\mathcal{G}^0[w^a])^0$. We set $\mathcal{G}'$ equal to the quotient of $(\mathcal{G}/\mathcal{G}^0[w^a])$ by

$$(\varpi^{-a}\alpha_1^0)(F_w^{n-h}/\mathcal{O}_{F,w}^{n-h}[(w^a g_1^0)^{-1}]) + \alpha_1^{\mathrm{et}}((F_w/\mathcal{O}_{F,w})^h [(g_1^{\mathrm{et}})^{-1}]).$$

This does not depend on the choice of ϖ. By corollary II.2.5 we see that the composite of $\varpi^{-a}\alpha_1^0$ with

$$\varpi^a g_1^0 : (w^{-t'}\mathcal{O}_{F,w}/\mathcal{O}_{F,w})^{n-h} \hookrightarrow$$
$$(w^{-t}\mathcal{O}_{F,w}/\mathcal{O}_{F,w})^{n-h}/((F_w/\mathcal{O}_{F,w})^{n-h}[(w^a g_1^0)^{-1}])$$

gives a Drinfeld $w^{t'}$-structure on $(\mathcal{G}')^0$. This Drinfeld $w^{t'}$-structure is also independent of the choice of ϖ.

It is tedious but straightforward to check that this does define an action. We see that $(p^{-2}, p^{-1}, \ldots, p^{-1})$ acts in the same way as $p \in G(\mathbb{A}^{\infty,p})$ and so acts invertibly on the inverse system. Thus this definition can be extended to the whole of $\mathbb{Q}_p^\times \times (GL_{n-h}(F_w) \times GL_h(F_w))^+ \times \prod_{i=2}^r (B_{w_i}^{\mathrm{op}})^\times$. (A less tedious argument is to use the compatibility described below with the action of $G(\mathbb{A}^{\infty,p}) \times \mathbb{Q}_p^\times \times P_M(F_w) \times \prod_{i=2}^r (B_{w_i}^{\mathrm{op}})^\times$ on the inverse system of the $X_{U^p,m,M}^\wedge$.) Note that $\widetilde{\mathrm{Fr}_\varpi^*}^{f_1(n-h)}$ maps $(I_{U^p,m}^{(h)})^\wedge(t)$ to itself and defines a lifting of $(\mathrm{Fr}^*)^{f_1(n-h)}$, which is analogous to the canonical lifting of Frobenius in the theory of elliptic modular curves.

Now fix homomorphisms $j^{\mathrm{et}} : \Lambda_{11} \twoheadrightarrow \mathcal{O}_{F,w}^h$ and $j^0 : \Lambda_{11} \twoheadrightarrow \mathcal{O}_{F,w}^{n-h}$ such that $j^0 \oplus j^{\mathrm{et}}$ is an isomorphism. Let $M = \ker j^{\mathrm{et}}$. These choices define a Levi component $L_{(j^0,j^{\mathrm{et}})} \subset P_M$, i.e. the elements of P_M which also preserve $\ker j^0$. They also induce an isomorphism

$$
\begin{array}{ccc}
(j^0, j^{\mathrm{et}})_* : L_{(j^0,j^{\mathrm{et}})} & \xrightarrow{\sim} & GL_{n-h}(F_w) \times GL_h(F_w) \\
g & \longmapsto & (j^0 \circ g \circ (j^0)^{-1}, j^{\mathrm{et}} \circ g \circ (j^{\mathrm{et}})^{-1}).
\end{array}
$$

If ϖ is a uniformiser in $\mathcal{O}_{F,w}$, we will define a morphism

$$
(j^0, j^{\mathrm{et}}, \varpi)^* : (I_{U^p,m}^{(h)})^\wedge(m_1) \longrightarrow X_{U^p,m,M}^\wedge
$$

which extends the morphism

$$
(j^{\mathrm{et}})^* : I_{U^p,m}^{(h)} \longrightarrow \overline{X}_{U^p,m,M}.
$$

To define such a morphism it suffices (by lemma III.4.9) to specify a deformation of the pair

$$
(\mathcal{G}^{(p^{m_1 f_1(n-h)})}, F^{m_1 f_1(n-h)} \circ \alpha_1^{\mathrm{et}} \circ j^{\mathrm{et}})/I_{U^p,m}^{(h)}
$$

to $(I_{U^p,m}^{(h)})^\wedge(m_1)$. As a deformation of $\mathcal{G}^{(p^{m_1 f_1(n-h)})}$ we take $\mathcal{G}/\mathcal{G}^0[w^{m_1}]$. Then we have the identification

$$
\begin{array}{ccc}
\mathcal{G}^0[w^{m_1}] \times \mathcal{G}^{\mathrm{et}}[w^{m_1}] & \xrightarrow{\sim} & (\mathcal{G}/\mathcal{G}^0[w^{m_1}])[w^{m_1}] \\
(x, y) & \longmapsto & \varpi^{-m_1} x + y.
\end{array}
$$

As a deformation of $F^{m_1 f_1(n-h)} \circ \alpha_1^{\mathrm{et}} \circ j^{\mathrm{et}}$ we take $(\alpha_1^0 \circ j^0) \oplus (\alpha_1^{\mathrm{et}} \circ j^{\mathrm{et}})$. (Note that over $I_{U^p,m}^{(h)}$ we are identifying $\mathcal{G}^{(p^{m_1 f_1(n-h)})}$ and $\mathcal{G}/\mathcal{G}^0[w^{m_1}]$ so that $F^{m_1 f_1(n-h)} : \mathcal{G} \to \mathcal{G}^{(p^{m_1 f_1(n-h)})}$ corresponds to the natural projection $\mathcal{G} \to \mathcal{G}/\mathcal{G}^0[w^{m_1}]$.)

Lemma IV.1.2   *1. The morphism $(j^0, j^{\mathrm{et}}, \varpi)^*$ is an isomorphism.*

2. *Suppose that $g \in G(\mathbb{A}^{\infty,p}) \times \mathbb{Q}_p^\times \times L_{(j^0,j^{et})} \times \prod_{i=2}^r (B_{w_i}^{op})^\times$ and suppose that $(j^0, j^{et})_*(g) \in G(\mathbb{A}^{\infty,p}) \times \mathbb{Q}_p^\times \times (GL_{n-h}(F_w) \times GL_h(F_w))^+ \times \prod_{i=2}^r (B_{w_i}^{op})^\times$. Suppose also that $(U^p)' \supset g^{-1} U^p g$ and that for each i we have $m_i >> m_i'$. Then*

$$
\begin{array}{ccc}
(I_{U^p,m}^{(h)})^\wedge(m_1) & \xrightarrow{(j^0,j^{et})_*(g)\widetilde{\mathrm{Fr}^*_\varpi}^{f_1(n-h)(m_1-m_1')}} & (I_{(U^p)',m'}^{(h)})^\wedge(m_1') \\
\downarrow & & \downarrow \\
X_{U^p,m,M}^\wedge & \xrightarrow{\ g\ } & X_{(U^p)',m',M}^\wedge
\end{array}
$$

commutes, where the vertical maps are $(j^0, j^{et}, \varpi)^*$.

Proof: The second part is formal. To prove the first part we will verify that $X_{U^p,m,M}^\wedge$ has the same universal property as $(I_{U^p,m,M}^{(h)})^\wedge(m_1)$. That is we will show that if $\mathcal{X}$ is a locally noetherian $\mathcal{O}_{F,w}$-formal scheme, if $p = 0$ on $\mathcal{X}^{red}$, if $\mathcal{H}/\mathcal{X}$ is a Barsotti-Tate $\mathcal{O}_{F,w}$-module, if $f : \mathcal{X}^{red} \to \overline{X}_{U^p,m,M}$ is a morphism under which $\mathcal{G}$ pulls back to $(\mathcal{H}/\mathcal{H}^0[w^{m_1}])|_{\mathcal{X}^{red}}$ and if γ is a Drinfeld w^{m_1}-structure on $\mathcal{H}^0/\mathcal{X}$, then there is a unique extension of f to a morphism $\widetilde{f} : \mathcal{X} \to (I_{U^p,m}^{(h)})^\wedge$ under which $\mathcal{G}$ pulls back to $\mathcal{H}/\mathcal{H}^0[w^{m_1}]$ and the canonical Drinfeld w^t-structure $\alpha_1 \circ j^0$ on $\mathcal{G}^0/X_{U^p,m,M}^\wedge$ pulls back to $\varpi^{-m_1}\gamma$. To see that $X_{U^p,m,M}^\wedge$ has this universal property we use lemma III.4.9 and note that there is a natural bijection between

- Drinfeld w^{m_1}-structures γ on $\mathcal{H}^0/\mathcal{X}$

- and Drinfeld w^{m_1}-structures $\delta : w^{-m_1}\Lambda_{11}/\Lambda_{11} \to (\mathcal{H}/\mathcal{H}^0[w^{m_1}])[w^{m_1}]$ over $\mathcal{X}$ which restrict to $\alpha_1 \circ j^{et}$ on $\mathcal{X}^{red}$.

This bijection sends δ to $\varpi^{m_1}\delta \circ (j^0|_{w^{-m_1}M/M}^{-1})$ and γ to $\varpi^{-m_1}\gamma \circ j^0 + \widetilde{\alpha}_1^{et}$, where $\widetilde{\alpha}_1^{et} : (w^{-m_1}\Lambda_{11}/\Lambda_{11}) \twoheadrightarrow \mathcal{H}^{et}[w^{m_1}]$ is the unique lifting over $\mathcal{X}$ of the pullback from $X_{U^p,m,M}^\wedge$ to $\mathcal{X}^{red}$ of $\alpha_1 : (w^{-m_1}\Lambda_{11}/\Lambda_{11}) \twoheadrightarrow \mathcal{G}[w^{m_1}]^{et}$. $\square$

We will let $\Phi^i(t)/I_{U^p,m}^{(h)} \times_{\mathrm{Spec}\,k(w)} \mathrm{Spec}\,k(w)^{ac}$ denote the formal vanishing cycles for $I_{U^p,m}^{(h)} \subset (I_{U^p,m}^{(h)})^\wedge(t)$. (If $t = m_1$ then it follows from the last lemma that $(I_{U^p,m}^{(h)})^\wedge(t)$ is isomorphic to the completion of a proper scheme of finite type over $\mathcal{O}_{F,w}$ along a locally closed subscheme of the special fibre. In general $(I_{U^p,m}^{(h)})^\wedge(t)$ is etale locally isomorphic to $(I_{U^p,m'}^{(h)})^\wedge(t)$ with $m_1' = t$. Thus $\Phi^i(t)/I_{U^p,m}^{(h)} \times_{\mathrm{Spec}\,k(w)} \mathrm{Spec}\,k(w)^{ac}$ is well defined.)

Note that, if $U^p \supset (U^p)'$ and for each i we have $m_i \leq m_i'$, then the restriction of $\Phi^i(t)/I_{U^p,m}^{(h)} \times_{\mathrm{Spec}\,k(w)} \mathrm{Spec}\,k(w)^{ac}$ to $I_{(U^p)',m'}^{(h)} \times_{\mathrm{Spec}\,k(w)} \mathrm{Spec}\,k(w)^{ac}$ is canonically isomorphic to

$$
\Phi^i(t)/I_{(U^p)',m'}^{(h)} \times_{\mathrm{Spec}\,k(w)} \mathrm{Spec}\,k(w)^{ac}
$$

(see [Berk3]).

Suppose that x is a closed point of $I^{(h)}_{U^p,m} \times \operatorname{Spec} k(w)^{\mathrm{ac}}$ and suppose that $j_x : \Sigma_{n-h} \xrightarrow{\sim} \mathcal{G}^0_x$. Then we obtain a natural map

$$j^*_x : ((I^{(h)}_{U^p,m})^{\wedge}(t) \times_{\operatorname{Spf}\mathcal{O}_{F,w}} \operatorname{Spf}\mathcal{O}_{\widehat{F}^{\mathrm{nr}}_w})^{\wedge}_x \longrightarrow \operatorname{Spf} R_{F_w,n-h,t},$$

and hence a homomorphism

$$(j^*_x)^* : \Psi^j_{F_w,l,n-h,t} \longrightarrow \Phi^j(t)_x.$$

Lemma IV.1.3

$$(j^*_x)^* : \Psi^j_{F_w,l,n-h,t} \xrightarrow{\sim} \Phi^j(t)_x.$$

Proof: We will let $\operatorname{Spf} R(\mathcal{G}_x)$ (resp. $\operatorname{Spf} R_t(\mathcal{G}^0_x)$) denote the universal deformation space for $\mathcal{G}_x$ (resp. $\mathcal{G}^0_x$ with its (unique) Drinfeld w^t level structure). Then we have

$$((I^{(h)}_{U^p,m})^{\wedge}(t) \times_{\operatorname{Spf}\mathcal{O}_{F,w}} \operatorname{Spf}\mathcal{O}_{\widehat{F}^{\mathrm{nr}}_w})^{\wedge}_x \cong \operatorname{Spf} R(\mathcal{G}_x) \times_{\operatorname{Spf} R_0(\mathcal{G}^0_x)} \operatorname{Spf} R_t(\mathcal{G}^0_x)$$
$$\xrightarrow{\sim} \operatorname{Hom}(T\mathcal{G}_x, \widetilde{\Sigma}_{F_w,n-h}) \times_{\operatorname{Spf} R_{F_w,n-h}} \operatorname{Spf} R_{F_w,n-h,t}.$$

As $\operatorname{Hom}(T\mathcal{G}_x, \widetilde{\Sigma}_{F_w,n-h})$ and $\operatorname{Spf} R_{F_w,n-h}$ are formally smooth we see that

$$\operatorname{Hom}(T\mathcal{G}_x, \widetilde{\Sigma}_{F_w,n-h}) \times_{\operatorname{Spf} R_{F_w,n-h}} \operatorname{Spf} R_{F_w,n-h,t} \longrightarrow \operatorname{Spf} R_{F_w,n-h,t}$$

induces an isomorphism on vanishing cycles (see lemma I.5.6). The lemma follows. $\square$

The inverse system of sheaves

$$\Phi^i(t)/I^{(h)}_{U^p,m} \times_{\operatorname{Spec} k(w)} \operatorname{Spec} k(w)^{\mathrm{ac}}$$

has an action of $G(\mathbb{A}^{\infty,p}) \times \mathbb{Q}^\times_p \times (GL_{n-h}(F_w) \times GL_h(F_w))^+ \times \prod^r_{i=2}(B^{\mathrm{op}}_{w_i})^\times \times W_{F_w}$ in the following sense. If

$$(g,\sigma) \in G(\mathbb{A}^{\infty,p}) \times \mathbb{Q}^\times_p \times (GL_{n-h}(F_w) \times GL_h(F_w))^+ \times \prod^r_{i=2}(B^{\mathrm{op}}_{w_i})^\times \times W_{F_w}$$

and if $[g] : I^{(h)}_{U^p,m} \to I^{(h)}_{(U^p)',m'}$ then for $t >> t'$ we get a natural map

$$(g,\sigma) : ([g] \times (\operatorname{Frob}^{w(\sigma)}_w)^*)^* \Phi(t') \longrightarrow \Phi(t)$$

on $I^{(h)}_{U^p,m} \times \operatorname{Spec} k(w)^{\mathrm{ac}}$.

We now wish to describe the action of (g,σ) on stalks. Thus let x be a closed point of $I^{(h)}_{U^p,m} \times \operatorname{Spec} k(w)^{\mathrm{ac}}$ and let $y = ([g] \times (\operatorname{Frob}^{w(\sigma)}_w)^*)x$, a closed point of $I^{(h)}_{(U^p)',m'} \times \operatorname{Spec} k(w)^{\mathrm{ac}}$. Suppose also that $j_x : \Sigma_{n-h} \overset{\sim}{\to} \mathcal{G}^0_x$. If $\delta \in D^\times_{F_w,n-h}$ and if $w(\det \delta) = w(\det g^0_w) - w(\sigma)$ then we will define

$$([g] \times (\operatorname{Frob}^{w(\sigma)}_w)^* \times \delta)(j_x) : \Sigma_{F_w,n-h} \overset{\sim}{\to} \mathcal{G}^0_y.$$

To do so it suffices to give an isomorphism

$$\Sigma^{(p^{f_1(w(\det g^0_w)-w(\sigma)))}}_{F_w,n-h} \overset{\sim}{\to} \mathcal{G}^0_x.$$

We simply take $j_x \circ \delta$. We will write simply

$$j_y = ([g] \times (\operatorname{Frob}^{w(\sigma)}_w)^* \times \delta)(j_x),$$

but recall that it depends on a choice of δ. We see that, for $t >> t'$,

$$
\begin{array}{ccc}
((I^{(h)}_{U^p,m})^\wedge(t) \times_{\operatorname{Spf}\mathcal{O}_{F,w}} \operatorname{Spf}\mathcal{O}_{\widehat{F}^{\mathrm{nr}}_w})^\wedge_x & \overset{j^*_x}{\longrightarrow} & \operatorname{Spf} R_{F_w,n-h,t} \\
\downarrow & & \downarrow \\
((I^{(h)}_{U^p,m})^\wedge(t') \times_{\operatorname{Spf}\mathcal{O}_{F,w}} \operatorname{Spf}\mathcal{O}_{\widehat{F}^{\mathrm{nr}}_w})^\wedge_y & \overset{j^*_y}{\longrightarrow} & \operatorname{Spf} R_{F_w,n-h,t'}
\end{array}
$$

commutes (the left vertical arrow being $g \times (\operatorname{Frob}^{w(\sigma)}_w)^*$ and the right vertical arrow (g^0_w, δ)). Hence

$$
\begin{array}{ccc}
\Phi^j(t)_x & \overset{g \times \sigma}{\longleftarrow} & \Phi^j(t')_y \\
\uparrow & & \uparrow \\
\Psi^j_{F_w,l,n-h,t} & \overset{(g^0_w,\delta,\sigma)}{\longleftarrow} & \Psi^j_{F_w,l,n-h,t'}
\end{array}
$$

also commutes (the left vertical arrow being $(j^*_x)^*$ and the right vertical arrow being $(j^*_y)^* = ([g] \times (\operatorname{Frob}^{w(\sigma)}_w)^* \times \delta)(j_x)^{**}$).

If we set

$$H^i_c(I^{(h)}, \Phi^j \otimes \mathcal{L}_\xi) = \varinjlim_{U^p,m,t} H^i_c(I^{(h)}_{U^p,m} \times_{\operatorname{Spec} k(w)} \operatorname{Spec} k(w)^{\mathrm{ac}}, \Phi^j(t) \otimes \mathcal{L}_\xi),$$

then $H^i_c(I^{(h)}, \Phi^j \otimes \mathcal{L}_\xi)$ becomes a smooth

$$G(\mathbb{A}^{\infty,p}) \times \mathbb{Q}^\times_p \times (GL_{n-h}(F_w) \times GL_h(F_w))^+ \times \prod_{i=2}^r (B^{\mathrm{op}}_{w_i})^\times \times W_{F_w}$$

module. Then we have the following lemma.

Lemma IV.1.4   *1. If Γ_t denotes the kernel of the homomorphism*

$$GL_{n-h}(\mathcal{O}_{F,w}) \twoheadrightarrow GL_{n-h}(\mathcal{O}_{F,w}/w^t)$$

then for $t \leq t'$ we have a canonical isomorphism

$$\Phi^i(t) \overset{\sim}{\longrightarrow} \Phi^i(t')^{\Gamma_t}.$$

2. The action of the semigroup

$$G(\mathbb{A}^{\infty,p}) \times \mathbb{Q}_p^\times \times (GL_{n-h}(F_w) \times GL_h(F_w))^+ \times \prod_{i=2}^{r} (B_{w_i}^{\mathrm{op}})^\times \times W_{F_w}$$

on $H_c^i(I^{(h)}, \Phi^j \otimes \mathcal{L}_\xi)$ is admissible/continuous.

Proof: The first part follows by calculating on stalks using isomorphisms $(j_x^*)^*$ and the above compatibility. The second part follows from the first. □

We also have the following lemma which we will prove in the next section.

Lemma IV.1.5 $\widetilde{\mathrm{Fr}_\varpi^*}^{f_1(n-h)}$ *acts invertibly on each*

$$H_c^i(I_{U^p,m}^{(h)} \times_{\mathrm{Spec}\,k(w)} \mathrm{Spec}\,k(w)^{\mathrm{ac}}, \Phi^j(t) \otimes \mathcal{L}_\xi).$$

We are now in a position to prove the following result.

Proposition IV.1.6   *1. The action of the semi-group*

$$G(\mathbb{A}^{\infty,p}) \times \mathbb{Q}_p^\times \times (GL_{n-h}(F_w) \times GL_h(F_w))^+ \times \prod_{i=2}^{r} (B_{w_i}^{\mathrm{op}})^\times \times W_{F_w}$$

on $H_c^i(I^{(h)}, \Phi^j \otimes \mathcal{L}_\xi)$ extends uniquely to an admissible/continuous action of

$$G(\mathbb{A}^{\infty,p}) \times \mathbb{Q}_p^\times \times GL_{n-h}(F_w) \times GL_h(F_w) \times \prod_{i=2}^{r} (B_{w_i}^{\mathrm{op}})^\times \times W_{F_w}.$$

2. If we fix a triple $(j^0, j^{\mathrm{et}}, \varpi)$ as above then we get an isomorphism of smooth

$$G(\mathbb{A}^{\infty,p}) \times \mathbb{Q}_p^\times \times GL_{n-h}(F_w) \times GL_h(F_w) \times \prod_{i=2}^{r} (B_{w_i}^{\mathrm{op}})^\times \times W_{F_w}$$
$$\cong G(\mathbb{A}^{\infty,p}) \times \mathbb{Q}_p^\times \times L_{(j^0, j^{\mathrm{et}})} \times \prod_{i=2}^{r} (B_{w_i}^{\mathrm{op}})^\times \times W_{F_w}$$

modules

$$H_c^i(I^{(h)}, \Phi^j \otimes \mathcal{L}_\xi) \xrightarrow{\sim} H_c^i(\overline{X}_M, \Phi^j \otimes \mathcal{L}_\xi).$$

3. The unipotent radical of $P_M(F_w)$ acts trivially on

$$H_c^i(\overline{X}_M, \Phi^j \otimes \mathcal{L}_\xi).$$

Proof: The first part follows from the lemma IV.1.5. For the second part we consider the maps

$$\widetilde{\mathrm{Fr}_\varpi^*}^{f_1(n-h)m_1} \circ ((j^0, j^{\mathrm{et}}, \varpi)^*)^{-1} : X_{U^p,m,M}^\wedge \longrightarrow (I_{U^p,m}^{(h)})^\wedge(m_1)$$

and the induced maps

$$(((j^0, j^{\mathrm{et}}, \varpi)^*)^{-1})^* \circ (\widetilde{\mathrm{Fr}_\varpi^*}^{f_1(n-h)m_1})^*$$

from

$$H_c^i(I_{U^p,m}^{(h)} \times_{\mathrm{Spec}\, k(w)} \mathrm{Spec}\, k(w)^{\mathrm{ac}}, \Phi^j(m_1) \otimes \mathcal{L}_\xi)$$

to

$$H_c^i(\overline{X}_{U^p,m,M} \times_{\mathrm{Spec}\, k(w)} \mathrm{Spec}\, k(w)^{\mathrm{ac}}, \Phi^j \otimes \mathcal{L}_\xi).$$

Combining lemmas IV.1.2 and IV.1.4 we see that these latter maps are isomorphisms. Again by lemma IV.1.2 they are compatible as U^p and m vary and give in the limit an isomorphism

$$H_c^i(I^{(h)}, \Phi^j \otimes \mathcal{L}_\xi) \xrightarrow{\sim} H_c^i(\overline{X}_M, \Phi^j \otimes \mathcal{L}_\xi).$$

That this isomorphism is compatible with

$$G(\mathbb{A}^{\infty,p}) \times \mathbb{Q}_p^\times \times GL_{n-h}(F_w) \times GL_h(F_w) \times \prod_{i=2}^r (B_{w_i}^{\mathrm{op}})^\times \times W_{F_w} \cong$$
$$\cong G(\mathbb{A}^{\infty,p}) \times \mathbb{Q}_p^\times \times L_{(j^0, j^{\mathrm{et}})} \times \prod_{i=2}^r (B_{w_i}^{\mathrm{op}})^\times \times W_{F_w}$$

actions again follows from lemma IV.1.2.

The final part follows from the second and from lemma I.2.1. □

Corollary IV.1.7 *For $h = 0, \ldots, n-1$ choose homomorphisms $j_h^0 : \Lambda_{11} \to$*
$\to \mathcal{O}_{F,w}^{n-h}$ and $j_h^{\mathrm{et}} : \Lambda_{11} \twoheadrightarrow \mathcal{O}_{F,w}^h$ such that $j_h^0 \oplus j_h^{\mathrm{et}}$ is an isomorphism. Let
$M_h = \ker j_h^{\mathrm{et}}$. Then we have an equality of virtual $G(\mathbb{A}^\infty) \times W_{F_w}$-modules

$$[H(X, \mathcal{L}_\xi)^{\mathbb{Z}_p^\times}] = \sum_{h,j,i} (-1)^{n-1+i+j} [\mathrm{Ind}_{P_{M_h}(F_w)}^{(B_w^{\mathrm{op}})^\times} H_c^i(I^{(h)}, \Phi^j \otimes \mathcal{L}_\xi)].$$

IV.2 Igusa varieties of the second kind

We now come to the slightly less obvious generalisation of Igusa curves. More precisely we define the *Igusa variety of the second kind,*

$$J_{U^p,m,s}^{(h)} = J^{(s)}(\mathcal{G}^0 / I_{U^p,m}^{(h)})$$

(see section II.2). Then $J_{U^p,m,s}^{(h)}/I_{U^p,m}^{(h)} \times \operatorname{Spec} k(w)^{\mathrm{ac}}$ is Galois with group $(\mathcal{O}_{D_{F_w,n-h}}/w^s)^\times$ (acting on the right).

We will describe an action of the semigroup

$$G(\mathbb{A}^{\infty,p}) \times \mathbb{Q}_p^\times \times (\mathbb{Z} \times GL_h(F_w))^+ \times \prod_{i=2}^r (B_{w_i}^{\mathrm{op}})^\times \times D_{F_w,n-h}^\times$$

on the inverse system of the $J_{U^p,m,s}^{(h)}$. Consider an element

$$(g^p, g_{p,0}, c, g_w^{\mathrm{et}}, g_{w_i}, \delta)$$

in

$$G(\mathbb{A}^{\infty,p}) \times \mathbb{Q}_p^\times \times (\mathbb{Z} \times GL_h(F_w))^+ \times \prod_{i=2}^r (B_{w_i}^{\mathrm{op}})^\times \times D_{F_w,n-h}^\times.$$

Choose $a \in F_w^\times$ with $c + (n-h) > (n-h)w(a) \geq c$. If $(U^p)' \supset (g^p)^{-1} U^p g^p$ and if for all i we have $m_i >> m_i'$, then we let

$$(g^p, g_{p,0}, c, g_w^{\mathrm{et}}, g_{w_i}, \delta) : I_{U^p,m}^{(h)} \times \operatorname{Spec} k(w)^{\mathrm{ac}} \longrightarrow I_{(U^p)',m'}^{(h)} \times \operatorname{Spec} k(w)^{\mathrm{ac}}$$

be the map

$$(g^p, g_{p,0}, c, g_w^{\mathrm{et}}, g_{w_i}) \times (\operatorname{Frob}_w^*)^{c-w(\det \delta)}.$$

We will extend this to a compatible series of morphisms

$$(g^p, g_{p,0}, c, g_w^{\mathrm{et}}, g_{w_i}, \delta) : J_{U^p,m,s}^{(h)} \longrightarrow J_{(U^p)',m',s'}^{(h)}$$

for $s > s'$. For this it suffices to give compatible isomorphisms

$$((\operatorname{Frob}_w^*)^{c-w(\det \delta)} \Sigma_{F_w,n-h})[w^{s'}] \overset{\sim}{\to} ((g^p, g_{p,0}, c, g_w^{\mathrm{et}}, g_{w_i})^* \mathcal{G}^0)[w^{s'}]$$

over $J_{U^p,m,s}^{(h)}$. First note that $a^{-1}\delta$ gives an isomorphism

$$a^{-1}\delta : (\operatorname{Frob}_w^*)^{c-w(\det \delta)} \Sigma_{F_w,n-h} \overset{\sim}{\to} \Sigma_{F_w,n-h}/\Sigma_{F_w,n-h}[aF^{-f_1 c}].$$

Also note that

$$a : \mathcal{G}^0/\mathcal{G}^0[aF^{-cf_1}] \overset{\sim}{\to} (g^p, g_{p,0}, c, g_w^{\mathrm{et}}, g_{w_i})^* \mathcal{G}^0.$$

Thus for our isomorphism

$$((\operatorname{Frob}_w^*)^{c-w(\det \delta)} \Sigma_{F_w,n-h})[w^{s'}] \overset{\sim}{\to} ((g^p, g_{p,0}, c, g_w^{\mathrm{et}}, g_{w_i})^* \mathcal{G}^0)[w^{s'}]$$

we may simply take $a^{-1}\delta$, followed by the map induced by the universal isomorphism

$$\Sigma_{F_w,n-h}[w^s] \overset{\sim}{\to} \mathcal{G}^0[w^s]$$

over $J_{U^p,m,s}^{(h)}$, in turn followed by a. It is straightforward but tedious to check this is independent of the choice of a and does define an action.

Note that the element $(\widetilde{\mathrm{Fr}^*}^{f_1}, 1)$ simply acts as $(\mathrm{Fr}^*)^{f_1}$. (So for instance on $I_{U^p,m}^{(h)} \times \mathrm{Spec}\, k(w)^{\mathrm{ac}}$ it acts as $(\mathrm{Fr}^*)^{f_1} = (\mathrm{Fr}^*)^{f_1} \times (\mathrm{Frob}_w^*)^{-1}$.)

We will be most interested in the part of this action which is an action of $k(w)^{\mathrm{ac}}$-schemes. To this end define

$$(D_{F_w,n-h}^\times \times GL_h(F_w))^+$$

to be the set of elements

$$(\delta, \gamma) \in D_{F_w,n-h}^\times \times GL_h(F_w)$$

such that $(w(\det \delta), \gamma) \in (\mathbb{Z} \times GL_h(F_w))^+$. Also set

$$G^{(h)}(\mathbb{A}^\infty) = G(\mathbb{A}^{\infty,p}) \times \mathbb{Q}_p^\times \times D_{F_w,n-h}^\times \times GL_h(F_w) \times \prod_{i=2}^r (B_{w_i}^{\mathrm{op}})^\times$$

and

$$G^{(h)}(\mathbb{A}^\infty)^+ = G(\mathbb{A}^{\infty,p}) \times \mathbb{Q}_p^\times \times (D_{F_w,n-h}^\times \times GL_h(F_w))^+ \times \prod_{i=2}^r (B_{w_i}^{\mathrm{op}})^\times.$$

If we embed $G^{(h)}(\mathbb{A}^\infty)^+$ into

$$G(\mathbb{A}^{\infty,p}) \times \mathbb{Q}_p^\times \times (\mathbb{Z} \times GL_h(F_w))^+ \times \prod_{i=2}^r (B_{w_i}^{\mathrm{op}})^\times \times D_{F_w,n-h}^\times$$

by sending $(g^p, g_{p,0}, \delta, g_w^{\mathrm{et}}, g_{w_i})$ to $(g^p, g_{p,0}, w(\det \delta), g_w^{\mathrm{et}}, g_{w_i}, \delta)$. In this way $G^{(h)}(\mathbb{A}^\infty)$ acts on the inverse system of the $J_{U^p,m,s}^{(h)}$ over $k(w)^{\mathrm{ac}}$. We note that this action is compatible with $w \circ \det : D_{F_w,n-h}^\times \longrightarrow \mathbb{Z}$ and the action of

$$G(\mathbb{A}^{\infty,p}) \times \mathbb{Q}_p^\times \times (\mathbb{Z} \times GL_h(F_w))^+ \times \prod_{i=2}^r (B_{w_i}^{\mathrm{op}})^\times$$

on the inverse system of the $I_{U^p,m}^{(h)}$.

If ρ is an irreducible admissible representation of $D^{\times}_{F_w, n-h}$ over $\overline{\mathbb{Q}}_l^{ac}$ we get a lisse etale sheaf $\mathcal{F}_\rho / J^{(h)}_{U^p, m, s}$ coming from the restriction of ρ to

$$\ker(\mathcal{O}^{\times}_{D_{F_w, n-h}} \longrightarrow (\mathcal{O}_{D_{F_w, n-h}}/w^s)^{\times}).$$

If $(g^p, g_{p,0}, c, g_w^{et}, g_{w_i}, \delta)$ in

$$G(\mathbb{A}^{\infty, p}) \times \mathbb{Q}_p^{\times} \times (\mathbb{Z} \times GL_h(F_w))^+ \times \prod_{i=2}^r (B_{w_i}^{op})^{\times} \times D^{\times}_{F_w, n-h}$$

defines a map

$$J^{(h)}_{U^p, m, s} \longrightarrow J^{(h)}_{(U^p)', m', s'}$$

then we obtain a morphism

$$(g^p, g_{p,0}, c, g_w^{et}, g_{w_i}, \delta) : (g^p, g_{p,0}, c, g_w^{et}, g_{w_i}, \delta)^*(\mathcal{F}_\rho \otimes \mathcal{L}_\xi) \longrightarrow \mathcal{F}_\rho \otimes \mathcal{L}_\xi.$$

Note moreover that

$$\ker(\mathcal{O}^{\times}_{D_{F_w, n-h}} \longrightarrow (\mathcal{O}_{D_{F_w, n-h}}/w^s)^{\times})$$

acts trivially on $\mathcal{F}_\rho \otimes \mathcal{L}_\xi$ over $J^{(h)}_{U^p, m, s}$.

We remark that as a sheaf (without the action of any groups) $\mathcal{F}_\rho$ only depends on ρ up to twists by unramified characters. Thus $\mathcal{F}_\rho \cong \mathcal{F}_{\rho'}$ for some ρ' with finite image. The representation ρ' can be thought of a representation of the fundamental group of $I^{(h)}_{U^p, m}$ and hence defines a lisse etale sheaf over $I^{(h)}_{U^p, m}$ which has base change to $I^{(h)}_{U^p, m} \times \operatorname{Spec} k(w)^{ac}$ isomorphic to $\mathcal{F}_\rho$.

We will set

$$H_c^i(I^{(h)}, \mathcal{F}_\rho \otimes \mathcal{L}_\xi) = \varinjlim_{U^p, m} H_c^i(I^{(h)}_{U^p, m} \times \operatorname{Spec} k(w)^{ac}, \mathcal{F}_\rho \otimes \mathcal{L}_\xi).$$

Then $H_c^i(I^{(h)}, \mathcal{F}_\rho \otimes \mathcal{L}_\xi)$ has an admissible action of

$$G(\mathbb{A}^{\infty, p}) \times \mathbb{Q}_p^{\times} \times (\mathbb{Z} \times GL_h(F_w))^+ \times \prod_{i=2}^r (B_{w_i}^{op})^{\times} \times (D^{\times}_{F_w, n-h}/\mathcal{O}^{\times}_{D_{F_w, n-h}})$$

and the element $(\widetilde{\operatorname{Fr}^*}^{f_1}, 1)$ acts trivially. Thus this action factors through the surjection from

$$G(\mathbb{A}^{\infty, p}) \times \mathbb{Q}_p^{\times} \times (\mathbb{Z} \times GL_h(F_w))^+ \times \prod_{i=2}^r (B_{w_i}^{op})^{\times} \times (D^{\times}_{F_w, n-h}/\mathcal{O}^{\times}_{D_{F_w, n-h}})$$

to $G^{(h)}(\mathbb{A}^\infty)/\mathcal{O}^\times_{D_{F_w,n-h}}$ which sends

$$(g^p, g_{p,0}, c, g_w^{\text{et}}, g_{w_i}, \delta)$$

to

$$(g^p, g_{p,0}p^{f_1(w(\det\delta)-c)}, \delta, g_w^{\text{et}}, g_{w_i}).$$

Thus we may consider $H^i_c(I^{(h)}, \mathcal{F}_\rho \otimes \mathcal{L}_\xi)$ as an admissible $G^{(h)}(\mathbb{A}^\infty)$-module. This is compatible with the action of $G^{(h)}(\mathbb{A}^\infty)^+$ on the inverse system of the $J^{(h)}_{U^p,m,s}$ over $k(w)^{\text{ac}}$.

Similarly the action of $\mathcal{O}^\times_{D_{F_w,n-h}}$ on $\Psi^j_{F_w,l,n-h,t}$ (resp. $\Psi^j_{F_w,l,n-h,t}[\rho]$, where ρ is an irreducible admissible representation of $D^\times_{F_w,n-h}$) and on the inverse system of the $J^{(h)}_{U^p,m,s}$ defines a sheaf $\mathcal{F}(\Psi^j_{F_w,l,n-h,t})$ (resp. $\mathcal{F}(\Psi^j_{F_w,l,n-h,t}[\rho])$) on each $J^{(h)}_{U^p,m,s}$. If $(g^p, g_{p,0}, c, g_w^{\text{et}}, g_{w_i}, \delta)$ in

$$G(\mathbb{A}^{\infty,p}) \times \mathbb{Q}_p^\times \times (\mathbb{Z} \times GL_h(F_w))^+ \times \prod_{i=2}^r (B_{w_i}^{\text{op}})^\times \times D^\times_{F_w,n-h}$$

defines a map

$$J^{(h)}_{U^p,m,s} \longrightarrow J^{(h)}_{(U^p)',m',s'},$$

and if

$$(\delta, \gamma, \sigma) \in A_{F_w,n-h}$$

defines a map

$$\Psi^j_{F_w,l,n-h,t'} \longrightarrow \Psi^j_{F_w,l,n-h,t}$$

then we obtain a morphism $(g^p, g_{p,0}, c, g_w^{\text{et}}, g_{w_i}, (\delta, \gamma, \sigma))$:

$$(g^p, g_{p,0}, c, g_w^{\text{et}}, g_{w_i}, \delta)^*(\mathcal{F}(\Psi^j_{F_w,l,n-h,t'}) \otimes \mathcal{L}_\xi) \longrightarrow \mathcal{F}(\Psi^j_{F_w,l,n-h,t}) \otimes \mathcal{L}_\xi$$

(resp.

$$(g^p, g_{p,0}, c, g_w^{\text{et}}, g_{w_i}, \delta)^*(\mathcal{F}(\Psi^j_{F_w,l,n-h,t'}[\rho]) \otimes \mathcal{L}_\xi) \to \mathcal{F}(\Psi^j_{F_w,l,n-h,t}[\rho]) \otimes \mathcal{L}_\xi.)$$

Note moreover that

- $\ker(\mathcal{O}^\times_{D_{F_w,n-h}} \to (\mathcal{O}_{D_{F_w,n-h}}/w^s)^\times)$ acts trivially on $\mathcal{F}(\Psi^j_{F_w,l,n-h,t}) \otimes \mathcal{L}_\xi$ and $\mathcal{F}(\Psi^j_{F_w,l,n-h,t}[\rho]) \otimes \mathcal{L}_\xi$ over $J^{(h)}_{U^p,m,s}$,

- that if $\Gamma_t = \ker(GL_{n-h}(\mathcal{O}_{F,w}) \to GL_{n-h}(\mathcal{O}_{F,w}/w^t))$ then

$$\mathcal{F}(\Psi^j_{F_w,l,n-h,t}) \xrightarrow{\sim} \mathcal{F}(\Psi^j_{F_w,l,n-h,t'})^{\Gamma_t}$$

and

$$\mathcal{F}(\Psi^j_{F_w,l,n-h,t}[\rho]) \xrightarrow{\sim} \mathcal{F}(\Psi^j_{F_w,l,n-h,t'}[\rho])^{\Gamma_t},$$

- and that

$$\mathcal{F}(\Psi^j_{F_w,l,n-h,t}) = \bigoplus_{\rho} \mathcal{F}(\Psi^j_{F_w,l,n-h,t}[\rho]),$$

where ρ runs over a set representatives of the inertial equivalence classes of representations of $D^\times_{F_w,n-h}$.

We will set $H^i_c(I^{(h)}, \mathcal{F}(\Psi^j) \otimes \mathcal{L}_\xi)$ equal to

$$\varinjlim_{U^p,m,t} H^i_c(I^{(h)}_{U^p,m} \times \operatorname{Spec} k(w)^{\mathrm{ac}}, \mathcal{F}(\Psi^j_{F_w,l,n-h,t}) \otimes \mathcal{L}_\xi)$$

and $H^i_c(I^{(h)}, \mathcal{F}(\Psi^j[\rho]) \otimes \mathcal{L}_\xi)$ equal to

$$\varinjlim_{U^p,m,t} H^i_c(I^{(h)}_{U^p,m} \times \operatorname{Spec} k(w)^{\mathrm{ac}}, \mathcal{F}(\Psi^j_{F_w,l,n-h,t}[\rho]) \otimes \mathcal{L}_\xi).$$

Then $H^i_c(I^{(h)}, \mathcal{F}(\Psi^j) \otimes \mathcal{L}_\xi)$ and $H^i_c(I^{(h)}, \mathcal{F}(\Psi^j[\rho]) \otimes \mathcal{L}_\xi)$ have admissible actions of

$$G(\mathbb{A}^{\infty,p}) \times \mathbb{Q}_p^\times \times (\mathbb{Z} \times GL_h(F_w))^+ \times \prod_{i=2}^r (B^{\mathrm{op}}_{w_i})^\times \times (A_{F_w,n-h}/\mathcal{O}^\times_{D_{F_w,n-h}})$$

i.e. of

$$G(\mathbb{A}^{\infty,p}) \times \mathbb{Q}_p^\times \times (\mathbb{Z} \times GL_h(F_w))^+ \times \prod_{i=2}^r (B^{\mathrm{op}}_{w_i})^\times \times GL_{n-h}(F_w) \times W_{F_w}.$$

Again the element $(\widetilde{\operatorname{Fr}^*}^{f_1}, 1)$ acts trivially. We see that

$$H^i_c(I^{(h)}, \mathcal{F}(\Psi^j) \otimes \mathcal{L}_\xi) = \bigoplus_{\rho} H^i_c(I^{(h)}, \mathcal{F}(\Psi^j[\rho]) \otimes \mathcal{L}_\xi)$$

where again ρ runs over a set of representatives of the inertial equivalence classes of representations of $D^\times_{F_w,n-h}$.

One also has that

$$\mathcal{F}_\rho \otimes \Psi^j_{F_w,l,n-h,t}(\rho) \xrightarrow{\sim} \bigoplus_{\delta \in \Delta[\rho]} \delta^* \mathcal{F}(\Psi^j[\rho]).$$

Hence

$$H_c^i(I^{(h)}, \mathcal{F}_\rho \otimes \mathcal{L}_\xi) \otimes \Psi_{F_w,l,n-h}^j(\rho) \cong H_c^i(I^{(h)}, \mathcal{F}(\Psi^j[\rho]) \otimes \mathcal{L}_\xi)^{e[\rho]},$$

where the action of

$$(g^p, g_{p,0}, c, g_w^{\mathrm{et}}, g_{w_i}, \gamma, \sigma)$$

on $H_c^i(I^{(h)}, \mathcal{F}(\Psi^j[\rho]) \otimes \mathcal{L}_\xi)^{e[\rho]}$ is compatible with the action of

$$(g^p, g_{p,0}p^{f_1(w(\det \gamma) - w(\sigma) - c)}, \delta, g_w^{\mathrm{et}}, g_{w_i}) \otimes (\gamma, \sigma)$$
$$\in G^{(h)}(\mathbb{A}^{\infty,p}) \times GL_{n-h}(F_w) \times W_{F_w}$$

on $H_c^i(I^{(h)}, \mathcal{F}_\rho \otimes \mathcal{L}_\xi) \otimes \Psi_{F_w,l,n-h}^j(\rho)$, where $\delta \in D_{F_w,n-h}^\times$ satisfies

$$w(\det \delta) = w(\det \gamma) - w(\sigma).$$

We record this as the following lemma.

Lemma IV.2.1 *We have a natural isomorphism*

$$H_c^i(I^{(h)}, \mathcal{F}(\Psi^j) \otimes \mathcal{L}_\xi)^{n-h}$$
$$\cong \bigoplus_\rho (H_c^i(I^{(h)}, \mathcal{F}_\rho \otimes \mathcal{L}_\xi) \otimes \Psi_{F_w,l,n-h}^j(\rho))^{(n-h)/e[\rho]}$$

under which the action of

$$(g^p, g_{p,0}, c, g_w^{\mathrm{et}}, g_{w_i}, \gamma, \sigma)$$

on $H_c^i(I^{(h)}, \mathcal{F}(\Psi^j) \otimes \mathcal{L}_\xi)$ is compatible with the action of

$$(g^p, g_{p,0}p^{f_1(w(\det \gamma) - w(\sigma) - c)}, \delta, g_w^{\mathrm{et}}, g_{w_i}) \otimes (\gamma, \sigma)$$
$$\in G^{(h)}(\mathbb{A}^{\infty,p}) \times GL_{n-h}(F_w) \times W_{F_w}$$

on $H_c^i(I^{(h)}, \mathcal{F}_\rho \otimes \mathcal{L}_\xi) \otimes \Psi_{F_w,l,n-h}^j(\rho)$, where $\delta \in D_{F_w,n-h}^\times$ satisfies

$$w(\det \delta) = w(\det \gamma) - w(\sigma).$$

By a closed point x_∞ of $J_{U^p,m,\infty}^{(h)}$ we shall mean a compatible system of closed points x_s of $J_{U^p,m,s}^{(h)}$ as s varies. If x is a closed point of $I_{U^p,m}^{(h)} \times \mathrm{Spec}\, k(w)^{\mathrm{ac}}$ then the following are equivalent

- the choice of an isomorphism $j : \Sigma_{F_w,n-h} \overset{\sim}{\to} \mathcal{G}_x^0$

- and the choice of a closed point x_∞ of $J_{U^p,m,\infty}^{(h)}$ above x.

We will write j_{x_∞} for the isomorphism corresponding to x_∞. Note that if

$$(g^p, g_{p,0}, c, g_w^{\mathrm{et}}, g_{w_i}) \in G(\mathbb{A}^{\infty,p}) \times \mathbb{Q}_p^\times \times (\mathbb{Z} \times GL_h(F_w))^+ \times \prod_{i=2}^r (B_{w_i}^{\mathrm{op}})^\times,$$

if $\delta \in D_{F_w, n-h}^\times$ and if $e = c - w(\det \delta)$ then

$$((g^p, g_{p,0}, c, g_w^{\mathrm{et}}, g_{w_i}) \times (\mathrm{Frob}_w^e)^* \times \delta)(j_{x_\infty}) = j_{(g^p, g_{p,0}, c, g_w^{\mathrm{et}}, g_{w_i}, \delta) x_\infty}.$$

If x_∞ is a closed point of $J_{U^p, m, \infty}^{(h)}$ above a closed point x of $I_{U^p, m}^{(h)} \times \mathrm{Spec}\, k(w)^{\mathrm{ac}}$ then we set

$$\mathcal{F}(\Psi_{F_w, l, n-h, t}^j)_{x_\infty} = \lim_{\to s} \mathcal{F}(\Psi_{F_w, l, n-h, t}^j)_{x_s}.$$

We then have canonical isomorphisms (the composite depending on the choice of x_∞ above x)

$$\mathcal{F}(\Psi_{F_w, l, n-h, t}^j)_x \xrightarrow{\sim} \mathcal{F}(\Psi_{F_w, l, n-h, t}^j)_{x_\infty} \xrightarrow{\sim} \Psi_{F_w, l, n-h, t}^j.$$

Moreover for $(g^p, g_{p,0}, c, g_w^{\mathrm{et}}, g_{w_i}, (\delta, \gamma, \sigma))$ in

$$G(\mathbb{A}^{\infty,p}) \times \mathbb{Q}_p^\times \times (\mathbb{Z} \times GL_h(F_w))^+ \times \prod_{i=2}^r (B_{w_i}^{\mathrm{op}})^\times \times A_{F_w, n-h},$$

for

$$y = ((g^p, g_{p,0}, c, g_w^{\mathrm{et}}, g_{w_i}) \times (\mathrm{Frob}_w^*)^{c-w(\det \delta)}) x$$

and for

$$y_\infty = (g^p, g_{p,0}, c, g_w^{\mathrm{et}}, g_{w_i}, \delta) x_\infty$$

we have a commutative diagram

$$
\begin{array}{ccc}
\mathcal{F}(\Psi_{F_w, l, n-h, t'}^j)_y & \xrightarrow{(g^p, g_{p,0}, c, g_w^{\mathrm{et}}, g_{w_i}, \gamma, \sigma)} & \mathcal{F}(\Psi_{F_w, l, n-h, t}^j)_x \\
\downarrow & & \downarrow \\
\mathcal{F}(\Psi_{F_w, l, n-h, t'}^j)_{y_\infty} & \xrightarrow{(g^p, g_{p,0}, c, g_w^{\mathrm{et}}, g_{w_i}, (\delta, \gamma, \sigma))} & \mathcal{F}(\Psi_{F_w, l, n-h, t}^j)_{x_\infty} \\
\downarrow & & \downarrow \\
\Psi_{F_w, l, n-h, t'}^j & \xrightarrow{(\delta, \gamma, \sigma)} & \Psi_{F_w, l, n-h, t}^j.
\end{array}
$$

The following proposition is of key importance for us. In the original version of this work we reduced its proof to an abstract result on formal vanishing cycles which Vladimir Berkovich kindly proved for us. This result with Berkovich's proof is reproduced in his appendix to this book. However, we have found it simpler to also incorporate Berkovich's argument directly into this section. Thus the latter half of the proof given below is entirely due to Berkovich.

Proposition IV.2.2 *There is a unique homomorphism*

$$\kappa : \Phi^j(t) \longrightarrow \mathcal{F}(\Psi^j_{F_w,l,n-h,t})$$

over $I^{(h)}_{U^p,m} \times \operatorname{Spec} k(w)^{\mathrm{ac}}$ *such that if* x_∞ *is a closed point of* $J^{(h)}_{U^p,m,\infty}$ *above a closed point* x *of* $I^{(h)}_{U^p,m} \times \operatorname{Spec} k(w)^{\mathrm{ac}}$, *then*

$$\kappa : \Phi^j(t)_x \longrightarrow \mathcal{F}(\Psi^j_{F_w,l,n-h,t})_x$$

coincides with the inverse of the composite

$$\mathcal{F}(\Psi^j_{F_w,l,n-h,t})_x \longrightarrow \mathcal{F}(\Psi^j_{F_w,l,n-h,t})_{x\infty} \longrightarrow \Psi^j_{F_w,l,n-h,t} \xrightarrow{(j^*_{x\infty})^*} \Phi^j(t)_x.$$

Proof: Because of the uniqueness it suffices to work locally on $I^{(h)}_{U^p,m} \times \operatorname{Spec} k(w)^{\mathrm{ac}}$. Thus let $W \subset I^{(h)}_{U^p,m} \times \operatorname{Spec} k(w)^{\mathrm{ac}}$ be an open subset such that $W^\wedge$, the restriction of $(I^{(h)}_{U^p,m})^\wedge(t) \times_{\operatorname{Spf} \mathcal{O}_{F,w}} \operatorname{Spf} \mathcal{O}_{\widehat{F}^{\mathrm{nr}}_w}$ to W, is affine. For any positive integer s let W_s denote the pullback of W to $J^{(h)}_{U^p,m,s}$ and let $W^\wedge_s/W^\wedge$ denote the unique etale cover with reduced subschemes W_s/W.

Choose a lifting of $W/k(w)^{\mathrm{ac}}$ to a smooth scheme of finite type $Y/\mathcal{O}_{\widehat{F}^{\mathrm{nr}}_w}$. (If $W = \operatorname{Spec} k(w)^{\mathrm{ac}}[T_1,\ldots,T_m]/(f_1,\ldots,f_s)$ then let

$$Y = \operatorname{Spec} \mathcal{O}_{\widehat{F}^{\mathrm{nr}}_w}[T_1,\ldots,T_m]/(\widetilde{f}_1,\ldots,\widetilde{f}_s),$$

where $\widetilde{f}_i$ lifts f_i.) Set

$$Y^\wedge = Y^\wedge_W \times_{\operatorname{Spf} \mathcal{O}_{\widehat{F}^{\mathrm{nr}}_w}} \operatorname{Spf} R_{F_w,n-h,t},$$

and more generally

$$Y^\wedge_s = (Y^\wedge_W)_s \times_{\operatorname{Spf} \mathcal{O}_{\widehat{F}^{\mathrm{nr}}_w}} \operatorname{Spf} R_{F_w,n-h,t},$$

where $(Y^\wedge_W)_s/Y^\wedge_W$ is the unique etale cover with reduced subschemes W_s/W. Also set

$$Y^\wedge\{N\} = Y^\wedge_W \times_{\operatorname{Spf} \mathcal{O}_{\widehat{F}^{\mathrm{nr}}_w}} \operatorname{Spf} R_{F_w,n-h,t}/\mathfrak{m}^N_{F_w,n-h,t},$$

and more generally

$$Y^\wedge_s\{N\} = (Y^\wedge_W)_s \times_{\operatorname{Spf} \mathcal{O}_{\widehat{F}^{\mathrm{nr}}_w}} \operatorname{Spf} R_{F_w,n-h,t}/\mathfrak{m}^N_{F_w,n-h,t}.$$

Note also that the formal vanishing cycles for $Y^\wedge_s$ are just the constant sheaves $\Psi^j_{F_w,l,n-h,t}$.

Choose $N(s)$ as in lemma II.2.6 for the triple F_w, $n - h$, t. Then the action of $(1 + w^s \mathcal{O}_{D_{F_w}, n-h})$ on $\Sigma_{F_w, n-h}$ lifts to an action on

$$\widetilde{\Sigma}_{F_w, n-h} \times_{R_{F_w, n-h, t}} R_{F_w, n-h, t} / \mathfrak{m}_{R_{F_w, n-h, t}}^{N(s)}.$$

Moreover for any positive integer $a \geq t$ the action of $(1 + w^{s-t+a} \mathcal{O}_{D_{F_w}, n-h})$ on

$$\widetilde{\Sigma}_{F_w, n-h}[w^a] \times_{R_{F_w, n-h, t}} R_{F_w, n-h, t} / \mathfrak{m}_{R_{F_w, n-h, t}}^{N(s)}$$

is trivial. Thus $(1 + w^s \mathcal{O}_{D_{F_w}, n-h})/(1 + w^{s+a-t} \mathcal{O}_{D_{F_w}, n-h})$ acts diagonally on

$$\widetilde{\Sigma}_{F_w, n-h, t}[w^a] \times Y^{\wedge}_{s+a-t}\{N(s)\}.$$

As $Y^{\wedge}_{s+a-t}\{N(s)\}/Y^{\wedge}_s\{N(s)\}$ is finite, etale and Galois with Galois group

$$(1 + w^s \mathcal{O}_{D_{F_w}, n-h})/(1 + w^{s+a-t} \mathcal{O}_{D_{F_w}, n-h}),$$

the quotient is a finite, flat group scheme $\mathcal{H}^0_s[w^a]/Y^{\wedge}_s\{N(s)\}$. The direct system of the $\mathcal{H}^0[w^a]$ define a formal Barsotti-Tate $\mathcal{O}_{F,w}$-module $\mathcal{H}^0_s$ over $Y^{\wedge}_s\{N(s)\}$. Note that $\mathcal{H}^0_s[w^t] \cong \widetilde{\Sigma}_{F_w, n-h, t}[w^t]$, and so $\mathcal{H}^0_s$ inherits a Drinfeld w^t-structure. Also note that over W_{s+a-t} we have an isomorphism

$$\mathcal{H}^0_s[w^a] \cong \Sigma_{F_w, n-h}[w^a] \overset{\sim}{\to} \mathcal{G}^0[w^a].$$

The composite isomorphism descends to an isomorphism over W_s. Thus we see that over W_s we get a canonical isomorphism between $\mathcal{G}^0$ and $\mathcal{H}^0_s$. More generally for $s' > s$ we get a canonical isomorphism between the pull back of $\mathcal{H}^0_s$ to $Y^{\wedge}_{s'}\{N(s)\}$ and the restriction of $\mathcal{H}^0_{s'}$ to $Y^{\wedge}_{s'}\{N(s)\}$.

Let $\mathcal{H}^{\mathrm{et}}_s/Y^{\wedge}_s$ denote the unique lifting of $\mathcal{G}^{\mathrm{et}}/W_s$ to an ind-etale Barsotti-Tate $\mathcal{O}_{F,w}$-module. By corollary II.1.10 we can recursively find an extension of Barsotti-Tate $\mathcal{O}_{F,w}$-modules

$$(0) \longrightarrow \mathcal{H}^0_s \longrightarrow \mathcal{H}_s \longrightarrow \mathcal{H}^{\mathrm{et}}_s \longrightarrow (0)$$

over $Y^{\wedge}_s\{N(s)\}$ which restricts over $Y^{\wedge}_s\{N(s-1)\}$ to the pullback

$$(0) \longrightarrow \mathcal{H}^0_{s-1} \longrightarrow \mathcal{H}_{s-1} \longrightarrow \mathcal{H}^{\mathrm{et}}_{s-1} \longrightarrow (0).$$

We will simply write $\mathcal{H}$ for any $\mathcal{H}_s$.

From the universal property of $(I^{(h)}_{U^p, m})^{\wedge}(t)$ we obtain a unique morphism of formal schemes over $\mathrm{Spf}\, \mathcal{O}_{\widehat{F}^{\mathrm{nr}}_w}$,

$$\kappa^*_s : Y^{\wedge}_s\{N(s)\} \longrightarrow (I^{(h)}_{U^p, m})^{\wedge}(t) \times \mathrm{Spf}\, \mathcal{O}_{\widehat{F}^{\mathrm{nr}}_w}$$

which extends the natural map $W_s \to I_{U^p,m}^{(h)} \times \operatorname{Spec} k(w)^{\mathrm{ac}}$ and such that $\mathcal{G}$ pulls back to $\mathcal{H}$ and the universal Drinfeld w^t-structure on $\mathcal{G}^0$ pulls back to the Drinfeld w^t-structure we have just defined on $\mathcal{H}^0$. From the definition of $W_s^\wedge$ and lemma I.5.8 we see that κ_s^* lifts to a unique morphism

$$\kappa_s^* : Y_s^\wedge\{N(s)\} \longrightarrow W_s^\wedge$$

which extends the identity map on W_s and such that $\mathcal{G}$ pulls back to $\mathcal{H}$ and the universal Drinfeld w^t-structure on $\mathcal{G}^0$ pulls back to the Drinfeld w^t-structure we have just defined on $\mathcal{H}^0$. Note that

$$
\begin{array}{ccc}
Y_s^\wedge\{N(s)\} & \xrightarrow{\kappa_s^*} & W_s^\wedge \\
\uparrow & & \uparrow \\
Y_{s+1}^\wedge\{N(s)\} & \xrightarrow{\kappa_{s+1}^*} & W_{s+1}^\wedge
\end{array}
$$

commutes. Note also that if x is a closed point of W_s and if x_∞ is a closed point of $J_{U^p,m,\infty}^{(h)}$ lying above x then

$$
\begin{array}{ccc}
(Y_s^\wedge\{N(s)\})_x^\wedge & \xrightarrow{\kappa_s^*} & (W_s^\wedge)_x^\wedge \\
\downarrow & & \downarrow \\
\operatorname{Spf} R_{F_w,n-h,t}/\mathfrak{m}_{R_{F_w,n-h,t}}^{N(s)} & \hookrightarrow & \operatorname{Spf} R_{F_w,n-h,t}
\end{array}
$$

commutes, where the right hand vertical arrow is $j_{x_\infty}^*$.

The rest of the argument is entirely due to Vladimir Berkovich, to whom we are extremely grateful. By his lemma I.5.9 we see that for every positive integer N we can find a positive integer s such that $N(s) \geq N$ and a morphism

$$\kappa(N)^* : Y_s^\wedge \longrightarrow W_s^\wedge$$

such that for each closed point x of W_s and each closed point x_∞ of $J_{U^p,m,\infty}^{(h)}$ lying above x

$$
\begin{array}{ccc}
(Y_s^\wedge\{N\})_x^\wedge & \xrightarrow{\kappa(N)^*} & (W_s^\wedge)_x^\wedge \\
\downarrow & & \downarrow \\
\operatorname{Spf} R_{F_w,n-h,t}/\mathfrak{m}_{R_{F_w,n-h,t}}^{N} & \hookrightarrow & \operatorname{Spf} R_{F_w,n-h,t}
\end{array}
$$

commutes, where the right hand vertical arrow is $j_{x_\infty}^*$.

We will let $\Psi^j(\mathbb{Z}/l^r\mathbb{Z})$ denote the jth vanishing cycle sheaf constructed for the constant sheaf $\mathbb{Z}/l^r\mathbb{Z}$ on $\operatorname{Spf} R_{F_w,n-h,t}$ and $\Phi^j(\mathbb{Z}/l^r\mathbb{Z})$ the jth vanishing cycle sheaf constructed for the constant sheaf $\mathbb{Z}/l^r\mathbb{Z}$ on $W_s^\wedge$. Thus

$$\Psi_{F_w,l,n-h,t}^j \cong (\varprojlim_r \Psi^j(\mathbb{Z}/l^r\mathbb{Z})) \otimes_{\mathbb{Z}_l} \overline{\mathbb{Q}}_l^{\mathrm{ac}}$$

and

$$\Phi^j(t) \cong (\varprojlim_r \Phi^j(\mathbb{Z}/l^r\mathbb{Z})) \otimes_{\mathbb{Z}_l} \overline{\mathbb{Q}}_l^{ac}.$$

By theorem 4.1 of [Berk3], for any positive integer r we may choose a positive integer N such that any two morphisms of formal schemes over $\mathcal{O}_{\widehat{F}_w^{nr}}$,

$$\mathrm{Spf}\, R_{F_w,n-h,t}[[X_1,\ldots,X_h]] \longrightarrow \mathrm{Spf}\, R_{F_w,n-h,t}$$

which agree on $\mathrm{Spf}\,(R_{F_w,n-h,t}/\mathfrak{m}_{R_{F_w,n-h,t}}^N)[[X_1,\ldots,X_h]]$ induce the same map on vanishing cycles

$$\Psi^j(\mathbb{Z}/l^r\mathbb{Z}) \longleftarrow \Psi^j(\mathbb{Z}/l^r\mathbb{Z}).$$

As the formal completion of any $W_s^\wedge$ or $Y_s^\wedge$ at a closed point is isomorphic to $\mathrm{Spf}\, R_{F_w,n-h,t}[[X_1,\ldots,X_h]]$ we see that we can find a positive integer s (still perhaps depending on r) and a morphism of constructible sheaves over W_s,

$$\kappa : \Phi^j(\mathbb{Z}/l^r\mathbb{Z}) \longrightarrow \Psi^j(\mathbb{Z}/l^r\mathbb{Z}),$$

such that for any closed point x of W_s and any closed point x_∞ of $J_{U^p,m,\infty}^{(h)}$ above x the morphism

$$\kappa : \Phi^j(\mathbb{Z}/l^r\mathbb{Z})_x \longrightarrow \Psi^j(\mathbb{Z}/l^r\mathbb{Z})$$

coincides with the inverse of $(j_{x_\infty}^*)^*$. In particular we see that κ is an isomorphism.

Moreover if $\delta \in \mathcal{O}_{D_{F_w},n-h}^\times$ we see that the natural map $\delta^*\Phi^j(\mathbb{Z}/l^r\mathbb{Z}) \overset{\sim}{\to} \Phi^j(\mathbb{Z}/l^r\mathbb{Z})$ (which arises as $\Phi^j(\mathbb{Z}/l^r\mathbb{Z})$ is a pullback from W) corresponds under κ to the composite of the natural map $\delta^*\Psi^j(\mathbb{Z}/l^r\mathbb{Z}) \overset{\sim}{\to} \Psi^j(\mathbb{Z}/l^r\mathbb{Z})$ (which arises as $\Phi^j(\mathbb{Z}/l^r\mathbb{Z})$ is a pullback from $\mathrm{Spec}\, k(w)^{ac}$) with the automorphism δ of $\Psi^j(\mathbb{Z}/l^r\mathbb{Z})$ (which arises from the action of $\mathcal{O}_{D_{F_w},n-h}^\times$ on $\mathrm{Spf}\, R_{F_w,n-h,t}$). (This can be checked by working on stalks and using the commutativity of the diagram

$$
\begin{array}{ccc}
\Phi^j(\mathbb{Z}/l^r\mathbb{Z})_{\delta x} & \overset{(j_{\delta x_\infty}^*)^*}{\longrightarrow} & \Psi^j(\mathbb{Z}/l^r\mathbb{Z}) \\
\downarrow & & \downarrow \\
\Phi^j(\mathbb{Z}/l^r\mathbb{Z})_x & \overset{(j_{x_\infty}^*)^*}{\longrightarrow} & \Psi^j(\mathbb{Z}/l^r\mathbb{Z}),
\end{array}
$$

where the right hand vertical arrow is δ.)

We will let $\mathcal{F}(\Psi^j(\mathbb{Z}/l^r\mathbb{Z}))$ denote the etale sheaf over W obtained by descending $\Psi^j(\mathbb{Z}/l^r\mathbb{Z})/W_s$ by the diagonal action of $(\mathcal{O}_{D_{F_w},n-h}/w^s)^\times$. Thus

$$\mathcal{F}(\Psi_{F_w,l,n-h,t}^j) \cong (\varprojlim_r \mathcal{F}(\Psi^j(\mathbb{Z}/l^r\mathbb{Z}))) \otimes_{\mathbb{Z}_l} \overline{\mathbb{Q}}_l^{ac}.$$

We see that κ descends to an isomorphism

$$\kappa : \Phi^j(\mathbb{Z}/l^r\mathbb{Z}) \longrightarrow \mathcal{F}(\Psi^j(\mathbb{Z}/l^r\mathbb{Z}))$$

over W such that for any closed point x of W and any closed point x_∞ of $J^{(h)}_{U^p,m,\infty}$ above x the morphism

$$\kappa : \Phi^j(\mathbb{Z}/l^r\mathbb{Z})_x \longrightarrow \mathcal{F}(\Psi^j(\mathbb{Z}/l^r\mathbb{Z}))_x$$

coincides with the inverse of

$$\mathcal{F}(\Psi^j(\mathbb{Z}/l^r\mathbb{Z}))_x \overset{\sim}{\to} \mathcal{F}(\Psi^j(\mathbb{Z}/l^r\mathbb{Z}))_{x_s} \cong \Psi^j(\mathbb{Z}/l^r\mathbb{Z}) \overset{(j^*_{x_\infty})^*}{\longrightarrow} \Phi^j_x,$$

where x_s is the image in W_s of x_∞. By looking at stalks we see that the morphisms κ are compatible as r varies and hence we can glue the morphisms κ to give the map whose existence is asserted in the proposition. $\square$

The next two corollaries are checked by working on stalks.

Corollary IV.2.3 *The homomorphism κ in the proposition is an isomorphism.*

Corollary IV.2.4 *Suppose that $g = (g^p, g_{p,0}, g^0_w, g^{\mathrm{et}}_w, g_{w_i})$ is an element of*

$$G(\mathbb{A}^{\infty,p}) \times \mathbb{Q}^\times_p \times (GL_{n-h}(F_w) \times GL_h(F_w))^+ \times \prod_{i=2}^{r} (B^{\mathrm{op}}_{w_i})^\times,$$

and that

$$[g] : I^{(h)}_{U^p,m} \longrightarrow I^{(h)}_{(U^p)',m'}.$$

Suppose also that $(g^0_w, \delta, \sigma) \in A_{F_w,n-h}$ and that $t >> t'$. Then

$$
\begin{array}{ccc}
([g] \times (\mathrm{Frob}^{w(\sigma)}_w)^*)^* \Phi^j(t') & \overset{g \times \sigma}{\longrightarrow} & \Phi^j(t) \\
\downarrow & & \downarrow \\
([g] \times (\mathrm{Frob}^{w(\det g^0_w)-w(\det \delta)}_w)^*)^* \mathcal{F}(\Psi^j_{F_w,l,n-h,t'}) & \longrightarrow & \mathcal{F}(\Psi_{F_w,l,n-h,t})
\end{array}
$$

commutes, where the lower horizontal arrow is $[g] \times (g^0_w, \delta, \sigma)$ and the vertical arrows are induced by κ.

The next three corollaries follow easily from the previous one.

Corollary IV.2.5 *Under κ the homomorphism $\widetilde{\mathrm{Fr}^*_\varpi}^{f_1(n-h)} \times 1$ from*

$$H^i_c(I^{(h)}_{U^p,m} \times \mathrm{Spec}\, k(w)^{\mathrm{ac}}, \Phi^j(t) \otimes \mathcal{L}_\xi)$$

to

$$H^i_c(I^{(h)}_{U^p,m} \times \operatorname{Spec} k(w)^{\mathrm{ac}}, \Phi^j(t) \otimes \mathcal{L}_\xi)$$

corresponds to the homomorphism $1 \times (\varpi^{-1}, \varpi^{-1}, 1)$ *from*

$$H^i_c(I^{(h)}_{U^p,m} \times \operatorname{Spec} k(w)^{\mathrm{ac}}, \mathcal{F}(\Psi^j_{F_w,l,n-h,t}) \otimes \mathcal{L}_\xi)$$

to

$$H^i_c(I^{(h)}_{U^p,m} \times \operatorname{Spec} k(w)^{\mathrm{ac}}, \mathcal{F}(\Psi^j_{F_w,l,n-h,t}) \otimes \mathcal{L}_\xi).$$

Hence lemma IV.1.5 follows.

Corollary IV.2.6 κ *induces an isomorphism*

$$H^i_c(I^{(h)}, \Phi^j \otimes \mathcal{L}_\xi) \xrightarrow{\sim} H^i_c(I^{(h)}, \mathcal{F}(\Psi^j) \otimes \mathcal{L}_\xi).$$

Moreover if $g = (g^p, g_{p,0}, g^0_w, g^{\mathrm{et}}_w, g_{w_i})$ *is an element of*

$$G(\mathbb{A}^{\infty,p}) \times \mathbb{Q}_p^\times \times (GL_{n-h}(F_w) \times GL_h(F_w))^+ \times \prod_{i=2}^r (B^{\mathrm{op}}_{w_i})^\times$$

and if $\sigma \in W_{F_w}$, *then the action of* $g \times \sigma$ *on the left hand side corresponds to the action of* $[g] \times g^0_w \times \sigma$ *on the right hand side.*

Corollary IV.2.7 *The action of*

$$G(\mathbb{A}^{\infty,p}) \times \mathbb{Q}_p^\times \times (GL_{n-h}(F_w) \times GL_h(F_w))^+ \times \prod_{i=2}^r (B^{\mathrm{op}}_{w_i})^\times$$

on $H^i_c(I^{(h)}, \Phi^j \otimes \mathcal{L}_\xi)$ *extends uniquely to an action of*

$$G(\mathbb{A}^{\infty,p}) \times \mathbb{Q}_p^\times \times GL_{n-h}(F_w) \times GL_h(F_w) \times \prod_{i=2}^r (B^{\mathrm{op}}_{w_i})^\times.$$

Combining these with lemma IV.2.1 we get the following corollary.

Corollary IV.2.8 *We have an isomorphism*

$$H^i_c(I^{(h)}, \Phi^j \otimes \mathcal{L}_\xi)^{n-h} \xrightarrow{\sim} \bigoplus_\rho (H^i_c(I^{(h)}, \mathcal{F}_\rho \otimes \mathcal{L}_\xi) \otimes \Psi^j_{F_w,l,n-h}(\rho))^{(n-h)/e[\rho]},$$

where ρ *runs over irreducible admissible representations of* $D^\times_{F_w,n-h}$ *up to inertial equivalence. Moreover if* $(g^p, g_{p,0}, g^0_w, g^{\mathrm{et}}_w, g_{w_i}, \sigma)$ *is an element of*

$$G(\mathbb{A}^{\infty,p}) \times \mathbb{Q}_p^\times \times GL_{n-h}(F_w) \times GL_h(F_w) \times \prod_{i=2}^r (B^{\mathrm{op}}_{w_i})^\times \times W_{F_w},$$

then the action of $(g^p, g_{p,0}, g_w^0, g_w^{\mathrm{et}}, g_{w_i}, \sigma)$ on the left hand side corresponds to the action of

$$(g^p, g_{p,0}p^{-f_1 w(\sigma)}, w(\delta), g_w^{\mathrm{et}}, g_{w_i}) \times (g_w^0, \sigma)$$

on $H_c^i(I^{(h)}, \mathcal{F}_\rho \otimes \mathcal{L}_\xi) \otimes \Psi_{F_w,l,n-h}^j(\rho)$, where $\delta \in D_{F_w,n-h}^\times$ satisfies

$$w(\det \delta) = w(\det g_w^0) - w(\sigma).$$

We will let d_h denote the homomorphism

$$\begin{array}{ccc} GL_{n-h}(F_w) \times W_{F_w} & \longrightarrow & G^{(h)}(\mathbb{A}^\infty)/\mathcal{O}_{D_{F_w,n-h}}^\times \\ (g_w^0, \sigma) & \longmapsto & (1, p^{-f_1 w(\sigma)}, \delta, 1, 1), \end{array}$$

where

$$w(\det \delta) = w(\det g_w^0) - w(\sigma).$$

If we set

$$[H_c(I^{(h)}, \mathcal{F}_\rho \otimes \mathcal{L}_\xi)] = \sum_i (-1)^{h-i}[H_c^i(I^{(h)}, \mathcal{F}_\rho \otimes \mathcal{L}_\xi)],$$

then we can combine corollaries IV.1.7 and IV.2.8 to obtain the following theorem.

Theorem IV.2.9 (The first basic identity) *For each $h = 0, \ldots, n-1$ choose homomorphisms $j_h^0 : \Lambda_{11} \twoheadrightarrow \mathcal{O}_{F,w}^{n-h}$ and $j_h^{\mathrm{et}} : \Lambda_{11} \twoheadrightarrow \mathcal{O}_{F,w}^h$ such that $j_h^0 \oplus j_h^{\mathrm{et}}$ is an isomorphism. Let $M_h = \ker j_h^{\mathrm{et}}$. Then we have an equality of virtual $G(\mathbb{A}^\infty) \times W_{F_w}$-modules between $[H(X, \mathcal{L}_\xi)^{\mathbb{Z}_p^\times}]$ and*

$$\sum_{h=0}^{n-1} \sum_\rho e[\rho]^{-1} \mathrm{Ind}_{P_{M_h}(F_w)}^{(B_w^{\mathrm{op}})^\times}([H_c(I^{(h)}, \mathcal{F}_\rho \otimes \mathcal{L}_\xi)] *_{d_h} [\Psi_{F_w,l,n-h}(\rho)]),$$

where ρ runs over representatives of the inertial equivalence classes of irreducible admissible representations of $D_{F_w,n-h}^\times$

(See the discussion before lemma II.2.9 for the definition of d_n. See section I.2 for the definition of $*_{d_n}$.)

Chapter V

Counting Points

V.1 An application of Fujiwara's trace formula

Let $\varphi \in C_c^\infty(G^{(h)}(\mathbb{A}^\infty)^+/\mathbb{Z}_p^\times \times \mathcal{O}_{D_{F_w,n-h}}^\times)$. In this section we will find (subject to some restrictions) a formula for

$$\operatorname{tr}(\varphi|H_c(I^{(h)}, \mathcal{F}_\rho \otimes \mathcal{L}_\xi)).$$

Our main tool will be Fujiwara's trace formula (see [Fu]). This is a form of the Lefschetz trace formula for the cohomology with compact supports of smooth but not necessarily proper varieties over finite fields. This formula was conjectured by Deligne and had been proved modulo some sort of resolution of singularities by Pink (see [P]). However first we must say something about the $k(w)^{\mathrm{ac}}$ points on Igusa varieties of the second kind.

If U^p is an open compact subgroup of $G(\mathbb{A}^{\infty,p})$, if $m \in \mathbb{Z}_{\geq 0}^r$ and if $s \in \mathbb{Z}_{\geq 0}$ then set $U^p(m, s)$ equal to

$$U^p \times \mathbb{Z}_p^\times \times (1 + w^s \mathcal{O}_{D_{F_w,n-h}}) \times$$
$$\times (1 + w^{m_1} M_h(\mathcal{O}_{F,w})) \times \prod_{i=2}^r (1 + w_i^{m_i} \mathcal{O}_{B,w_i}^{\mathrm{op}}),$$

an open compact subgroup of $G^{(h)}(\mathbb{A}^\infty)$.

Also set

$$J^{(h)}(k(w)^{\mathrm{ac}}) = \varprojlim_{U^p,m,s} J_{U^p,m,s}^{(h)}(k(w)^{\mathrm{ac}}).$$

This set has a natural right action of $G^{(h)}(\mathbb{A}^\infty)^+$ and we see that

$$J_{U^p,m,s}^{(h)}(k(w)^{\mathrm{ac}}) = J^{(h)}(k(w)^{\mathrm{ac}})/U^p(m, s).$$

149

Also if $x \in J^{(h)}(k(w)^{\mathrm{ac}})$ then we define the stalk of $\mathcal{F}_\rho \otimes \mathcal{L}_\xi$ to be the direct limit of its stalks at the images of x in each $J^{(h)}_{U^p,m,s}$. For each such x there is a canonical isomorphism

$$(\mathcal{F}_\rho \otimes \mathcal{L}_\xi)_x \overset{\sim}{\to} \rho \otimes \xi$$

such that for any $g \in G^{(h)}(\mathbb{A}^\infty)^+$ we have a commutative diagram

$$
\begin{array}{ccc}
(\mathcal{F}_\rho \otimes \mathcal{L}_\xi)_{xg} & \overset{g}{\longrightarrow} & (\mathcal{F}_\rho \otimes \mathcal{L}_\xi)_x \\
\downarrow & & \downarrow \\
\rho \otimes \xi & \overset{\rho(g_w^0) \otimes \xi(g_l)}{\longrightarrow} & \rho \otimes \xi.
\end{array}
$$

As a first step we will see that when considering $k(w)^{\mathrm{ac}}$-points one can (as in characteristic zero) work with abelian varieties up to isogeny, rather than up to prime to p-isogeny. To this end we have the following lemma, whose proof is straightforward. (The main point being that if x is a closed geometric point of $J^{(h)}_{U^p,m,s}$ then $\mathcal{G}_x \cong \mathcal{G}_x^0 \times \mathcal{G}_x^{\mathrm{et}}$ (canonically).)

Lemma V.1.1   1. $J^{(h)}(k(w)^{\mathrm{ac}})$ is in bijection with equivalence classes of $(r+6)$-tuples

$$(A, \lambda, i, \eta^p, \eta_{p,0}, \eta_w^0, \eta_w^e, \eta_{w_i}),$$

  where

- $A/k(w)^{\mathrm{ac}}$ *is an abelian variety of dimension* dn^2;
- $\lambda : A \to A^\vee$ *is a polarisation;*
- $i : B \hookrightarrow \mathrm{End}\,(A) \otimes_{\mathbb{Z}} \mathbb{Q}$ *such that for all* $b \in B$ *we have* $\lambda \circ i(b) = i(b^*)^\vee \circ \lambda$;
- $\eta^p : V \otimes_{\mathbb{Q}} \mathbb{A}^{\infty,p} \overset{\sim}{\to} V^p A$ *is an isomorphism of* $B \otimes_{\mathbb{Q}} \mathbb{A}^{\infty,p}$-*modules under which the standard pairing,* $(\ ,\ )$, *on* V *corresponds to a* $(\mathbb{A}^{\infty,p})^\times$ *multiple of the* λ-*Weil pairing on* $V^p A$;
- $\eta_{p,0} : \mathbb{Q}_p \overset{\sim}{\to} \mathbb{Q}_p(1)$;
- $\eta_w^0 : \Sigma_{n-h} \longrightarrow \varepsilon A[w^\infty]^0$ *is an isogeny;*
- $\eta_w^e : F_w^h \overset{\sim}{\to} \varepsilon V_w A$;
- *for* $i > 1$, $\eta_{w_i} : \Lambda_i \otimes_{\mathbb{Z}_p} \mathbb{Q}_p \overset{\sim}{\to} V_{w_i} A$ *is an isomorphism of* B_{w_i}-*modules.*

Here we call two such $(r+6)$-*tuples,* $(A, \lambda, i, \eta^p, \eta_{p,0}, \eta_w^0, \eta_w^e, \eta_{w_i})$ *and* $(A', \lambda', i', (\eta^p)', \eta'_{p,0}, (\eta_w^0)', (\eta_w^e)', \eta'_{w_i})$, *equivalent if there is an isogeny* $\alpha : A \to A'$, $\gamma \in \mathbb{Q}^\times$ *and* $a \in \mathbb{Z}_p^\times$ *such that*

- $\gamma\lambda = \alpha^\vee \circ \lambda' \circ \alpha$,

- $\alpha \circ i(b) = i'(b) \circ \alpha$ *for all* $b \in B$,
- $(\eta^p)' \circ \alpha = \eta^p$,
- $\gamma a \eta_{p,0} = \eta'_{p,0}$,
- $(\eta_w^0)' \circ \alpha = \eta_w^0$,
- $(\eta_w^e)' \circ \alpha = \eta_w^e$,
- $(\eta_{w_i})' \circ \alpha = \eta_{w_i}$.

2. *Under this bijection the action of*

$$(g^p, g_{p,0}, g_w^0, g_w^e, g_{w_i}) \in G^{(h)}(\mathbb{A}^\infty)^+$$

on $J^{(h)}(k(w)^{\mathrm{ac}})$ *corresponds to the map which sends*

$$(A, \lambda, i, \eta^p, \eta_{p,0}, \eta_w^0, \eta_w^e, \eta_{w_i})$$

to

$$(A, \lambda, i, \eta^p \circ g^p, \eta_{p,0} \circ g_{p,0}, \eta_w^0 \circ g_w^0, \eta_w^e \circ g_w^e, \eta_{w_i} \circ g_{w_i}).$$

3. *In particular the action of* $G^{(h)}(\mathbb{A}^\infty)^+$ *on* $J^{(h)}(k(w)^{\mathrm{ac}})$ *extends to an action of* $G^{(h)}(\mathbb{A}^\infty)$.

We will let $\mathrm{PIC}^{(h)}$ denote the set of equivalence classes of triples (A, λ, i), where

- $A/k(w)^{\mathrm{ac}}$ is an abelian variety of dimension dn^2,

- $\lambda : A \to A^\vee$ is a polarisation, and

- $i : B \hookrightarrow \mathrm{End}\,(A) \otimes_{\mathbb{Z}} \mathbb{Q}$ such that for all $b \in B$ we have $\lambda \circ i(b) = i(b^*)^\vee \circ \lambda$;

for which there exist

- an isomorphism $V \otimes_{\mathbb{Q}} \mathbb{A}^{\infty,p} \overset{\sim}{\to} V^p A$ of $B \otimes_{\mathbb{Q}} \mathbb{A}^{\infty,p}$-modules under which the standard pairing, $(\ ,\ )$, on $V \otimes \mathbb{A}^{\infty,p}$ corresponds to a $(\mathbb{A}^{\infty,p})^\times$ multiple of the λ-Weil pairing on $V^p A$,

- an isogeny $\eta_w^0 : \Sigma_{n-h} \overset{\sim}{\to} \varepsilon A[w^\infty]^0$,

- an isomorphism $F_w^h \overset{\sim}{\to} \varepsilon V_w A$, and

- for each $j > 1$, an isomorphism $\Lambda_j \otimes_{\mathbb{Z}_p} \mathbb{Q}_p \overset{\sim}{\to} V_{w_j} A$ of B_{w_j}-modules.

Here, we call two triples (A, λ, i) and (A', λ', i') equivalent if there exists $\gamma \in \mathbb{Q}^\times$ and an isogeny $\alpha : A \to A'$ such that

- $\gamma\lambda = \alpha^\vee \circ \lambda' \circ \alpha$ and

- $\alpha \circ i(b) = i'(b) \circ \alpha$ for all $b \in B$.

There is a natural map

$$\pi : J^{(h)}(k(w)^{\mathrm{ac}}) \longrightarrow \mathrm{PIC}^{(h)}$$
$$(A, \lambda, i, \eta^p, \eta_{p,0}, \eta_w^0, \eta_w^e, \eta_{w_i}) \longmapsto (A, \lambda, i).$$

The group $G^{(h)}(\mathbb{A}^\infty)$ preserves and acts transitively on each fibre of π. To describe the fibres more precisely we need a little more notation.

Suppose $[(A, \lambda, i)] \in \mathrm{PIC}^{(h)}$. Let

- $C_{(A,\lambda,i)} = \mathrm{End}^0_B(A)$,

- $M_{(A,\lambda,i)}$ denote the centre of $C_{(A,\lambda,i)}$,

- $\ddagger_{(A,\lambda,i)}$ denote the λ-Rosati involution on $C_{(A,\lambda,i)}$, and

- $H_{(A,\lambda,i)}/\mathbb{Q}$ denote the algebraic group such that, for any $\mathbb{Q}$-algebra R, we have

$$H_{(A,\lambda,i)}(R) = \{g \in C_{(A,\lambda,i)} \otimes_\mathbb{Q} R : gg^{\ddagger_{(A,\lambda,i)}} \in R^\times\}.$$

We see that $M_{(A,\lambda,i)}$ is canonically determined by $[(A, \lambda, i)]$, and that both the pair $(C_{(A,\lambda,i)}, \ddagger_{(A,\lambda,i)})$ and the algebraic group $H_{(A,\lambda,i)}$ are determined by $[(A, \lambda, i)]$, but only up to $H_{(A,\lambda,i)}(\mathbb{Q})$-conjugacy.

The choice of some isomorphism $V \otimes_\mathbb{Q} \mathbb{A}^{\infty,p} \overset{\sim}{\to} V^p A$ of $B \otimes_\mathbb{Q} \mathbb{A}^{\infty,p}$-modules under which the standard pairing, $(\ ,\ )$, on $V \otimes \mathbb{A}^{\infty,p}$ corresponds to a $(\mathbb{A}^{\infty,p})^\times$ multiple of the λ-Weil pairing on $V^p A$, gives rise to an embedding

$$H_{(A,\lambda,i)} \times \mathbb{A}^{\infty,p} \hookrightarrow G \times \mathbb{A}^{\infty,p}.$$

Similarly the choice an isogeny $\Sigma_{n-h} \longrightarrow \varepsilon A[w^\infty]^0$ gives rise to a map

$$C_{(A,\lambda,i),w} \longrightarrow D_{F_w,n-h},$$

the choice of an isomorphism $F_w^h \overset{\sim}{\to} \varepsilon V_w A$ gives rise to a map

$$C_{(A,\lambda,i),w} \longrightarrow M_h(F_w),$$

and for each $j > 1$ the choice of an isomorphism of B_{w_j}-modules $\Lambda_j \otimes_{\mathbb{Z}_p} \mathbb{Q}_p \overset{\sim}{\to} V_{w_j} A$ gives rise to embeddings

$$C_{(A,\lambda,i),w_j} \hookrightarrow B_{w_j}^{\mathrm{op}}.$$

Thus we get an embedding

$$\iota_{(A,\lambda,i)} : H_{(A,\lambda,i)}(\mathbb{A}^\infty) \hookrightarrow G^{(h)}(\mathbb{A}^\infty)$$

which is canonical up to $G^{(h)}(\mathbb{A}^\infty)$-conjugacy. The following lemma is easy.

Lemma V.1.2 *As sets with right $G^{(h)}$-action we have*

$$\pi^{-1}[(A, \lambda, i)] \cong \iota_{(A,\lambda,i)}(H_{(A,\lambda,i)}(\mathbb{Q}))\backslash G^{(h)}(\mathbb{A}^\infty).$$

The choice of $\iota_{(A,\lambda,i)}$ (in its conjugacy class) and the isomorphism depend on the choice of a point in $\pi^{-1}[(A, \lambda, i)]$. If this point is varied then there exists $g \in G^{(h)}$ such that $\iota_{(A,\lambda,i)}$ changes by conjugation by g and the isomorphism changes by left translation by g.

In what follows we will often use z to denote an element of $\mathrm{PIC}^{(h)}$, and will write H_z (resp. M_z, etc) for $H_{(A,\lambda,i)}$ (resp. $M_{(A,\lambda,i)}$ etc) for some $(A, \lambda, i) \in z$.

Let $\varphi \in C_c^\infty(G^{(h)}(\mathbb{A}^\infty)^+/\mathbb{Z}_p^\times \times \mathcal{O}_{D_{F_w},n-h}^\times)$. Any such φ can be written as a finite sum of the form

$$\varphi = \sum_g a_g \mathrm{char}\,_{U^p(m,0)gU^p(m,0)}$$

for some fixed U^p and m (depending on φ). As always we can and will assume that U^p is sufficiently small.

By a fixed point of $[U^p(m,0)gU^p(m,0)]$ we will mean a point

$$x \in J^{(h)}(k(w)^{\mathrm{ac}})/(U^p(m,0) \cap gU^p(m,0)g^{-1})$$

such that $x = xg \in J^{(h)}(k(w)^{\mathrm{ac}})/U^p(m,0)$. This set appears to depend on g, not just on $U^p(m,0)gU^p(m,0)$, but if we replace g by u_1gu_2 (with $u_1, u_2 \in U^p(m,0)$) the two sets are in natural bijection via $x \mapsto xu_1^{-1}$. (If $u_1gu_2 = u_1'gu_2'$, with $u_1', u_2' \in U^p(m,0)$ as well, then $x \mapsto x(u_1')^{-1}$ gives the same map.) We will denote this set defined up to canonical bijection $\mathrm{Fix}([U^p(m,0)gU^p(m,0)])$.

Suppose x is such a fixed point. Choose $g \in U^p(m,0)gU^p(m,0)$ and a point $\widetilde{x} \in J^{(h)}(k(w)^{\mathrm{ac}})$ above

$$x \in J^{(h)}(k(w)^{\mathrm{ac}})/U^p(m,0) \cap gU^p(m,0)g^{-1}.$$

Then we see that

$$\widetilde{x}g = \widetilde{x}u$$

for some $u \in U^p(m,0)$. We have

$$\mathrm{tr}\,[U^p(m,0)gU^p(m,0)]|(\mathcal{F}_\rho \otimes \mathcal{L}_\xi)_x = \mathrm{tr}\,(\rho \otimes \xi)(gu^{-1})$$

(see page 433 of [Ko3]). (We remark that the right hand side is indeed independent of the various choices. First if we replace $\widetilde{x}$ by $\widetilde{x}v$ for some $v \in U^p(m,0) \cap gU^p(m,0)g^{-1}$ then gu^{-1} is replaced by $v^{-1}gu^{-1}v$ and so the value of the trace is unchanged. Secondly if we replace g by u_1gu_2 and $\widetilde{x}$ by $\widetilde{x}(u_1)^{-1}$ then gu^{-1} is replaced by $u_1gu^{-1}u_1^{-1}$ and again the value of the trace is unchanged.)

Again suppose that $x \in \text{Fix}([U^p(m,0)gU^p(m,0)])$ and again choose $g \in U^p(m,0)gU^p(m,0)$ and $\widetilde{x} \in J^{(h)}(k(w)^{\text{ac}})$ above

$$x \in J^{(h)}(k(w)^{\text{ac}})/U^p(m,0) \cap gU^p(m,0)g^{-1}.$$

Let $z = \pi(\widetilde{x})$. Then we can represent $\widetilde{x}$ by an element $y \in G^{(h)}(\mathbb{A}^\infty)$, and we see that

$$yg = \iota_z(a)yu$$

for some $a \in H_z(\mathbb{Q})$ and some $u \in U^p(m,0)$. We will show that the conjugacy class $[a]$ of a in $H_z(\mathbb{Q})$ depends only on x. We have to check independence of the following choices.

- We could postmultiply a by an element of

$$(\iota_z)^{-1}(yU^p(m,0)y^{-1}) \cap H_z(\mathbb{Q}).$$

  But as $H_z(\mathbb{R})$ is compact modulo the centre this intersection is a finite group and so as $U^p(m,0)$ is sufficiently small we see that

$$(\iota_z)^{-1}(yU^p(m,0)y^{-1}) \cap H_z(\mathbb{Q}) = \{1\}.$$

- We could replace y by $\iota_z(b)yv$ with $b \in H_z(\mathbb{Q})$ and $v \in U^p(m,0) \cap gU^p(m,0)g^{-1}$. In this case a is replaced by bab^{-1} and u is replaced by $v^{-1}u(g^{-1}vg)$.

- We could replace g by u_1gu_2 and y by yu_1^{-1} with $u_1, u_2 \in U^p(m,0)$. Then a remains unchanged and u is replaced by uu_2.

- We could preconjugate ι_z by $b \in H_z(\mathbb{Q})$ and postconjugate by $g' \in G^{(h)}(\mathbb{A}^\infty)$ while replacing y by $g'y$. Then u is unchanged and a is replaced by $b^{-1}ab$.

Thus we may write $[a(x)]$ for this conjugacy class. Notice that

$$\text{tr}\,[U^p(m,0)gU^p(m,0)]|(\mathcal{F}_\rho \otimes \mathcal{L}_\xi) = \text{tr}\,(\rho \otimes \xi)(\iota_z(a(x))),$$

because $gu^{-1} = y^{-1}\iota_z(a)y$.

Now we ask the converse question: given $a \in H_z(\mathbb{Q})$ how many points $x \in \text{Fix}([U^p(m,0)gU^p(m,0)])$ are there with $[a(x)] = [a]$? One may check that the answer is the cardinality of the double coset space

$$\#(\iota_z(H_z(\mathbb{Q}))\backslash X/U^p(m,0) \cap gU^p(m,0)g^{-1})$$

where

$$X = \{y \in G^{(h)}(\mathbb{A}^\infty) : y^{-1}\iota_z([a])y \cap gU^p(m,0) \neq \emptyset\}.$$

If $a, b \in H_z(\mathbb{Q})$ with both $y^{-1}\iota_z(a)y$ and $y^{-1}\iota_z(b)y \in gU^p(m,0)$ then $\iota_z(a^{-1}b) \in y^{-1}gUg^{-1}y$ and so (because U^p is sufficiently small and $H_z(\mathbb{R})$ is compact modulo its centre) we see that $a = b$. We deduce that the number of $x \in \mathrm{Fix}([U^p(m,0)gU^p(m,0)])$ with $[a(x)] = [a]$ is also given by

$$\#(\iota_z(Z_{H_z}(a)(\mathbb{Q}))\backslash X'/U^p(m,0) \cap gU^p(m,0)g^{-1})$$

where

$$X' = \{y \in G^{(h)}(\mathbb{A}^\infty) : y^{-1}\iota_z(a)y \in gU^p(m,0)\}.$$

A similar argument shows that for any $y \in G^{(h)}(\mathbb{A}^\infty)$ we have

$$\iota_z(Z_{H_z}(a)(\mathbb{Q}))y(U^p(m,0) \cap gU^p(m,0)g^{-1}) =$$
$$= \coprod_{b \in Z_{H_z}(a)(\mathbb{Q})} \iota_z(b)y(U^p(m,0) \cap gU^p(m,0)g^{-1}),$$

and so the number of $x \in \mathrm{Fix}([U^p(m,0)gU^p(m,0)])$ with $[a(x)] = [a]$ is also given by

$$\mathrm{vol}\,(U^p(m,0) \cap gU^p(m,0)g^{-1})^{-1}$$
$$\mathrm{vol}\,(\{y \in \iota_z(Z_{H_z}(a)(\mathbb{Q}))\backslash G^{(h)}(\mathbb{A}^\infty) : y^{-1}\iota_z(a)y \in gU^p(m,0)\}),$$

where we use any Haar measure on $G^{(h)}(\mathbb{A}^\infty)$ and where we use a Haar measure on $\iota_z(Z_{H_z}(a)(\mathbb{Q}))$ which gives each point volume 1. This can be rewritten

$$\mathrm{vol}\,(U^p(m,0) \cap gU^p(m,0)g^{-1})^{-1}$$
$$\mathrm{vol}\,(\iota_z(Z_{H_z}(a)(\mathbb{Q}))\backslash Z_{G^{(h)}(\mathbb{A}^\infty)}(a))O_{\iota_z(a)}^{G^{(h)}(\mathbb{A}^\infty)}(\mathrm{char}\,_{gU^p(m,0)}),$$

where again the measure on $\iota_z(Z_{H_z}(a)(\mathbb{Q}))$ gives each point volume 1 and where the Haar measures on the other groups are arbitrary as long as they are chosen consistently for each occurrence of a given group. This appears to depend on the choice of $g \in [U^p(m,0)gU^p(m,0)]$. Adding the formulas for g running over a set of representatives for $U^p(m,0)gU^p(m,0)/U^p(m,0)$ and dividing by

$$\#(U^p(m,0)gU^p(m,0)/U^p(m,0)) = [U^p(m,0) : U^p(m,0) \cap gU^p(m,0)g^{-1}],$$

we see that the number of $x \in \mathrm{Fix}([U^p(m,0)gU^p(m,0)])$ with $[a(x)] = [a]$ is also given by

$$\mathrm{vol}\,(U^p(m,0))^{-1}$$
$$\mathrm{vol}\,(\iota_z(Z_{H_z}(a)(\mathbb{Q}))\backslash Z_{G^{(h)}(\mathbb{A}^\infty)}(a))O_{\iota_z(a)}^{G^{(h)}(\mathbb{A}^\infty)}(\mathrm{char}\,_{U^p(m,0)gU^p(m,0)}).$$

We remark that this number may be infinite.

We will say that $\varphi \in C_c^\infty(G^{(h)}(\mathbb{A}^\infty)^+/\mathbb{Z}_p^\times \times \mathcal{O}_{D_{F_w},n-h}^\times)$ is *acceptable* if it can be written as a finite sum

$$\varphi = \sum_j \alpha_j \text{char}_{U^p(m,0)g_j U^p(m,0)}$$

with U^p sufficiently small and where

1. for $y \in \bigcup_j U^p(m,0)g_j U^p(m,0)$ the p-adic valuation of every eigenvalue of y_w^0 is strictly less than the p-adic valuation of every eigenvalue of y_w^e;

2. each $\text{Fix}([U^p(m,0)g_j U^p(m,0)])$ is a finite set;

3. and, for each j,

$$\sum_i (-1)^i \text{tr}\left([U^p(m,0)g_j U^p(m,0)]|_{H_c^i(I_{U^p,m}^{(h)} \times \text{Spec}\, k(w)^{\text{ac}}, \mathcal{F}_\rho \otimes \mathcal{L}_\xi))}\right)$$
$$= \sum_{x \in \text{Fix}([U^p(m,0)g_j U^p(m,0)])} \text{tr}\left([U^p(m,0)g_j U^p(m,0)]|(\mathcal{F}_\rho \otimes \mathcal{L}_\xi)_x\right).$$

Here $[U^p(m,0)g_j U^p(m,0)]$ defines a correspondence on $I_{U^p,m}^{(h)}$ which lifts to a natural cohomological correspondence on $\mathcal{F}_\rho \otimes \mathcal{L}_\xi$ over $(I_{U^p,m}^{(h)} \times \text{Spec}\, k(w)^{\text{ac}})$. It therefore acts on

$$H_c^i(I_{U^p,m}^{(h)} \times \text{Spec}\, k(w)^{\text{ac}}, \mathcal{F}_\rho \otimes \mathcal{L}_\xi).$$

(See [Fu].)

This definition is only useful if we have a good supply of acceptable functions φ. This is provided by the following lemma, whose key ingredient is Fujiwara's trace formula.

Lemma V.1.3 *Suppose that* $\varphi \in C_c^\infty(G^{(h)}(\mathbb{A}^\infty)^+/\mathbb{Z}_p^\times \times \mathcal{O}_{D_{F_w},n-h}^\times)$. *Fix* $\delta \in D_{F_w,n-h}^\times$ *with* $w(\det \delta) = 1$. *Then for* $N >> 0$ *the function*

$$\varphi|(1, p^{f_1 N}, \delta^N, 1, 1)$$

defined by

$$(\varphi|(1, p^{f_1 N}, \delta^N, 1, 1))(y) = \varphi(y(1, p^{f_1 N}, \delta^N, 1, 1))$$

is acceptable.

Proof: The first condition is easily checked. The latter two conditions follow from corollary 5.4.5 of [Fu] because

$$[U^p(m,0)g_i(1, p^{-f_1 N}, \delta^{-N}, 1, 1)U^p(m,0)] = (\text{Fr}^*)^{f_1 N}.[U^p(m,0)g_i U^p(m,0)]$$

in the notation of [Fu] (see the third paragraph of section IV.2). □

We now show how we can calculate the trace of an acceptable function on the cohomology of Igusa varieties of the first kind.

Lemma V.1.4 *Suppose that* $\varphi \in C_c^\infty(G^{(h)}(\mathbb{A}^\infty)^+/\mathbb{Z}_p^\times \times \mathcal{O}_{D_{F_w,n-h}}^\times)$ *is acceptable. Then*

$$\operatorname{tr}(\varphi|H_c(I^{(h)}, \mathcal{F}_\rho \otimes \mathcal{L}_\xi)) = (-1)^h \sum_{z \in \text{PIC}^{(h)}} \sum_{[a] \subset H_z(\mathbb{Q})}$$
$$\operatorname{vol}(\iota_z(Z_{H_z}(a)(\mathbb{Q})) \backslash Z_{G^{(h)}(\mathbb{A}^\infty)}(\iota_z(a))(\mathbb{A}^\infty)) O_{\iota_z(a)}^{G^{(h)}(\mathbb{A}^\infty)}(\varphi) \operatorname{tr}(\rho \otimes \xi)(\iota_z(a)).$$

This sum is finite and all the terms occurring are finite numbers. We may make any consistent choices for the Haar measures implicit in this formula as long as the Haar measure on $Z_{H_z^{\mathbb{A}^\vee}}(a)(\mathbb{Q})$ *gives points measure 1.*

Proof: If

$$\varphi = \sum_i \alpha_i \operatorname{char}_{U^p(m,0)g_i U^p(m,0)}$$

then

$$\operatorname{tr}(\varphi|H_c(I^{(h)}, \mathcal{F}_\rho \otimes \mathcal{L}_\xi)) = \operatorname{vol}(U^p(m,0)) \sum_{i,j} \alpha_i (-1)^{h-j}$$
$$\operatorname{tr}([U^p(m,0)g_i U^p(m,0)]|H_c^j(I_{U^p,m}^{(h)} \times \operatorname{Spec} k(w)^{\text{ac}}, \mathcal{F}_\rho \otimes \mathcal{L}_\xi)),$$

where we use the same Haar measure to compute $\operatorname{tr}\varphi$ as we do to compute $\operatorname{vol}(U^p(m,0))$. The lemma now follows from the definition of acceptable and from our previous calculations. □

V.2 Honda-Tate theory

In this section we will recall how the theory of Honda and Tate [Tat1] allows us to classify simple abelian varieties over $\mathbb{F}_p^{\text{ac}}$. We then apply this to begin to understand $\text{PIC}^{(h)}$.

By a CM field we will mean a number field M such that for any embedding $i: M \hookrightarrow \mathbb{C}$ the image iM is stable under complex conjugation and such that the automorphism c of M induced by complex conjugation on iM is independent of the embedding i. Equivalently either M is totally real or a totally imaginary quadratic extension of a totally real field. We will let $\mathbb{Q}[\mathfrak{P}_M]$ denote the $\mathbb{Q}$-vector space with basis the places of M above p. If $i: M \hookrightarrow N$ is a finite extension we get natural maps $i_*: \mathbb{Q}[\mathfrak{P}_M] \to \mathbb{Q}[\mathfrak{P}_N]$ induced by $x \mapsto \sum_{y|x} e_{y/x}y$ and $i^*: \mathbb{Q}[\mathfrak{P}_N] \to \mathbb{Q}[\mathfrak{P}_M]$ induced by $y \mapsto f_{y/x}x$ if $y|x$. If I is a fractional ideal of M then we set $[I] = \sum_x x(I)x \in \mathbb{Q}[\mathfrak{P}_M]$.

By a p-adic type for a CM field M we shall mean an element $\eta \in \mathbb{Q}[\mathfrak{P}_M]$ such that $\eta + c_*\eta = [p]$. We will call p-adic types $\eta \in \mathbb{Q}[\mathfrak{P}_M]$ and $\eta' \in \mathbb{Q}[\mathfrak{P}_{M'}]$ equivalent if there is a CM field M'', a p-adic type $\eta'' \in \mathbb{Q}[\mathfrak{P}_{M''}]$ and embeddings $i : M'' \hookrightarrow M$ and $i' : M'' \hookrightarrow M'$ such that $i_*(\eta'') = \eta$ and $i'_*(\eta'') = \eta'$. By a p-adic type we shall mean an equivalence class of p-adic types for various CM fields M. Then any p-adic type $\mathfrak{b}$ has a minimal representative (M, η) such that if $(M', \eta') \in \mathfrak{b}$ then there exists $i : M \hookrightarrow M'$ such that $\eta' = i_*\eta$. (To see this choose $(M', \eta') \in \mathfrak{b}$ with $M'/\mathbb{Q}$ Galois, let H denote the subgroup of $\sigma \in \mathrm{Gal}\,(M'/\mathbb{Q})$ such that $\sigma_*\eta' = \eta'$ and set $M = (M')^H$.) We call a p-adic type $\mathfrak{b}$ ordinary if for any $(M, \eta) \in \mathfrak{b}$ and any $x \in \mathfrak{P}_M$ we have $\eta_x\eta_{cx} = 0$.

Suppose that $q = p^r$ is a power of p. By a q-number one means an algebraic number π such that for any embedding $i : \mathbb{Q}[\pi] \hookrightarrow \mathbb{C}$ we have $\pi c(\pi) = q$. To any q-number π we associate a CM-type $\mathfrak{b}(\pi) = [(\mathbb{Q}[\pi], [\pi]/r)]$. To π Honda and Tate also associate a simple abelian variety $A_\pi/\mathbb{F}_q$. The simple factors of $A_\pi \times_{\mathbb{F}_q} \mathbb{F}_{q^s}$ are all isogenous to A_{π^s}. Thus all the simple factors of $A_\pi \times_{\mathbb{F}_q} \mathbb{F}_p^{\mathrm{ac}}$ are isogenous: we will denote them $A'_\pi/\mathbb{F}_p^{\mathrm{ac}}$. We see that A'_π is isogenous to A'_{π^s}. If $\mathfrak{b}(\pi) = \mathfrak{b}(\pi')$ then we can find positive integers n and n' such that (up to Galois conjugation) $[\pi^n] = [(\pi')^{n'}]$. Then (after replacing π by a suitable Galois conjugate) $\pi^n/(\pi')^{n'}$ is a unit with all archimedean absolute values 1 and hence a root of unity. Thus we can find a positive integer m such that π^{nm} and $(\pi')^{n'm}$ are Galois conjugate. Thus $A_{\pi^{nm}}$ is isogenous to $A_{(\pi')^{n'm}}$ and so A'_π is isogenous to $A'_{\pi'}$. This allows us to write $A_{\mathfrak{b}(\pi)}$ for $A'_\pi/\mathbb{F}_p^{\mathrm{ac}}$. If η is is any p-adic type for a CM field M then for some positive integer r we have that $r\eta \in \mathbb{Z}[\mathfrak{P}_M]$ and hence for a second positive integer h we have that $rh\eta = [\alpha]$ for some $\alpha \in \mathcal{O}_M$. Then $\beta = \alpha c(\alpha)/p^{rh}$ is a unit in the ring of integers of the maximally totally real subfield M^+ of M. Thus $\pi = \alpha^2\beta^{-1}$ is a p^{2hr}-number and $\mathfrak{b}(\pi) = [(M, \eta)]$. Thus to any p-adic type we can associate a well defined isogeny class of simple abelian varieties $A_\mathfrak{b}/\mathbb{F}_p^{\mathrm{ac}}$. It follows easily from the theory of Honda and Tate that this gives a bijection between p-types and isogeny classes of simple abelian varieties over $\mathbb{F}_p^{\mathrm{ac}}$. The following further results also follow easily from their theory.

1. If $\mathfrak{b}$ is a p-adic type with minimal representative (M, η) then the endomorphism algebra $\mathrm{End}^0(A_\mathfrak{b})$ is the division algebra with centre M and invariants

   - $\mathrm{inv}_x\mathrm{End}^0(A_\mathfrak{b}) = 1/2$ if x is real;
   - $\mathrm{inv}_x\mathrm{End}^0(A_\mathfrak{b}) = \eta_x f_{x/p}$ if $x|p$;
   - $\mathrm{inv}_x\mathrm{End}^0(A_\mathfrak{b}) = 0$ otherwise.

2. If $\mathfrak{b}$ is a p-adic type with minimal representative (M, η) then $\dim A_\mathfrak{b} = [M : \mathbb{Q}][\mathrm{End}^0(A_\mathfrak{b}) : M]^{1/2}/2$.

3. If $\mathfrak{b}$ is a p-adic type with minimal representative (M, η) and if x is a place of M above p then $A_{\mathfrak{b}}[x^\infty]$ has height $[M_x : \mathbb{Q}_p][\text{End}^0(A_{\mathfrak{b}}) : M]^{1/2}$ and its Newton polygon has pure slope $\eta_x/e_{x/p}$.

Now fix a CM field L. If M/L and M'/L are CM field extensions and if η and η' are p-adic types for M and M' respectively then we will call them equivalent over L if there is a CM field extension M''/L, a p-adic type $\eta'' \in \mathbb{Q}[\mathfrak{P}_{M''}]$ and embeddings $i : M'' \hookrightarrow M$ and $i' : M'' \hookrightarrow M$ over L such that $i_*(\eta'') = \eta$ and $i'_*(\eta'') = \eta'$. By a p-adic type over L we shall mean an L-equivalence class of p-adic types for various CM fields M/L. Then any p-adic type $\mathfrak{b}$ has a minimal representative (M, η) such that if $(M', \eta') \in \mathfrak{b}$ then there exists $i : M \hookrightarrow M'$ over L such that $\eta' = i_*\eta$.

Now let C be a division algebra with centre L. We will consider the category of pairs (A, i) up to isogeny, where $A/\mathbb{F}_p^{\text{ac}}$ is an abelian variety and $i : C \hookrightarrow \text{End}^0(A/\mathbb{F}_p^{\text{ac}})$. As in section 3 of [Ko3] we can use the results of the last paragraph to describe the simple objects of this category. They are in bijection with p-adic types over L. If $\mathfrak{b}$ is such a type we will let $(A_{\mathfrak{b}}, i_{\mathfrak{b}})$ denote the corresponding simple object. We have the following additional properties.

1. If $\mathfrak{b}$ is a p-adic type over L with minimal representative (M, η) then $\text{End}_C(A_{\mathfrak{b}})$ is the division algebra with centre M and invariants

   - $\text{inv}_x \text{End}_C(A_{\mathfrak{b}}) = 1/2 - \text{inv}_x(C \otimes_L M)$ if x is real;
   - $\text{inv}_x \text{End}_C(A_{\mathfrak{b}}) = \eta_x f_{x/p} - \text{inv}_x(C \otimes_L M)$ if $x|p$;
   - $\text{inv}_x \text{End}_C(A_{\mathfrak{b}}) = -\text{inv}_x(C \otimes_L M)$ otherwise.

2. If $\mathfrak{b}$ is a p-adic type over L with minimal representative (M, η) then $\dim A_{\mathfrak{b}} = [M : \mathbb{Q}][C : L]^{1/2}[\text{End}_B(A_{\mathfrak{b}}) : M]^{1/2}/2$.

3. If $\mathfrak{b}$ is a p-adic type over L with minimal representative (M, η) and if x is a place of M above p then $A_{\mathfrak{b}}[x^\infty]$ has height

$$[M_x : \mathbb{Q}_p][C : L]^{1/2}[\text{End}(A_{\mathfrak{b}}) : M]^{1/2}$$

and its Newton polygon has pure slope $\eta_x/e_{x/p}$.

Finally in this section we return to the specific case of interest to us and take a first step towards a concrete description of $\text{PIC}^{(h)}$.

Lemma V.2.1 *1. There is a bijection between isogeny classes of pairs (A, i) where $A/k(w)^{\text{ac}}$ is an abelian variety of dimension dn^2 and $i : B \hookrightarrow \text{End}^0(A)$ such that*

   - *$A[w_j^\infty]$ is ind-etale for $j > 1$,*

- $A[w^\infty]^e$ *has height* nh, *and*
- $A[w^\infty]^0$ *has dimension* n

and pairs $(M, \tilde{w})$ *where*

- M/F *is a CM field extension which is embeddable into* B *over* F,

- $\tilde{w}$ *is a place of* M *above* w *such that*

$$[M_{\tilde{w}} : F_w]n = [M : F](n - h),$$

- *there is no intermediate field* $M \supset N \supset F$ *such that* $\tilde{w}|_N$ *is inert in* M.

2. *Suppose that* (A, i) *and* $(M, \tilde{w})$ *correspond. Then* $C = \mathrm{End}_B^0(A)$ *is the division algebra with centre* M *and invariants as follows.*

- *If* $x \nmid ww^c$ *is a place of* M *then* $\mathrm{inv}_x(C) = \mathrm{inv}_x(B^{\mathrm{op}} \otimes_F M)$.
- *If* $x|ww^c$ *is a place of* M *not dividing* $\tilde{w}\tilde{w}^c$ *then* $\mathrm{inv}_x(C) = 0$.
- $\mathrm{inv}_{\tilde{w}}(C) = [M : F]/n$.
- $\mathrm{inv}_{\tilde{w}^c}(C) = -[M : F]/n$.

Moreover

$$\dim_M(\mathrm{End}_B(A) \otimes_{\mathbb{Z}} \mathbb{Q}) = (n/[M : F])^2$$

and

$$A[w^\infty]^0 = A[\tilde{w}^\infty].$$

Proof: We will first show that any A as in the first part of the lemma is a simple object in the category of abelian varieties up to isogeny with B-action. To this end choose a simple factor A' of A in the category of abelian varieties up to isogeny with B-action. Suppose that A' corresponds to a p-adic type over F with minimal representative (M, η). Choose a place x of F such that B_x is a division algebra and such that $x \nmid u^c$. Let y be a place of M above x. If $x|p$ then (as $x|u$ but $x \neq w$) $A[y^\infty]$ is etale and hence $\eta_y = 0$. In any case we see that, if $C' = \mathrm{End}_B^0(A')$, then

$$\mathrm{inv}_y(C') = -[M_y : F_w]\mathrm{inv}_x(B),$$

and so

$$[C' : M]^{1/2} \geq n/[M : F].$$

Thus

$$\dim A' \geq dn^2,$$

which implies $A = A'$, as desired.

Thus let A correspond to a p-adic type over F with minimal representative (M, η). Note that if y is a place of M dividing u but not dividing w then $A[y^\infty]$ is etale and hence $\eta_y = 0$. Moreover as $A[w^\infty]^0$ is a simple object in the category of Barsotti-Tate groups up to isogeny with B_w-action we see that there is a unique place $\tilde{w}$ of M above w with $\eta_{\tilde{w}} \neq 0$. Then $A[w^\infty]^0 = A[\tilde{w}^\infty]$ and as this Barsotti-Tate group has height $[F_w : \mathbb{Q}_p]n(n-h)$ we see that

$$[M_{\tilde{w}} : F_w][C : M]^{1/2} = n - h.$$

Also as the Newton polygon of $A[w^\infty]^0$ has Newton polygon which is pure of slope $1/([F_w : \mathbb{Q}_p](n-h))$ we see that

$$\eta_{\tilde{w}} = e_{\tilde{w}/w}/((n-h)f_{w/p}).$$

Hence we see that

- inv $_{\tilde{w}}C = [M_{\tilde{w}} : F_w]/(n-h) = [C : M]^{-1/2}$,

- inv $_{\tilde{w}^c}C = -[M_{\tilde{w}} : F_w]/(n-h) = -[C : M]^{-1/2}$,

- and for any other place x of M we have

$$\text{inv }_x C = -[M_x : F_x]\text{inv }_x(B).$$

As

$$dn^2 = \dim A = [M : \mathbb{Q}]n[C : M]^{1/2}/2$$

we see that

$$n = [M : F][C : M]^{1/2}$$

and hence that

$$(n-h)[M : F] = n[M_{\tilde{w}} : F_w].$$

Because we can find an extension $N : M$ such that N splits B and $[N : F]^2 = [B : F]$, we see that M embeds in B over F. Finally as (M, η) was chosen minimal we see that there is no intermediate field $M \supset N \supset F$ such that $\tilde{w}|_N$ is inert in M.

Conversely if $(M, \tilde{w})$ is as in the theorem we consider the p-adic type over F, (M, η) where

- $\eta_{\tilde{w}} = e_{\tilde{w}/w}/((n-h)f_{w/p})$,

- $\eta_x = 0$ for any other place $x|u$ of M,

- $\eta_{\tilde{w}^c} = e_{\tilde{w}/p}(1 - 1/((n-h)[F_w : \mathbb{Q}_p]))$,

- $\eta_x = e_{x/p}$ for any other place $x|u^c$ of M.

Let (A, i) be the corresponding abelian variety with B-action. Again using the results of Honda-Tate recalled at the start of this section we see that (A, i) has the properties listed in the lemma. $\square$

Lemma V.2.2 *Suppose that* $\varphi \in C_c^\infty(G^{(h)}(\mathbb{A}^\infty)^+/\mathbb{Z}_p^\times \times \mathcal{O}_{D_{F_w},n-h}^\times)$ *is acceptable, then*

$$\text{tr}\,(\varphi|H_c(I^{(h)}, \mathcal{F}_\rho \otimes \mathcal{L}_\xi)) = (-1)^h \sum_{z \in \text{PIC}^{(h)}} \sum_{[a] \subset H_z(\mathbb{Q})}$$
$$\text{vol}\,(Z_{H_z}(a)(\mathbb{Q})\backslash Z_{H_z}(a)(\mathbb{A}^\infty))O_{\iota_z(a)}^{G^{(h)}(\mathbb{A}^\infty)}(\varphi)\text{tr}\,(\rho \otimes \xi)(\iota_z(a)).$$

This sum is finite and all the terms occurring are finite numbers. For each non-zero term we have $F(a) \supset M_z$ *and*

$$Z_{G^{(h)}(\mathbb{A}^\infty)}(\iota_z(a)) = \iota_z(Z_{H_z}(a)(\mathbb{A}^\infty)).$$

We may make any consistent choices for the Haar measures implicit in this formula as long as the Haar measure on $Z_{H_z}(a)(\mathbb{Q})$ *gives points measure 1 and as long as the measures used on* $Z_{G^{(h)}(\mathbb{A}^\infty)}(\iota_z(a))$ *and* $Z_{H_z}(a)(\mathbb{A}^\infty)$ *correspond under* ι_z.

Proof: By lemma V.1.4 it suffices to check that if $O_{\iota_z^{AV}(a)}^{G^{(h)}(\mathbb{A}^\infty)}(\varphi) \neq 0$ then $F(a) \supset M_z$ (which in turn implies that

$$Z_{G^{(h)}(\mathbb{A}^\infty)}(\iota_z(a)) = \iota_z(Z_{H_z}(a)(\mathbb{A}^\infty))).$$

If $O_{\iota_z(a)}^{G^{(h)}(\mathbb{A}^\infty)}(\varphi) \neq 0$ then the p-adic valuation of every eigenvalue of $\iota_z(a)_w^0$ is strictly less than the p-adic valuation of every eigenvalue of $\iota_z(a)_w^e$ (because φ is acceptable). Thus there is a constant κ such that a place x of $M_z(a) \subset C_z$ above w divides $\tilde{w}$ if and only if

$$|a|_x^{1/[M_z(a)_x:F_w]} > \kappa.$$

Hence if x is a place of $M_z(a)$ above $\tilde{w}$ then any other place of $M_z(a)$ above $x|_{F(a)}$ also lies above $\tilde{w}$. Let $N/(F(a) \cap M_z)$ denote the normal closure of $M_z(a)/F(a) \cap M$. Then N has the same property that two places x and x' of N above the same place of $F(a)$ either both lie above $\tilde{w}$ or neither lies above $\tilde{w}$. Fix a place x of N above $\tilde{w}$ and let Δ denote the decomposition group for x in $\mathrm{Gal}\,(N/F(a)\cap M_z)$. Let $\sigma \in \mathrm{Gal}\,(N/F(a)\cap M_z)$. Then some (resp. all) places of N above the place $\sigma(x)|_{F(a)}$ lie above $\widetilde{w}$ if and only if

$$(\mathrm{Gal}\,(N/F(a))\sigma\Delta) \cap (\mathrm{Gal}\,(N/M_z)\Delta) \neq \emptyset$$

(resp.

$$(\mathrm{Gal}\,(N/F(a))\sigma\Delta) \subset (\mathrm{Gal}\,(N/M_z)\Delta)\ ).$$

Thus we see that $\mathrm{Gal}\,(N/M_z)\Delta$ is a union of double cosets of the form

$$\mathrm{Gal}\,(N/F(a))\sigma\Delta,$$

i.e.

$$\mathrm{Gal}\,(N/F(a))\mathrm{Gal}\,(N/M_z)\Delta = \mathrm{Gal}\,(N/M_z)\Delta.$$

As $\mathrm{Gal}\,(N/(F(a)\cap M_z))$ is generated by $\mathrm{Gal}\,(N/F(a))$ and $\mathrm{Gal}\,(N/M_z)$ we see that

$$\mathrm{Gal}\,(N/(F(a)\cap M_z)) = \mathrm{Gal}\,(N/M_z)\Delta.$$

This translates into $\widetilde{w}|_{F(a)\cap M_z}$ being inert in M_z. By the minimality of $(M_z, \widetilde{w})$ we conclude that $F(a) \supset M_z$. $\square$

V.3   Polarisations I

In this section we will explain how to add polarisations to the picture to get a better understanding of $\mathrm{PIC}^{(h)}$, as well as H_z and ι_z for any $z \in \mathrm{PIC}^{(h)}$. We will do this by employing techniques from Galois cohomology as in [Ko3] and [Ko4]. We start with a general lemma describing polarisations on certain abelian varieties.

Lemma V.3.1 *Let L be an imaginary quadratic field, C a a finite dimensional semi-simple L-algebra and $\dagger$ a positive involution on C which (necessarily) restricts to c on L. Let k be an algebraically closed field and A/k an abelian variety with a (fixed) embedding $C \hookrightarrow \mathrm{End}^0(A)$. We will call a polarisation of A a $\dagger$-polarisation if the corresponding Rosati involution takes C to itself and acts on C via $\dagger$. We call two polarisations λ and λ', C-equivalent if there exists $\gamma \in \mathbb{Q}^\times$ and $\alpha \in \mathrm{End}^0_C(A)$ such that*

$$\lambda = \gamma\alpha^\vee\lambda'\alpha.$$

There exists a $\dagger$-polarisation λ_0 on A. Let $\ddagger_0$ denote the λ_0-Rosati involution on $\mathrm{End}^0_C(A)$ and define an algebraic group $H_0/\mathbb{Q}$ by setting

$$H_0(R) = \{g \in \mathrm{End}^0_C(A) \otimes_{\mathbb{Q}} R :\ gg^{\ddagger_0} \in R^\times\}$$

for any $\mathbb{Q}$*-algebra* R. *There is a natural bijection between* C*-equivalence classes of* $\dagger$*-polarisations on* A *and*

$$\ker(H^1(\mathbb{Q}, H_0) \longrightarrow H^1(\mathbb{R}, H_0)).$$

If a $\dagger$*-polarisation* λ *corresponds to a cohomology class* ϕ *and if* S *is a finite set of places of* $\mathbb{Q}$ *containing* ∞ *and the characteristic of* k *(if this is positive), then the difference between the* $C \otimes \mathbb{A}^S$*-module* $V^S A$ *with* λ_0*-Weil pairing and the* λ*-Weil pairing (as* $\dagger$*-Hermitian alternating pairings up to equivalence) is measured by the image of* ϕ *in* $H^1(\mathbb{Q}, H_0(\overline{\mathbb{A}}^S))$.

Proof: The first assertion is lemma 9.2 of [Ko3]. Any $\dagger$-polarisation λ on A is of the form $\lambda_0 \beta$ for some $\beta \in \operatorname{End}_C^0(A)^\times$ with $\beta^{\ddagger_0} = \beta$ and $\beta = \delta^{\ddagger_0}\delta$ for some $\delta \in \operatorname{End}_C^0(A) \otimes_\mathbb{Q} \mathbb{R}$. Conversely any such β does give rise to a $\dagger$-polarisation. Two such elements β and β' give rise to C-equivalent polarisations if there exist $\gamma \in \mathbb{Q}^\times$ and $\alpha \in \operatorname{End}_C^0(A)^\times$ with $\beta = \gamma \alpha^{\ddagger_0}\beta'\alpha$. As in section I.6 we see that $\beta \in \operatorname{End}_C^0(A)^\times \cap \operatorname{End}_C^0(A)^{\ddagger_0 = 1}$ up to this equivalence are in bijection with

$$H^1(L/\mathbb{Q}, H_0(L)) \cong H^1(\mathbb{Q}, H_0).$$

The condition that $\beta = \delta^{\ddagger_0}\delta$ for some $\delta \in \operatorname{End}_C^0(A) \otimes_\mathbb{Q} \mathbb{R}$ is equivalent to the corresponding cohomology class mapping to zero in $H^1(\mathbb{R}, H_0)$. The lemma follows. $\square$

It will be convenient to define an equivalence relation which we will call *near equivalence* and denote $\sim$ on $\operatorname{PIC}^{(h)}$ as follows. We set $[(A, \lambda, i)] \sim [(A', \lambda', i')]$ if and only if

- there exists an isogeny $\alpha : A \to A'$ such that $\alpha \circ i(b) = i'(b) \circ \alpha$ for all $b \in B$, and

- the $B \otimes_F M_{[(A,\lambda,i)]} \otimes \mathbb{A}^{\infty,p}$-module $V^p A$ with the λ-Weil pairing is equivalent to $V^p A'$ with the λ'-Weil pairing.

Lemma V.3.2 *Suppose that* $[(A, i, \lambda)]$ *and* $[(A', i', \lambda')]$ *are nearly equivalent. Then there is an isomorphism*

$$H_{(A,i,\lambda)} \xrightarrow{\sim} H_{(A',i',\lambda')}$$

which is canonical up to composition with conjugation by an element of the group $H_{(A,i,\lambda)}(\mathbb{Q})$. *After base change to* $\mathbb{A}^{\infty,p}$ *this isomorphism is induced by some equivalence of* $B \otimes_F M_{[(A,\lambda,i)]} \otimes \mathbb{A}^{\infty,p}$*-modules with non-degenerate* $* \otimes c$*-Hermitian pairings between* $V^p A$ *and* $V^p A'$. *In particular it is compatible with* $\iota_{(A,i,\lambda)}$ *and* $\iota_{(A',i',\lambda')}$ *up to* $G^{(h)}(\mathbb{A}^\infty)$*-conjugacy.*

Proof: We can find an isogeny

$$\alpha : A \longrightarrow A'$$

such that

- $i'(x) = \alpha i(x) \alpha^{-1}$ for all $x \in B \otimes_F M$,

- and $\lambda' = \alpha^{-\vee} \lambda \gamma \alpha^{-1}$ for some $\gamma \in C_{(A,i,\lambda)}^{\ddagger(A,i,\lambda)=1}$ with $\gamma = \delta^{\ddagger(A,i,\lambda)} \delta$ for some $\delta \in C_{(A,i,\lambda)} \otimes_{\mathbb{Q}} \mathbb{R}$.

Then $(\gamma, 1) \in C_{(A,i,\lambda)}^{\times} \times E^{\times}$ represents an element of

$$\ker(H^1(E/\mathbb{Q}, H_{(A,i,\lambda)}(E)) \longrightarrow H^1(\mathbb{Q}, H_{(A,i,\lambda)}(\overline{\mathbb{A}}))).$$

Let $Z_{(A,i,\lambda)}$ denote the centre of $H_{(A,i,\lambda)}$. As

- $H^1(E, H_{(A,i,\lambda)}) = (0)$,

- $H^1(E, Z_{(A,i,\lambda)}) = (0)$,

- and $\ker^1(\mathbb{Q}, Z_{(A,i,\lambda)}) \overset{\sim}{\to} \ker^1(\mathbb{Q}, H_{(A,i,\lambda)})$ (by the same argument used on pages 393 and 394 of [Ko3]),

we deduce that

$$\ker(H^1(E/\mathbb{Q}, H_{(A,i,\lambda)}(E)) \longrightarrow H^1(\mathbb{Q}, H_{(A,i,\lambda)}(\overline{\mathbb{A}})))$$

equals

$$\ker(H^1(E/\mathbb{Q}, Z_{(A,i,\lambda)}(E)) \longrightarrow H^1(E/\mathbb{Q}, Z_{(A,i,\lambda)}(\mathbb{A}_E))).$$

Thus we can find $(\delta, \mu) \in C_{(A,i,\lambda)}^{\times} \times E^{\times}$ such that

$$(\mu^{-c} \delta^{\ddagger(A,i,\lambda)} \gamma \delta, \mu/\mu^c) \in M_{(A,\lambda,i)}^{\times} \times E^{\times}.$$

Note that we may take $\mu = 1$. Then replacing α by $\alpha \delta$ we see that without loss of generality we may suppose that $\gamma \in M_{(A,\lambda,i)}^{\times}$ (and hence that γ is a totally positive element of $M_{(A,\lambda,i)}^{+}$).

It follows (from $\gamma \in M_{(A,\lambda,i)}$) that α induces an isomorphism $C_{(A,i,\lambda)} \overset{\sim}{\to} C_{(A',i',\lambda')}$ ($x \mapsto \alpha x \alpha^{-1}$) which takes $\ddagger_{(A,i,\lambda)}$ to $\ddagger_{(A',i',\lambda')}$ and hence induces an isomorphism $H_{(A,i,\lambda)} \overset{\sim}{\to} H_{(A',i',\lambda')}$. Any two such isomorphisms must differ by conjugation by an element of $H_{(A,i,\lambda)}(\mathbb{Q})$.

We may choose $\mu \in (\mathbb{A}^{\infty,p})^{\times}$ and $\delta \in (M_{(A,i,\lambda)} \otimes \mathbb{A}^{\infty,p})^{\times}$ such that $\gamma = \mu^{-1} \delta^c \delta$. Suppose that $\beta : V^p A \overset{\sim}{\to} V^p A'$ is an isomorphism of

$B \otimes_F M_{(A,i,\lambda)} \otimes_{\mathbb{Q}} \mathbb{A}^{\infty,p}$-modules with alternating pairings up to $(\mathbb{A}^{\infty,p})^\times$-multiples. Then we see that

$$\beta^{-1}(V^p \alpha)\delta^{-1} \in H_{(A,i,\lambda)}(\mathbb{A}^{\infty,p}).$$

Altering β on the right by an element of $H_{(A,i,\lambda)}(\mathbb{A}^{\infty,p})$, we may suppose that $\beta = (V^p \alpha)\delta^{-1}$ and hence that β and $V^p \alpha$ induce the same isomorphism

$$H_{(A,i,\lambda)} \times_{\mathbb{Q}} \mathbb{A}^{\infty,p} \xrightarrow{\sim} H_{(A',i',\lambda')} \times_{\mathbb{Q}} \mathbb{A}^{\infty,p}.$$

the second part of the lemma follows. $\square$

Thus for $z \in \mathrm{PIC}^{(h)}/\sim$ we have

- a CM field $M_z \supset F$ and a distinguished place $\tilde{w}$ of M_z above w such that
    - M_z is embeddable into B over F,
    - $[M_{z,\tilde{w}} : F_w]n = [M_z : F](n-h)$,
    - there is no intermediate field $M_z \supset N \supset F$ such that $\tilde{w}|_N$ is inert in M_z,

- and an algebraic group $H_z/\mathbb{Q}$ (canonical up to $H_z(\mathbb{Q})$-conjugation) and an embedding $\iota_z : H_z(\mathbb{A}^\infty) \hookrightarrow G^{(h)}(\mathbb{A}^\infty)$ well defined up to $G^{(h)}(\mathbb{A}^\infty)$-conjugacy.

Let $\nu_z : H_z \longrightarrow \mathbb{G}_m$ denote the multiplier character (so that $\nu_z(x) = x^{\ddagger_z}x$). Also let $Z_{H_z}(a)(\mathbb{A})^1$ (resp. $Z_{H_z}(a)(\mathbb{R})^1$) denote the kernel of

$$|\nu_z| : Z_{H_z}(a)(\mathbb{A}) \longrightarrow \mathbb{R}^\times_{>0}$$

(resp. the kernel of

$$|\nu_z|_{\mathbb{R}} : Z_{H_z}(a)(\mathbb{R}) \longrightarrow \mathbb{R}^\times_{>0}).$$

Let $\mathrm{FP}^{(h)}_{\mathrm{AV}}$ denote the set of pairs $(z, [a])$ where $z \in \mathrm{PIC}^{(h)}/\sim$ and $[a]$ is a $H_z(\mathbb{A})$-conjugacy class of elements $a \in H_z(\mathbb{Q})$ with $F(a) \supset M_z$.

Lemma V.3.3 *Suppose that* $\varphi \in C_c^\infty(G^{(h)}(\mathbb{A}^\infty)^+/\mathbb{Z}_p^\times \times \mathcal{O}_{D_{F_w},n-h}^\times)$ *is acceptable, then*

$$\mathrm{tr}\,(\varphi|H_c(I^{(h)}, \mathcal{F}_\rho \otimes \mathcal{L}_\xi)) = (-1)^h \kappa_B \sum_{(z,[a]) \in \mathrm{FP}^{(h)}_{\mathrm{AV}}} O^{G^{(h)}(\mathbb{A}^\infty)}_{\iota_z(a)}(\varphi)$$
$$\mathrm{vol}\,(Z_{H_z}(a)(\mathbb{R})^1)^{-1} \mathrm{tr}\,(\rho \otimes \xi)(\iota_z(a)),$$

where the Haar measures on $Z_{H_z}(a)(\mathbb{R})^1$ *and*

$$Z_{G^{(h)}(\mathbb{A}^\infty)}(\iota_z(a)) \cong Z_{H_z}(a)(\mathbb{A}^\infty)$$

are chosen to be compatible with Tamagawa measure on $Z_{H_z}(a)(\mathbb{A})^1$ and the exact sequence

$$\{1\} \to Z_{H_z}(a)(\mathbb{R})^1 \to Z_{H_z}(a)(\mathbb{A})^1 \to Z_{G^{(h)}(\mathbb{A}^\infty)}(\iota_z(a)) \to \{1\}.$$

(Recall that $\kappa_B = 2$ if $[B : \mathbb{Q}]/2$ is odd and $\kappa_B = 1$ otherwise.)

Proof: From lemma V.3.1 we see that the near equivalence class of $z \in$ PIC$^{(h)}$ has cardinality $\# \ker^1(\mathbb{Q}, H_z)$. Thus we can rewrite the equality of lemma V.2.2 as

$$\text{tr}\,(\varphi|H_c(I^{(h)}, \mathcal{F}_\rho \otimes \mathcal{L}_\xi)) = (-1)^h \sum_{z \in \text{PIC}^{(h)}/\sim} \sum_{[a] \subset H_z(\mathbb{Q})} \# \ker^1(\mathbb{Q}, H_z)$$
$$\text{vol}\,(Z_{H_z}(a)(\mathbb{Q})\backslash Z_{H_z}(a)(\mathbb{A}^\infty))O^{G^{(h)}(\mathbb{A}^\infty)}_{\iota_z(a)}(\varphi)\text{tr}\,(\rho \otimes \xi)(\iota_z(a)),$$

where the second sum is only over those conjugacy classes $[a]$ for which $F(a) \supset M_z$.

We have an exact sequence

$$\{1\} \to Z_{H_z}(a)(\mathbb{R})^1 \to Z_{H_z}(a)(\mathbb{Q})\backslash Z_{H_z}(a)(\mathbb{A})^1 \to$$
$$\to Z_{H_z}(a)(\mathbb{Q})\backslash Z_{H_z}(a)(\mathbb{A}^\infty) \to \{1\},$$

and hence $\text{vol}\,(Z_{H_z}(a)(\mathbb{Q})\backslash Z_{H_z}(a)(\mathbb{A}^\infty))$ equals

$$\text{vol}\,(Z_{H_z}(a)(\mathbb{Q})\backslash Z_{H_z}(a)(\mathbb{A}))\text{vol}\,(Z_{H_z}(a)(\mathbb{R})^1)^{-1}.$$

Moreover if we use Tamagawa measure on $Z_{H_z}(a)(\mathbb{A})^1$ then the main theorem of [Ko5] tells us that

$$\text{vol}\,(Z_{H_z}(a)(\mathbb{Q})\backslash Z_{H_z}(a)(\mathbb{A}^\infty))$$
$$= \#A(Z_{H_z}(a)(\mathbb{Q}))(\# \ker^1(\mathbb{Q}, Z_{H_z}(a)))^{-1}\text{vol}\,(Z_{H_z}(a)(\mathbb{R})^1)^{-1}$$
$$= \kappa_B(\# \ker^1(\mathbb{Q}, Z_{H_z}(a)(\mathbb{Q})))^{-1}\text{vol}\,(Z_{H_z}(a)(\mathbb{R})^1)^{-1}.$$

(See the introduction to [Ko5] and formula 4.2.2 of [Ko1], and compute $A(Z_{H_z}(a)(\mathbb{Q}))$ directly from the definition. Note that $[B : \mathbb{Q}]/2$ is even if and only if $[Z_{C_z}(a) : \mathbb{Q}]/2$ is.) Thus the right hand side of the above trace formula can be rewritten

$$(-1)^h \kappa_B \sum_{z \in \text{PIC}^{(h)}/\sim} \sum_{[a] \subset H_z(\mathbb{Q})} (\# \ker^1(\mathbb{Q}, H_z)/\# \ker^1(\mathbb{Q}, Z_{H_z}(a)))$$
$$O^{G^{(h)}(\mathbb{A}^\infty)}_{\iota_z(a)}(\varphi)\text{vol}\,(Z_{H_z}(a)(\mathbb{R})^1)^{-1}\text{tr}\,(\rho \otimes \xi)(\iota_z(a)),$$

where the second sum is only over those conjugacy classes $[a]$ with $F(a) \supset M_z$ and where we use measures on $Z_{H_z}(a)(\mathbb{R})^1$ and $Z_{G^{(h)}(\mathbb{A}^\infty)}(\iota_z(a))$ compatible with Tamagawa measure on $Z_{H_z}(a)(\mathbb{A})^1$ and the exact sequence

$$\{1\} \to Z_{H_z}(a)(\mathbb{R})^1 \to Z_{H_z}(a)(\mathbb{A})^1 \to Z_{G^{(h)}(\mathbb{A}^\infty)}(\iota_z(a)) \to \{1\}.$$

Now suppose that a and $a' \in H_z(\mathbb{Q})$ are conjugate under $H_z(\mathbb{A})$. Then $Z_{H_z}(a)$ and $Z_{H_z}(a')$ are inner forms of each other which become isomorphic over $\mathbb{A}$. Moreover the Tamagawa measures on $Z_{H_z}(a)(\mathbb{A})^1$ and $Z_{H_z}(a')(\mathbb{A})^1$ agree under this isomorphism (use the definition of Tamagawa measure and the discussion in paragraph two on page 631 of [Ko3]). Thus

$$O^{G^{(h)}(\mathbb{A}^\infty)}_{\iota_z(a)}(\varphi)\mathrm{vol}\,(Z_{H_z}(a)(\mathbb{R})^1)^{-1} = O^{G^{(h)}(\mathbb{A}^\infty)}_{\iota_z(a')}(\varphi)\mathrm{vol}\,(Z_{H_z}(a')(\mathbb{R})^1)^{-1}.$$

The number of $H_z(\mathbb{Q})$-conjugacy classes in the $H_z(\mathbb{A})$-conjugacy class of a in $H_z(\mathbb{Q})$ is

$$\#\ker(\ker^1(\mathbb{Q}, Z_{H_z}(a)) \longrightarrow \ker^1(\mathbb{Q}, H_z)).$$

For the moment let Z (resp. Z') denote the centre of H_z (resp. $Z_{H_z}(a)$). Then we have homomorphisms

$$\ker^1(\mathbb{Q}, Z) \longrightarrow \ker^1(\mathbb{Q}, Z') \longrightarrow \ker^1(\mathbb{Q}, Z_{H_z}(a)) \longrightarrow \ker^1(\mathbb{Q}, H_z),$$

where as on pages 393 and 394 of [Ko3] we see that

$$\ker^1(\mathbb{Q}, Z') \xrightarrow{\sim} \ker^1(\mathbb{Q}, Z_{H_z}(a))$$

and

$$\ker^1(\mathbb{Q}, Z) \xrightarrow{\sim} \ker^1(\mathbb{Q}, H_z).$$

Thus

$$\#\ker(\ker^1(\mathbb{Q}, Z_{H_z}(a)) \longrightarrow \ker^1(\mathbb{Q}, H_z))$$

equals

$$\#\ker^1(\mathbb{Q}, Z_{H_z}(a))/\#\ker^1(\mathbb{Q}, H_z).$$

In particular this number only depends on a up to $H_z(\mathbb{A})$-conjugacy. The lemma follows. $\square$

V.4   Polarisations II

The next lemma is presumably the analogue in our language of "the vanishing of the Kottwitz invariant". It is similar to a result of Th.Zink in [Zi].

Lemma V.4.1 *Fix an embedding* $\tau : F \hookrightarrow \mathbb{C}$. *Suppose that M is a CM-field extension of F, that $\widetilde{w}$ is a place of M above w such that*

$$[M_{\widetilde{w}} : F_w]n = [M : F](n - h)$$

and such that there is no intermediate field $M \supset N \supset F$ with $\widetilde{w}|_N$ is inert in M. Let $A/k(w)^{\mathrm{ac}}$ be an abelian variety of dimension dn^2 and $i : B \otimes_F M \hookrightarrow \mathrm{End}^0(A)$ an embedding such that

- *M is the centre of $\mathrm{End}_B^0(A)$,*

- *$A[w^\infty]^0 = A[\widetilde{w}^\infty]$ has dimension n,*

- *$A[w^\infty]^e$ has height nh,*

- *and $A[w_i^\infty]$ is ind-etale for all $i > 1$.*

Then we can find

- *a $* \otimes c$-polarisation $\lambda_0 : A \to A^\vee$ (with respect to $B \otimes_F M$), and*

- *a finitely generated $B \otimes M$-module W_0 together with an non-degenerate alternating pairing*

$$\langle \ , \ \rangle_0 : W_0 \times W_0 \longrightarrow \mathbb{Q}$$

which is $* \otimes c$-Hermitian

such that

1. *there is an isomorphism of $B \otimes_F M \otimes \mathbb{A}^{\infty,p}$-modules*

$$W_0 \otimes \mathbb{A}^{\infty,p} \overset{\sim}{\to} V^p A$$

which takes $\langle \ , \ \rangle_0$ to an $(\mathbb{A}^{\infty,p})^\times$-multiple of the λ_0-Weil pairing on $V^p A$,

2. *and there is an isomorphism of $B \otimes \mathbb{R}$-modules*

$$W_0 \otimes \mathbb{R} \overset{\sim}{\to} V \otimes \mathbb{R}$$

which takes $\langle \ , \ \rangle_0$ to a $\mathbb{R}^\times$ multiple of our standard pairing $(\ , \ )_\tau$ on $V \otimes \mathbb{R}$.

Proof: By lemma 9.2 of [Ko3] there is a polarisation $\lambda_0 : A \to A^\vee$ such that the λ_0-Rosati involution preserves $B \otimes_F M$ and acts on it as $* \otimes c$. Let $C = \mathrm{End}_B(A) \otimes_{\mathbb{Z}} \mathbb{Q}$ and let $\ddagger_0$ denote the λ_0-Rosati involution restricted to C. We see from lemma V.2.1 that C is a division algebra with centre M and that $\ddagger_0|_M = c$.

The first step of the proof will be to show that, up to isogeny, we can lift A with its $B \otimes_F M$-action and its polarisation over $\mathcal{O}_{F_w^{\mathrm{ac}}}$.

Let M^+ denote the maximal totally real subfield of M. For $a | [C : M]^{1/2}$ we will let $X_a \subset C$ denote the locally closed M^+-subvariety of semi-simple elements δ such that $\delta^{\ddagger_0} = \delta$ and the characteristic polynomial of δ over M is an ath, but no higher, power. (Note that if a monic polynomial over M^+ is an ath power over $(M^+)^{\mathrm{ac}}$ it is already one over M^+.) Then

$$C^{\ddagger_0=1} = \coprod X_a(M^+).$$

Computing over $(M^+)^{\mathrm{ac}}$ we see that

$$\dim X_a = [C : M] + (a - [C : M]/a).$$

As $C^{\ddagger_0=1}$ is an affine space over M^+ of dimension $[C : M]$ and as for $a \neq [C : M]^{1/2}$ we have $\dim X_a < [C : M]$, we see that we can find an element

$$\delta \in C^{\ddagger_0=1} - \coprod_{a \neq [C:M]^{1/2}} X_a(M^+).$$

Set $N = M(\delta)$. Then N is a maximal subfield of C, which is preserved by $\ddagger_0$, which is a CM-field and which satisfies $\ddagger_0|_N = c$. Set $N^+ = M^+(\delta)$ the maximal totally real subfield of N.

We see that N also splits B (as it does so locally at all places of F by lemma V.2.1). Choose an isomorphism $\alpha : B \otimes_F N \xrightarrow{\sim} M_n(N)$ and let $*_\alpha$ denote the involution $\alpha \circ (* \otimes c) \circ \alpha^{-1}$ on $M_n(N)$. If $x \in M_n(N)$ we will define $x' \in M_n(M)$ by $(x')_{ij} = x_{ji}^c$. Then we have $x^{*_\alpha} = ax'a^{-1}$ for some $a \in GL_n(N)$ with $a'a^{-1} \in N^\times$. We see that $(a'a^{-1})(a'a^{-1})^c = (a')^{-1}(a'a^{-1})a = 1$ and so by Hilbert's theorem 90 $a'a^{-1} = \gamma/\gamma^c$ for some $\gamma \in N^\times$. Replacing a by γa we see that we may suppose that $a' = a$. If we compose α with conjugation by b in $GL_n(N)$ then we change a to bab'. Thus by suitable choice of α we may suppose that

$$\alpha(x^{* \otimes c}) = a\alpha(x)'a^{-1}$$

for some diagonal matrix $a \in GL_n(N^+)$, with diagonal entries $a_1, \ldots, a_n$ say. As $* \otimes c$ is a positive involution on $M_n(N)$ we see that $a_i a_j$ is totally positive for all i and j. Thus multiplying a by a scalar we may suppose that each a_i is totally positive.

Considering $\varepsilon \in M_n(N)$, let $A_1 = \varepsilon A$ and let $j : A_1 \to A$ denote the tautological map. Let $i_1 : N \hookrightarrow \mathrm{End}^0(A_1)$ be the map induced by i and let $\lambda_1 = j^\vee \circ \lambda_0 \circ j$, a polarisation on A_1. If for $i = 1, \ldots, n$ we let ε_i denotes the element of $M_n(N)$ which has a 1 in the ith entry in the first column and zeroes elsewhere, then we get an isogeny

$$(\varepsilon_1 \circ j) + \cdots + (\varepsilon_n \circ j) : A_1^n \longrightarrow A.$$

Then the diagram

$$
\begin{array}{ccc}
A & \xrightarrow{\;\lambda\;} & A^\vee \\
\uparrow & & \downarrow \\
A_1^n & \xrightarrow{\;\oplus\lambda_1\circ(a_i/a_1)\;} & A_1^n
\end{array}
$$

commutes. Note that, as the centraliser of $B \otimes_F N$ in $B \otimes_F \operatorname{End}_B^0(A)$ is N, the centraliser of $i_1(N)$ in $\operatorname{End}^0(A_1)$ is just $i_1(N)$ itself.

Now (up to isogeny) we can lift A_1 to an abelian scheme $\widetilde{A}_1/\mathcal{O}_{F_w^{\mathrm{ac}}}$ in such a way that the action i_1 of N lifts to an action $\widetilde{i}_1$ of N on $\widetilde{A}_1$. (See [Tat1].) Choose a polarisation $\mu : \widetilde{A}_1 \times F_w^{\mathrm{ac}} \to \widetilde{A}_1^\vee \times F_w^{\mathrm{ac}}$ for which the Rosati involution induces c on $\widetilde{i}_1(N)$ (see lemma 9.2 of [Ko3]). It extends to a homomorphism $\mu : \widetilde{A}_1 \to \widetilde{A}_1^\vee$, which is again a polarisation (see [Ko3], page 392). Let $\overline{\mu}$ denote the pullback of μ to A_1. As λ_1 and $\overline{\mu}$ both induce c on N and as $i_1(N)$ is its own centraliser in $\operatorname{End}^0(A_1)$ we see that $\lambda_1 = \overline{\mu} \circ x$ for some $x \in N$. As λ_1 and $\overline{\mu}$ are both polarisations, x must in fact be a totally positive element of N^+. Then $\widetilde{\lambda}_1 = \mu \circ x$ is a polarisation of $\widetilde{A}_1$ which reduces to λ_1. Set $\widetilde{A} = (\widetilde{A}_1)^n$, $\widetilde{i} = M_n(\widetilde{i}_1 \circ \alpha) : B \otimes_F M \to \operatorname{End}^0(\widetilde{A})$ and $\widetilde{\lambda}_0 = \oplus_{j=1}^n \widetilde{\lambda}_1 \circ \widetilde{i}(a_j/a_1)$.

Because $\operatorname{Lie}\widetilde{A}_1 \otimes_{\mathcal{O}_{F_w^{\mathrm{ac}}}} F_w^{\mathrm{ac}}$ is a $F \otimes F_w^{\mathrm{ac}} \cong (F_w^{\mathrm{ac}})^{\operatorname{Hom}(F, F_w^{\mathrm{ac}})}$-module, we get a decomposition

$$
\operatorname{Lie}\widetilde{A}_1 \otimes_{\mathcal{O}_{F_w^{\mathrm{ac}}}} F_w^{\mathrm{ac}} \cong \bigoplus_{\sigma \in \operatorname{Hom}(F, F_w^{\mathrm{ac}})} (\operatorname{Lie}\widetilde{A}_1)_\sigma .
$$

We also have a decomposition

$$
\operatorname{Lie}\widetilde{A}_1 \otimes_{\mathcal{O}_{F_w^{\mathrm{ac}}}} F_w^{\mathrm{ac}} \cong \operatorname{Lie}\widetilde{A}_1[p^\infty] \otimes_{\mathcal{O}_{F_w^{\mathrm{ac}}}} F_w^{\mathrm{ac}} \cong \bigoplus_x \operatorname{Lie}\widetilde{A}_1[x^\infty] \otimes_{\mathcal{O}_{F_w^{\mathrm{ac}}}} F_w^{\mathrm{ac}},
$$

where x runs over places of F above p. Then

$$
\operatorname{Lie}\widetilde{A}_1[x^\infty] \otimes_{\mathcal{O}_{F_w^{\mathrm{ac}}}} F_w^{\mathrm{ac}} \cong \bigoplus_\sigma (\operatorname{Lie}\widetilde{A}_1)_\sigma ,
$$

where now σ runs over embeddings $\sigma : F \hookrightarrow F_w^{\mathrm{ac}}$ such that x is the induced place of F. Let $\operatorname{Hom}(F, F_w^{\mathrm{ac}})^+$ denote those embeddings which induce the place u on E, so that $\operatorname{Hom}(F, F_w^{\mathrm{ac}}) = \operatorname{Hom}(F, F_w^{\mathrm{ac}})^+ \coprod \operatorname{Hom}(F, F_w^{\mathrm{ac}})^+ \circ c$. Then we deduce that there is a unique $\sigma_0 \in \operatorname{Hom}(F, F_w^{\mathrm{ac}})^+$ such that $(\operatorname{Lie}\widetilde{A}_1)_{\sigma_0} \neq (0)$. Moreover σ_0 induces w on F and $(\operatorname{Lie}\widetilde{A}_1)_{\sigma_0} \cong F_w^{\mathrm{ac}}$.

We can find an embedding $\kappa : F_w^{\mathrm{ac}} \hookrightarrow \mathbb{C}$ such that $\kappa \circ \sigma_0 = \tau$, our distinguished embedding $F \hookrightarrow \mathbb{C}$. Set

$$
W_0 = H_1((\widetilde{A} \times_{\operatorname{Spec}\mathcal{O}_{F_w^{\mathrm{ac}}}, \kappa} \operatorname{Spec}\mathbb{C})(\mathbb{C}), \mathbb{Q}).
$$

This is a $B \otimes_F M$-module with an alternating pairing (coming from $\tilde{\lambda}_0$) which is $* \otimes c$-Hermitian for the action of $B \otimes_F M$. We see at once that $W_0 \otimes \mathbb{A}^{\infty,p}$ is equivalent to $V^p A$ as a $B \otimes_F M \otimes_{\mathbb{Q}} \mathbb{A}^{\infty,p}$-module with $* \otimes c$-Hermitian $\mathbb{A}^{\infty,p}$-alternating pairing.

It remains to show that $W_0 \otimes \mathbb{R}$ has invariants (see section I.6) $(1, n-1)$ at τ and $(0, n)$ at any other embedding $F^+ \hookrightarrow \mathbb{R}$. Note first that the invariants of $W_0 \otimes \mathbb{R}$ are the same as the invariants of the $F \otimes_{\mathbb{Q}} \mathbb{R}$-module with c-Hermitian alternating pairing $\langle \ , \ \rangle_1$ on $W_1 \otimes_{\mathbb{Q}} \mathbb{R}$, where

$$W_1 = H_1((\tilde{A}_1 \times_{\mathrm{Spec}\, \mathcal{O}_{F_w^{\mathrm{ac}},\kappa}} \mathrm{Spec}\, \mathbb{C})(\mathbb{C}), \mathbb{Q}),$$

and $\langle \ , \ \rangle_1$ comes from $\tilde{\lambda}_1$. As an F-module we see that $W_1 \cong F^n$. Also

$$W_1 \otimes_{\mathbb{Q}} \mathbb{R} \cong (\mathrm{Lie}\, \tilde{A}_1) \otimes_{\mathcal{O}_{F_w^{\mathrm{ac}},\kappa}} \mathbb{C},$$

and so it is an $F \otimes_{\mathbb{Q}} \mathbb{C}$-module, not simply an $F \otimes_{\mathbb{Q}} \mathbb{R}$-module. Corresponding to $F \otimes_{\mathbb{Q}} \mathbb{C} \cong \mathbb{C}^{\mathrm{Hom}\,(F,\mathbb{C})}$ we get a decomposition

$$W_1 \otimes_{\mathbb{Q}} \mathbb{R} \cong \bigoplus_{\tau' \in \mathrm{Hom}\,(F,\mathbb{C})} (W_1 \otimes_{\mathbb{Q}} \mathbb{R})_{\tau'}.$$

As $W_1 \otimes_{\mathbb{Q}} \mathbb{R} \cong (F \otimes_{\mathbb{Q}} \mathbb{R})^n$ we see that for all τ',

$$\dim_{\mathbb{C}}(W_1 \otimes_{\mathbb{Q}} \mathbb{R})_{\tau'} + \dim_{\mathbb{C}}(W_1 \otimes_{\mathbb{Q}} \mathbb{R})_{\tau' \circ c} = n.$$

On the other hand

- $(W_1 \otimes_{\mathbb{Q}} \mathbb{R})_{\kappa \circ \sigma_0} \cong \mathbb{C}$,

- but if $\tau' \neq \kappa \circ \sigma_0$ while $\tau'|_E = (\kappa \circ \sigma_0|_E$ then $(W_1 \otimes_{\mathbb{Q}} \mathbb{R})_{\tau'} = (0)$.

As the alternating form on $(W_1 \otimes_{\mathbb{Q}} \mathbb{R})$ is $c \otimes c$-Hermitian for the action of $F \otimes_{\mathbb{Q}} \mathbb{C}$ we see that

$$W_1 \otimes_{\mathbb{Q}} \mathbb{R} = \bigoplus_{\tau} (W_1 \otimes_{\mathbb{Q}} \mathbb{R})_{\tau'}$$

is an orthogonal direct sum for this alternating form.

For a suitable choice of $i \in \mathbb{C}$ a square root of -1 the c-symmetric form $(W_1 \otimes_{\mathbb{Q}} \mathbb{R}) \times (W_1 \otimes_{\mathbb{Q}} \mathbb{R}) \to \mathbb{C}$ given by

$$x \times y \longmapsto \langle ix, y \rangle_1 + i \langle x, y \rangle_1$$

is positive definite. Choose $\sqrt{-1} \in E \otimes_{\mathbb{Q}} \mathbb{R}$ such that $\kappa \circ \sigma_0(\sqrt{-1}) = -i$. If $\tau'|_E = c \circ \kappa \circ \sigma_0|_E$ then

$$x \times y \longmapsto \langle \sqrt{-1}x, y \rangle_1 + \sqrt{-1} \langle x, y \rangle_1$$

is positive definite on $(W_1 \otimes_{\mathbb{Q}} \mathbb{R})_{\tau'}$, while if $\tau'|_E = \kappa \circ \sigma_0|_E$ then it is negative definite on $(W_1 \otimes_{\mathbb{Q}} \mathbb{R})_{\tau'}$. It follows that the invariants of $W_0 \otimes_{\mathbb{Q}} \mathbb{R}$ coincide with those of $(\ ,\ )_\tau$ on $V \otimes_{\mathbb{Q}} \mathbb{R}$ and so these are equivalent as $B \otimes_{\mathbb{Q}} \mathbb{R}$-modules with $*$-Hermitian $\mathbb{R}$-alternating pairings up to $\mathbb{R}^\times$-multiples. $\square$

Keep the notation of the last lemma. Let $\ddagger_0$ denote the λ_0-Rosati involution on $C_0 = \mathrm{End}_B^0(A)$. Let $H_0^{\mathrm{AV}}/\mathbb{Q}$ denote the reductive algebraic group such that for any $\mathbb{Q}$-algebra R the R-points of H_0^{AV} are the set of $g \in C_0 \otimes_{\mathbb{Q}} R$ such that $g^{\ddagger_0} g \in R^\times$. Also let $D_0 = \mathrm{End}_{B \otimes M}(W_0)$, so that D_0 is isomorphic to the centraliser of M in B^{op}, and let $\dagger_0$ denote the involution on $B^{\mathrm{op}} = \mathrm{End}_B(W_0)$ and on D_0 induced by $\langle\ ,\ \rangle_0$. Let $H_0^{\mathrm{LA}}/\mathbb{Q}$ (resp. $G_0/\mathbb{Q}$) denote the reductive algebraic group such that for any $\mathbb{Q}$-algebra R the R-points of H_0^{LA} are the set of $g \in D_0 \otimes_{\mathbb{Q}} R$ such that $g^{\dagger_0} g \in R^\times$ (resp. $g \in B^{\mathrm{op}} \otimes_{\mathbb{Q}} R$ such that $g^{\dagger_0} g \in R^\times$). Thus $H_0^{\mathrm{LA}} \subset G_0$ and we have a natural isomorphism

$$H_0^{\mathrm{AV}} \times \mathbb{A}^{\infty,p} \xrightarrow{\sim} H_0^{\mathrm{LA}} \times \mathbb{A}^{\infty,p}.$$

There is also an isomorphism $\psi : H_0^{\mathrm{AV}} \times \mathbb{Q}^{\mathrm{ac}} \xrightarrow{\sim} H_0^{\mathrm{LA}} \times \mathbb{Q}^{\mathrm{ac}}$ such that

- over $\overline{\mathbb{A}}^{\infty,p}$ the above natural isomorphism and ψ differ by an inner automorphism, and hence that

- for any $\sigma \in \mathrm{Gal}\,(\mathbb{Q}^{\mathrm{ac}}/\mathbb{Q})$, ψ and $\sigma(\psi)$ differ by an inner automorphism

(see lemma I.6.2). Also let ϕ_0 denote the class in $H^1(\mathbb{Q}, G_0)$ which represents the difference between $(V, (\ ,\ )_{\beta_\tau})$ and $(W_0, \langle\ ,\ \rangle_0)$.

Let $\tau : F \hookrightarrow \mathbb{C}$. We see that there are natural bijections between the following sets.

1. $G_\tau(\mathbb{Q})$-conjugacy classes of F-embeddings $j : M \hookrightarrow B^{\mathrm{op}}$ such that $\#_\tau \circ j = j \circ c$.

2. Equivalence classes of $B \otimes_F M$-modules with a $* \otimes c$-Hermitian $\mathbb{Q}$-alternating form which are equivalent to V as B-modules with $*$-Hermitian $\mathbb{Q}$-alternating pairing.

3. The preimage of ϕ_0 under $H^1(\mathbb{Q}, H_0^{\mathrm{LA}}) \longrightarrow H^1(\mathbb{Q}, G_0)$.

(See section I.7 for the definition of G_τ and $\#_\tau$. The map from the first to the second set sends j to $(V, (\ ,\ )_{\beta_\tau})$ considered as a $B \otimes_F M$-module via $\mathrm{Id} \otimes j : B \otimes_F M \to B \otimes_F B^{\mathrm{op}}$.)

Lemma V.4.2 *This bijection induces a bijection between the following sets.*

1. $G_\tau(\mathbb{A})$-conjugacy classes of F-embeddings $j : M \hookrightarrow B^{\mathrm{op}}$ such that $\#_\tau \circ j = j \circ c$.

2. The preimage of ϕ_0 under $H^1(\mathbb{Q}, H_0^{\mathrm{LA}}(\overline{\mathbb{A}})) \longrightarrow H^1(\mathbb{Q}, G_0(\overline{\mathbb{A}}))$.

Proof: We see that $G(\mathbb{A})$-conjugacy classes of F-embeddings $j : M \hookrightarrow B^{\mathrm{op}}$ such that $\#_\tau \circ j = j \circ c$ are in bijection with classes $x \in H^1(\mathbb{Q}, H_0^{\mathrm{LA}}(\overline{\mathbb{A}}))$ which lift to a class $y \in H^1(\mathbb{Q}, H_0^{\mathrm{LA}})$ mapping to $\phi_0 \in H^1(\mathbb{Q}, G_0)$. Any such x certainly maps to ϕ_0 in $H^1(\mathbb{Q}, G_0(\overline{\mathbb{A}}))$. So we must show that if $x \in H^1(\mathbb{Q}, H_0^{\mathrm{LA}}(\overline{\mathbb{A}}))$ maps to ϕ_0 in $H^1(\mathbb{Q}, G_0(\overline{\mathbb{A}}))$, then we can lift x to such a y.

Let $A(G_0)$ and $A(H_0^{\mathrm{LA}})$ be the groups defined in section 2.1 of [Ko2]. Then according to proposition 2.6 of [Ko2] we have a commutative diagram with exact rows

$$
\begin{array}{ccccccc}
(0){\to}\ker^1(\mathbb{Q}, H_0^{\mathrm{LA}}) & \to & H^1(\mathbb{Q}, H_0^{\mathrm{LA}}) & \to & H^1(\mathbb{Q}, H_0^{\mathrm{LA}}(\overline{\mathbb{A}})) & \to & A(H_0^{\mathrm{LA}}) \\
\downarrow & & \downarrow & & \downarrow & & \downarrow \\
(0){\to} \ker^1(\mathbb{Q}, G_0) & \to & H^1(\mathbb{Q}, G_0) & \to & H^1(\mathbb{Q}, G_0(\overline{\mathbb{A}})) & \to & A(G_0).
\end{array}
$$

The lemma now follows from a diagram chase, because of the following two observations.

1. $\ker^1(\mathbb{Q}, H_0^{\mathrm{LA}}) \twoheadrightarrow \ker^1(\mathbb{Q}, G_0)$.

2. $A(H_0^{\mathrm{LA}}) \hookrightarrow A(G_0)$.

The first of these follows because letting Z_0 denote the centre of G_0 the composite homomorphism

$$
\ker^1(\mathbb{Q}, Z_0) \longrightarrow \ker^1(\mathbb{Q}, H_0^{\mathrm{LA}}) \longrightarrow \ker^1(\mathbb{Q}, G_0)
$$

is an isomorphism by the argument on pages 393 and 394 of [Ko3]. The second follows by direct computation from the definitions. In fact, if $[F^+ : \mathbb{Q}][B : F]^{1/2} = [M^+ : \mathbb{Q}][D_0 : M]^{1/2}$ is odd then $A(G_0) = (0)$ and $A(H_0^{\mathrm{LA}}) = (0)$. If on the other hand $[F^+ : \mathbb{Q}][B : F]^{1/2} = [M^+ : \mathbb{Q}][D_0 : M]^{1/2}$ is even then the natural homomorphism $A(H_0^{\mathrm{LA}}) \to A(G_0)$ is the unique isomorphism

$$
\mathbb{Z}/2\mathbb{Z} \xrightarrow{\sim} \mathbb{Z}/2\mathbb{Z}.
$$

□

There is also a natural bijection between the following two sets.

1. $B \otimes_F M$-equivalence classes of $* \otimes c$-polarisations $\lambda : A \to A^\vee$ such that there is an equivalence of $B \otimes \mathbb{A}^{\infty,p}$-modules with $*$-Hermitian $\mathbb{A}^{\infty,p}$-alternating pairings between $V^p A$ with the λ-Weil pairing and $V \otimes \mathbb{A}^{\infty,p}$ with our standard pairing (,).

2. Those elements of $\ker(H^1(\mathbb{Q}, H_0^{\mathrm{AV}}) \longrightarrow H^1(\mathbb{R}, H_0^{\mathrm{AV}}))$ which map to ϕ_0 in $H^1(\mathbb{Q}, G_0(\overline{\mathbb{A}}^{\infty,p}))$.

Lemma V.4.3 *There are bijections between the following sets.*

1. *Near equivalence classes of $* \otimes c$-polarisations $\lambda : A \to A^\vee$ such that there is an equivalence of $B \otimes \mathbb{A}^{\infty,p}$-modules with $*$-Hermitian $\mathbb{A}^{\infty,p}$-alternating pairing between $V^p A$ with its λ-Weil pairing and $V \otimes \mathbb{A}^{\infty,p}$ with our standard pairing $(\ , \ )$.*

2. *The preimage of ϕ_0 under*

$$H^1(\mathbb{Q}, H_0^{\mathrm{AV}}(\overline{\mathbb{A}}^{\infty,p})) \cong H^1(\mathbb{Q}, H_0^{\mathrm{LA}}(\overline{\mathbb{A}}^{\infty,p})) \longrightarrow H^1(\mathbb{Q}, G_0(\overline{\mathbb{A}}^{\infty,p})).$$

Proof: We have to show that the preimage of ϕ_0 under the homomorphism

$$H^1(\mathbb{Q}, H_0^{\mathrm{AV}}(\overline{\mathbb{A}}^{\infty,p})) \cong H^1(\mathbb{Q}, H_0^{\mathrm{LA}}(\overline{\mathbb{A}}^{\infty,p})) \longrightarrow H^1(\mathbb{Q}, G_0(\overline{\mathbb{A}}^{\infty,p}))$$

is contained in the image of $\ker(H^1(\mathbb{Q}, H_0^{\mathrm{AV}}) \to H^1(\mathbb{R}, H_0^{\mathrm{AV}}))$. Suppose that $x \in H^1(\mathbb{Q}, H_0^{\mathrm{AV}}(\overline{\mathbb{A}}^{\infty,p}))$ maps to $\phi_0 \in H^1(\mathbb{Q}, G_0(\overline{\mathbb{A}}^{\infty,p}))$. Again by proposition 2.6 of [Ko3] we have an exact sequence

$$H^1(\mathbb{Q}, H_0^{\mathrm{AV}}) \longrightarrow H^1(\mathbb{Q}, H_0^{\mathrm{AV}}(\overline{\mathbb{A}}^{\infty,p})) \oplus H^1(\mathbb{R}, H_0^{\mathrm{AV}}) \longrightarrow A(H_0^{\mathrm{AV}}),$$

and so it suffices to check that x maps to zero under the map

$$H^1(\mathbb{Q}, H_0^{\mathrm{AV}}(\overline{\mathbb{A}}^{\infty,p})) \longrightarrow A(H_0^{\mathrm{AV}}).$$

By lemma 2.8 of [Ko3] we have a commutative diagram

$$
\begin{array}{ccccc}
H^1(\mathbb{Q}, H_0^{\mathrm{AV}}(\overline{\mathbb{A}}^{\infty,p})) & \xrightarrow{\sim} & H^1(\mathbb{Q}, H_0^{\mathrm{LA}}(\overline{\mathbb{A}}^{\infty,p})) & \longrightarrow & H^1(\mathbb{Q}, G_0(\overline{\mathbb{A}}^{\infty,p})) \\
\downarrow & & \downarrow & & \downarrow \\
A(H_0^{\mathrm{AV}}) & = & A(H_0^{\mathrm{LA}}) & \hookrightarrow & A(G_0).
\end{array}
$$

(The injectivity of the map $A(H_0^{\mathrm{LA}}) \to A(G_0)$ was explained in the proof of lemma V.4.2.) Thus it suffices to show that $\phi_0 \in H^1(\mathbb{Q}, G_0(\overline{\mathbb{A}}^{\infty,p}))$ maps to zero in $A(G_0)$. But $\phi_0 \in \ker(H^1(\mathbb{Q}, G_0) \to H^1(\mathbb{R}, G_0))$ by lemma V.4.1 and $H^1(\mathbb{Q}_p, G_0) = (0)$. Thus $\phi_0 \in H^1(\mathbb{Q}, G_0(\overline{\mathbb{A}}^{\infty,p}))$ has the same image in $A(G_0)$ as $\phi_0 \in H^1(\mathbb{Q}, G_0(\overline{\mathbb{A}}))$, i.e. 0 (by proposition 2.6 of [Ko3]). $\square$

If $\tau : F \hookrightarrow \mathbb{C}$, let $\mathrm{PHT}_\tau^{(h)}$ denote the set of triples $(M, \widetilde{w}, [j])$ where

- M is a CM-field extension of F,

- $\widetilde{w}$ is a place of M above w such that

$$[M_{\widetilde{w}} : F_w] n = [M : F](n - h),$$

- there is no intermediate field $M \supset N \supset F$ such that $\widetilde{w}|_N$ is inert in M,

- $[j]$ is a $G_\tau(\mathbb{A})$-conjugacy class of F-embeddings

$$j : M \hookrightarrow B^{\mathrm{op}}$$

such that $\#_\tau \circ j = j \circ c$.

Note that there is a natural bijection between $G_\tau(\mathbb{Q})$ conjugacy classes of F-embeddings

$$j : M \hookrightarrow B^{\mathrm{op}}$$

such that $\#_\tau \circ j = j \circ c$ and equivalence classes of $B \otimes_F M$-modules W with a non-degenerate $* \otimes c$-Hermitian alternating pairing $W \times W \to \mathbb{Q}$. (The equivalence sends j to $(V, (\ , \ )_{\beta_\tau})$ considered as an M-module via $j : M \hookrightarrow B^{\mathrm{op}}$. It depends on the choice of $(\ , \ )_{\beta_\tau}$.) If j, j' correspond to W, W' then j and j' are $G_\tau(\mathbb{A})$ conjugate if and only if $W \otimes \mathbb{A}$ and $W' \otimes \mathbb{A}$ are equivalent. For such a j let $(D_j, \dagger_j)$ denote the centraliser of the image of j and the restriction of $\#_\tau$ to j. Also let H_j denote the algebraic group over $\mathbb{Q}$ such that for any $\mathbb{Q}$-algebra R we have

$$H_j(R) = \{ g \in D_j \otimes_{\mathbb{Q}} R : \ g^{\dagger_j} g \in R^\times \}.$$

There is a natural embedding $\iota_j : H_j \hookrightarrow G_\tau$. Because the composite

$$\mathrm{ker}^1(\mathbb{Q}, H_j) \longrightarrow H^1(\mathbb{Q}, H_j / Z(H_j))$$

is trivial (see pages 393 and 394 of [Ko3]) we see that the pair $(D_j, \dagger_j)$, and hence the algebraic group H_j, only depends on the $G_\tau(\mathbb{A})$-conjugacy class of j (at least up to $H_j(\mathbb{Q})$-conjugacy). Moreover the embedding $\iota_j : H_j \hookrightarrow G_\tau$ only depends on the $G_\tau(\mathbb{A})$-conjugacy class of j up to $G_\tau(\mathbb{A})$-conjugacy. Thus if $y = (M, \widetilde{w}, [j]) \in \mathrm{PHT}_\tau^{(h)}$ we will write $(D_y^{\mathrm{LA}}, \dagger_y)$ (resp. $\iota_y^{\mathrm{LA}} : H_y^{\mathrm{LA}} \hookrightarrow G_\tau$) for $(D_j, \dagger_j)$ (resp. $\iota_j : H_j \hookrightarrow G_\tau$) for any $j \in [j]$. Then $(D_y^{\mathrm{LA}}, \dagger_y)$ and H_y^{LA} are determined uniquely up to $H_y^{\mathrm{LA}}(\mathbb{Q})$-isomorphism, and ι_y^{LA} is determined up to $G_\tau(\mathbb{A})$-conjugacy. We will also write W_y for the $B \otimes_F M$-module with a non-degenerate $* \otimes c$-Hermitian alternating pairing corresponding to some $j \in [j]$. Note that W_j is not determined up to equivalence, but $W_j \otimes \mathbb{A}$ is.

Also let $\mathrm{FP}_{\tau, \mathrm{LA}}^{(h)}$ be the set of pairs $(y, [a])$ where $y = (M, \widetilde{w}, [j]) \in \mathrm{PHT}_\tau^{(h)}$ and $[a]$ is a $H_y^{\mathrm{LA}}(\mathbb{A})$ conjugacy class of elements $a \in H_y^{\mathrm{LA}}(\mathbb{Q})$ such that $F(a) \supset M$ and a is elliptic in both $H_y^{\mathrm{LA}}(\mathbb{R})$ and $D_{y, \widetilde{w}}^\times$. If $(y, [a]) \in \mathrm{FP}_{\tau, \mathrm{LA}}^{(h)}$ we define a conjugacy class $[\widetilde{\iota}(y, [a])] \subset G^{(h)}(\mathbb{A}^\infty)$ as follows. Set

- $\widetilde{\imath}(y,[a])^p = \iota_y^{\mathrm{LA}}(a)^{\infty,p}$,

- $\widetilde{\imath}(a,\widetilde{w})_{p,0} = \nu(a) \in \mathbb{Q}_p^\times$,

- $\widetilde{\imath}(a,\widetilde{w})_{w_i} = a \in (B_{w_i}^{\mathrm{op}})^\times$ for $i > 1$,

- $\widetilde{\imath}(a,\widetilde{w})_w^0 = a \in F(a)_{\widetilde{w}}^\times \hookrightarrow D_{F_w,n-h}^\times$, and

- $\widetilde{\imath}(a,\widetilde{w})_w^e = a \in \prod_x F(a)_x^\times \subset GL_h(F_w)$, where the product is over places $x \neq \widetilde{w}$ of $F(a)$ which divide w.

Proposition V.4.4 *There is a natural surjection*

$$\mathcal{P}_\tau : \mathrm{PHT}_\tau^{(h)} \twoheadrightarrow \mathrm{PIC}^{(h)}/\sim$$

such that $\mathcal{P}_\tau^{-1}[(A,\lambda,i)]$ is identified with the set of embeddings from the centre of $\mathrm{End}_B^0(A)$ into $\mathbb{C}$ extending τ (unless $F^+ = \mathbb{Q}$ and $n = 2$ when the fibres each have only one element). (We write τ_y for the embedding associated to y.) If $\mathcal{P}_\tau(M,\widetilde{w},[j]) = [(A,\lambda,i)]$ then

- *M is the centre of $\mathrm{End}_B^0(A)$,*

- *$A[w^\infty]^0 = A[\widetilde{w}^\infty]$, and*

- *$V^p A$ with its λ-Weil pairing is equivalent to $V \otimes \mathbb{A}^{\infty,p}$ with the standard pairing $(\ ,\ )$ as $B \otimes_F M \otimes \mathbb{A}^{\infty,p}$-modules with $\mathbb{A}^{\infty,p}$-alternating pairings.*

Moreover if $y \in \mathrm{PHT}_\tau^{(h)}$ then there is an isomorphism

$$j_y : H_y^{\mathrm{LA}} \times \mathbb{A}^{\infty,p} \xrightarrow{\sim} H_{\mathcal{P}_\tau} \times \mathbb{A}^{\infty,p}$$

which is canonical up to $H_{\mathcal{P}_\tau(y)}(\mathbb{A}^{\infty,p})$-conjugacy and which is compatible with ι_y^{LA} and $\iota_{\mathcal{P}_\tau}$ (up to $G(\mathbb{A}^{\infty,p})$-conjugacy). Further there is an isomorphism

$$\psi : H_y^{\mathrm{LA}} \times \mathbb{Q}^{\mathrm{ac}} \xrightarrow{\sim} H_{\mathcal{P}_\tau} \times \mathbb{Q}^{\mathrm{ac}}$$

such that

- *for all $\sigma \in \mathrm{Gal}(\mathbb{Q}^{\mathrm{ac}}/\mathbb{Q})$ the automorphism $\psi^{-1}\sigma(\psi)$ of $H_y^{\mathrm{LA}} \times \mathbb{Q}^{\mathrm{ac}}$ is inner, and*

- *the automorphism $\psi^{-1}j_y$ of $H_y^{\mathrm{LA}} \times \overline{\mathbb{A}}^{\infty,p}$ is inner.*

Proof: This follows on combining lemmas I.6.2, V.2.1, V.4.1, V.4.2 and V.4.3 with the following observations. We keep the notation of lemma V.4.1. Firstly

$$H^1(\mathbb{Q}_p, H_0^{\mathrm{LA}}) = (0).$$

Secondly the fibre of ϕ_0 under

$$H^1(\mathbb{R}, H_0^{\mathrm{LA}}) \longrightarrow H^1(\mathbb{R}, G_0)$$

is naturally identified with equivalence classes of $B \otimes_F M \otimes \mathbb{R}$ modules with non-degenerate $* \otimes c$-Hermitian $\mathbb{R}$-alternating pairings which as $B \otimes \mathbb{R}$-modules with $*$-Hermitian $\mathbb{R}$-alternating pairings are equivalent to $V \otimes \mathbb{R}$ with the pairing $(\ ,\ )_\tau$, which in turn is naturally identified with the set of embeddings $M \hookrightarrow \mathbb{C}$ extending τ (unless $F^+ = \mathbb{Q}$ and $n = 2$, in which case the fibre has one element). (See the proof of lemma V.4.6 below for a similar argument carried out in slightly more detail.) $\square$

Corollary V.4.5 $J^{(0)}(k(w)^{\mathrm{ac}})$ *is non-empty.*

Proof: It suffices to see that $\mathrm{PHT}_\tau^{(0)}$ is non-empty. However it contains $(F, w, [j])$, where j is the canonical embedding. $\square$

Lemma V.4.6 *The map $\mathcal{P}_\tau$ extends to a surjective map*

$$\mathcal{P}_\tau : \mathrm{FP}_{\tau,\mathrm{LA}}^{(h)} \twoheadrightarrow \mathrm{FP}_{\mathrm{AV}}^{(h)}$$

such that

$$\mathcal{P}_\tau(y, [a]) = (\mathcal{P}_\tau(y), [a']),$$

where a' is an element of $H_{\mathcal{P}_\tau y}(\mathbb{Q})$ which is $H_{\mathcal{P}_\tau y}(\mathbb{A}^{\infty,p})$-conjugate to $j_y \iota_y^{\mathrm{LA}}(a)$. The fibre $\mathcal{P}_\tau^{-1}(z, [a'])$ contains $[F(a) : F]$ elements (unless $F^+ = \mathbb{Q}$ and $n = 2$ when it contains a unique element). Moreover

$$[\iota_{\mathcal{P}_\tau y} \mathcal{P}_\tau(y, [a])] = [\tilde{\iota}(y, [a])] \subset G^{(h)}(\mathbb{A}^\infty).$$

Proof: Fix $y = (M, \tilde{w}, [j]) \in \mathrm{PHT}_\tau^{(h)}$ and set $z = \mathcal{P}_\tau(y)$. We will give a natural map from conjugacy classes in $H_y^{\mathrm{LA}}(\mathbb{A})$, which are elliptic in $H_y^{\mathrm{LA}}(\mathbb{R})$ and in $D_{y,\tilde{w}}^\times$, to conjugacy classes in $H_z^{\mathrm{AV}}(\mathbb{A})$ as follows.

1. We use j_y to associate conjugacy classes in $H_y^{\mathrm{LA}}(\mathbb{A}^{\infty,p})$ and $H_z(\mathbb{A}^{\infty,p})$.

2. We use the isomorphism

$$\mathbb{Q}_p^\times \times \prod_{x|u,x\neq\widetilde{w}} D_{y,x}^\times \cong \mathbb{Q}_p^\times \times \prod_{x|u,x\neq\widetilde{w}} C_{z,x}^{\prime\times},$$

which is canonical up to conjugacy, to associate conjugacy classes in these two groups.

3. We use the natural identification of elliptic conjugacy classes in

$$D_{y,\widetilde{w}}^\times \cong GL_{n/[M:F]}(M_{\widetilde{w}})$$

with conjugacy classes in $C_{z,\widetilde{w}}^\times$. ($C_{z,\widetilde{w}}$ is a division algebra with centre $M_{\widetilde{w}}$ and dimension $(n/[M:F])^2$.)

4. We use the natural map from elliptic conjugacy classes in $H_y^{\mathrm{LA}}(\mathbb{R})$ to conjugacy classes in its compact mod centre inner form, $H_z(\mathbb{R})$. This map is surjective. (Elliptic conjugacy classes transfer to any inner form and in $H_z(\mathbb{R})$ stable conjugacy and conjugacy coincide because $H_z(\mathbb{R})$ is compact mod centre.)

We claim that under this map the preimage of $[a]$ has $[(M \otimes_{\mathbb{Q}} \mathbb{R})(a_\infty) : M \otimes_{\mathbb{Q}} \mathbb{R}]$ elements. (Unless $M^+ = \mathbb{Q}$ and $n = 2$ when it has only 1 element.) If $[a']$ is one such preimage then the other preimages are just the $[(a')^\infty \times a_\infty'']$ with a_∞'' stably conjugate to a_∞' in $H_y^{\mathrm{LA}}(\mathbb{R})$. Thus the preimages are in bijection with

$$\ker(H^1(\mathbb{R}, Z_{H_y^{\mathrm{LA}}}(a_\infty')) \longrightarrow H^1(\mathbb{R}, H_y^{\mathrm{LA}})).$$

Elements of this set also parametrise equivalence classes of $B \otimes_F ((M \otimes_{\mathbb{Q}} \mathbb{R})(a_\infty'))$-modules with a $* \otimes c$-Hermitian $\mathbb{R}$-alternating pairing which are equivalent as $B \otimes_F M \otimes_{\mathbb{Q}} \mathbb{R}$-modules with $* \otimes c$-Hermitian $\mathbb{R}$-alternating pairing to $V \otimes \mathbb{R}$ with its standard pairing, $(\ ,\ )_\tau$. These in turn are parametrised by sequences of pairs of integers $(a_{\tau'}, b_{\tau'})$ for $\tau' : (M^+ \otimes_{\mathbb{Q}} \mathbb{R})(a_\infty') \to \mathbb{R}$ where $(a_{\tau'}, b_{\tau'}) = (0, n/[(M \otimes_{\mathbb{Q}} \mathbb{R})(a_\infty') : F \otimes_{\mathbb{Q}} \mathbb{R}])$ for all but one such embedding τ_1 for which $\tau_1|_M = \tau_y$ and $(a_{\tau_1}, b_{\tau_1}) = (1, n/[(M \otimes_{\mathbb{Q}} \mathbb{R})(a_\infty') : F \otimes_{\mathbb{Q}} \mathbb{R}] - 1)$. (See proposition V.4.4 for the definition of τ_y.) Thus the preimage of $[a] \subset H_z^{\mathrm{AV}}(\mathbb{A})$ does have $[(M \otimes_{\mathbb{Q}} \mathbb{R})(a_\infty) : M \otimes_{\mathbb{Q}} \mathbb{R}]$ elements, unless $M^+ = \mathbb{Q}$ and $n = 2$ when it has only 1 element.

To complete the proof of the lemma it suffices to show that, if $[a] \subset H_y^{\mathrm{LA}}(\mathbb{A})$ maps to $[a'] \subset H_z^{\mathrm{AV}}(\mathbb{A})$, then $[a]$ contains an element of $H_y^{\mathrm{LA}}(\mathbb{Q})$ if and only if $[a']$ contains an element of $H_z(\mathbb{Q})$. This follows from theorem 6.6 of [Ko2] if we can verify that the group $\mathfrak{R}(I/\mathbb{Q})$ of that paper vanishes for every centraliser I of a semi-simple element in $H(\mathbb{Q})$ where $H/\mathbb{Q}$ is a quasi-split inner form of H_z (and hence of H_y^{LA}). (We remark that any element of $H_y^{\mathrm{LA}}(\mathbb{Q})$ or $H_z(\mathbb{Q})$ is conjugate over $\mathbb{Q}^{\mathrm{ac}}$ to an element of $H(\mathbb{Q})$, see for instance theorem 2.4.2 of [Lab].) This can be verified as in lemma 2 of [Ko4]. Alternatively one can note that it fits in an exact sequence

$$\ker^1(\mathbb{Q}, I) \to \ker^1(\mathbb{Q}, H) \to \mathrm{Hom}\,(\mathfrak{R}(I/\mathbb{Q}), \mathbb{Q}/\mathbb{Z}) \to A(I) \to A(H) \to (0)$$

(combine section 4.6 of [Ko2] with the definition of $A(H)$ and $A(I)$ and the isomorphism 4.2.2 of [Ko1]). If $[B : \mathbb{Q}]/2$ is odd then a direct calculation shows that $A(H) = A(I) = (0)$, while if $[B : \mathbb{Q}]/2$ is even then the morphism $A(I) \to A(H)$ is the unique isomorphism $\mathbb{Z}/2\mathbb{Z} \overset{\sim}{\to} \mathbb{Z}/2\mathbb{Z}$. On the other hand the map $\ker^1(\mathbb{Q}, I) \longrightarrow \ker^1(\mathbb{Q}, H)$ is surjective because (arguing as on pages 393 and 394 of [Ko3] we see that) the composite

$$\ker^1(\mathbb{Q}, Z(H)) \longrightarrow \ker^1(\mathbb{Q}, Z(I)) \longrightarrow \ker^1(\mathbb{Q}, I) \longrightarrow \ker^1(\mathbb{Q}, H)$$

(where $Z(H)$ (resp. $Z(I)$) denotes the centre of H (resp. I)) is an isomorphism. Thus $\mathfrak{R}(I/\mathbb{Q}) = (0)$. $\square$

We are now in a position to rewrite the trace formula of lemma V.3.3 in purely group theoretic terms, i.e. with no reference to abelian varieties. However we will get a slightly simpler formula if we first give a second description for $\mathrm{FP}_{\tau,\mathrm{LA}}^{(h)}$.

To this end let $\mathrm{FP}_\tau^{(h)}$ denote the set of equivalence classes of pairs $(a, \widetilde{w})$ where $a \in G_\tau(\mathbb{Q})$ is an element which is elliptic in $G_\tau(\mathbb{R})$ and where $\widetilde{w}$ is a place of the field $F(a)$ above w such that

$$(n - h)[F(a) : F] = n[F(a)_{\widetilde{w}} : F_w].$$

We consider two pairs $(a, \widetilde{w})$ and $(a', \widetilde{w}')$ equivalent if a and a' are conjugate by an element of $G(\mathbb{A})$ which induces an isomorphism $F(a)_w \overset{\sim}{\to} F(a')_w$ taking $\widetilde{w}$ to $\widetilde{w}'$. If $[(a, \widetilde{w})] \in \mathrm{FP}_\tau^{(h)}$ then we define a conjugacy class $[\iota(a, \widetilde{w})] \subset G^{(h)}(\mathbb{A}^\infty)$ as follows. We set

- $\iota(a, \widetilde{w})^p = a \in G(\mathbb{A}^{\infty, p})$,

- $\iota(a, \widetilde{w})_{p,0} = \nu(a) \in \mathbb{Q}_p^\times$,

- $\iota(a, \widetilde{w})_{w_i} = a \in (B_{w_i}^{\mathrm{op}})^\times$ for any $i > 1$,

- $\iota(a, \widetilde{w})_w^0 = a \in F(a)_{\widetilde{w}}^\times \hookrightarrow D_{F_w, n-h}^\times$, and

- $\iota(a, \widetilde{w})_w^e = a \in \prod_x F(a)_x^\times \subset GL_h(F_w)$, where the product is over places $x \neq \widetilde{w}$ of $F(a)$ which divide w.

Lemma V.4.7 *There is a bijection*

$$\phi : \mathrm{FP}_{\tau,\mathrm{LA}}^{(h)} \overset{\sim}{\longrightarrow} \mathrm{FP}_\tau^{(h)}$$

which sends $((M, \widetilde{w}, [j]), [a])$ *to* $[(\iota_z^{\mathrm{LA}}(a), \widetilde{w}')]$ *where* $\widetilde{w}'$ *is the unique place of* $M(a)$ *above the place* $\widetilde{w}$ *of* M. *Thus*

$$[\iota\phi(y, [a])] = [\widetilde{\iota}(y, [a])].$$

Proof: Suppose $(a, \widetilde{w})$ represents an element of $\mathrm{FP}_\tau^{(h)}$. Let M denote the minimal subfield of $F(a)$ which contains F and for which $\widetilde{w}|_M$ is inert in $F(a)$. (It is an exercise in the splitting of primes in number fields to show such a unique minimal subfield exists.) Let $j : M \hookrightarrow B^{\mathrm{op}}$ be the tautological embedding. Then $((M, \widetilde{w}|_M, [j]), [a])$ is a point of $\mathrm{FP}_{\tau, \mathrm{LA}}^{(h)}$ mapping to the class of $(a, \widetilde{w})$ and it is unique. The lemma follows. $\square$

Putting these results together with lemma V.3.3 we get the following formula for the trace of φ on $H_c(I^{(h)}, \mathcal{F}_\rho \otimes \mathcal{L}_\xi)$.

Proposition V.4.8 *Choose an embedding* $\tau : F \hookrightarrow \mathbb{C}$. *Suppose that* $\varphi \in C_c^\infty(G^{(h)}(\mathbb{A}^\infty)^+/\mathbb{Z}_p^\times \times \mathcal{O}_{D_{F_w, n-h}}^\times)$ *is acceptable. Then*

$$\mathrm{tr}\,(\varphi | H_c(I^{(h)}, \mathcal{F}_\rho \otimes \mathcal{L}_\xi)) = (-1)^h \kappa_B \sum_{y=[(a, \widetilde{w})] \in \mathrm{FP}_\tau^{(h)}}$$
$$[F(a) : F]^{-1} \mathrm{vol}\,(Z_{G_\tau}(a)(\mathbb{R})_0^1)^{-1} O_{\iota(y)}^{G^{(h)}(\mathbb{A}^\infty)}(\varphi) \mathrm{tr}\,(\rho \otimes \xi)(\iota(y)),$$

where $\kappa_B = 2$ *if* $[B : \mathbb{Q}]/2$ *is even and* $\kappa_B = 1$ *otherwise. (If* $F^+ = \mathbb{Q}$ *and* $n = 2$ *we must drop the* $[F(a) : F]^{-1}$ *term.)*

This sum is finite and all the terms occurring are finite numbers.

We choose measures on $Z_{G^{(h)}(\mathbb{A}^\infty)}(\iota(y))$ *and on* $Z_{G_\tau}(a)(\mathbb{R})_0^1$ *compatible with*

- *Tamagawa measure on* $Z_{G_\tau}(a)(\mathbb{A})^1$,

- *the exact sequence*

$$\{1\} \longrightarrow Z_{G_\tau}(a)(\mathbb{R})^1 \longrightarrow Z_{G_\tau}(a)(\mathbb{A})^1 \longrightarrow Z_{G_\tau}(a)(\mathbb{A}^\infty) \longrightarrow \{1\},$$

- *the association of measures on* $Z_{G_\tau}(a)(\mathbb{R})_0^1$ *and* $Z_G(a)(\mathbb{R})^1$ *(see page 631 of [Ko5]),*

- *the association of measures on*

$$Z_{(B_w^{\mathrm{op}})^\times}(a) \cong \prod_{x | w} GL_{n/[F(a):F]}(F(a)_x)$$

and on

$$Z_{D_{F_w, n-h}^\times \times GL_h(F_w)}(\iota(y)) \cong D_{F(a)_{\widetilde{w}}, (n-h)/[F(a)_{\widetilde{w}}:F_w]}^\times \times$$
$$\prod_{x | w, x \neq \widetilde{w}} GL_{h/([F(a):F] - [F(a)_{\widetilde{w}}:F_w])}(F(a)_x)$$

(see page 631 of [Ko5]),

- *and the isomorphism*

$$Z_{G^{(h)}(\mathbb{A}^\infty)}(\iota(y))$$

$$\cong Z_G(a)(\mathbb{A}^{\infty,p}) \times \mathbb{Q}_p^\times \times Z_{D_{F_w,n-h}^\times \times GL_h(F_w)}(\iota(y)) \times \prod_{i=2}^r Z_{(B_{w_i}^{\mathrm{op}})^\times}(a).$$

V.5   Some local harmonic analysis

In this section we will combine proposition V.4.8 with some local harmonic analysis to compute $H_c(I^{(h)}, \mathcal{F}_\rho \otimes \mathcal{L}_\xi)$ in terms of $H(X, \mathcal{L}_\xi)$. We will use results and notation from section I.3 without comment.

For $h = 0, \ldots, n-1$ let $P_h \subset GL_n$ denote the parabolic subgroup consisting of block upper triangular matrices with an $(n-h) \times (n-h)$-block in the top left hand corner and an $(h \times h)$-block in the bottom right hand corner. We will let P_h^{op} denote the opposite parabolic (lower triangular matrices with the same block structure) and L_h denote the Levi component $P_h \cap P_h^{\mathrm{op}} (\cong GL_{n-h} \times GL_h)$.

Fix an irreducible admissible representation ρ of $D_{F_w,n-h}^\times$. We will define a homomorphism

$$\mathrm{Red}_\rho^{(h)} : \mathrm{Groth}\,(GL_n(F_w)) \longrightarrow \mathrm{Groth}\,(D_{F_w,n-h}^\times/\mathcal{O}_{D_{F_w,n-h}}^\times \times GL_h(F_w))$$

as follows.

First we have a homomorphism

$$\mathrm{Groth}\,(GL_n(F_w)) \longrightarrow \mathrm{Groth}\,(GL_{n-h}(F_w) \times GL_h(F_w))$$

which sends $[\pi]$ to $[J_{N_h^{\mathrm{op}}}(\pi) \otimes \delta_{P_h}^{1/2}]$ (see section I.2). Secondly we have a homomorphism

$$\mathrm{Groth}\,(GL_{n-h}(F_w) \times GL_h(F_w))$$
$$\to \mathrm{Groth}\,(D_{F_w,n-h}^\times/\mathcal{O}_{D_{F_w,n-h}}^\times \times GL_h(F_w))$$

which sends $[\alpha \otimes \beta]$ to

$$\sum_\psi \mathrm{vol}\,(D_{F_w,n-h}^\times/F_w^\times)^{-1} \mathrm{tr}\,\alpha(\varphi_{\mathrm{JL}\,(\rho^\vee \otimes \psi)})[\psi \otimes \beta],$$

where ψ runs over characters of $D_{F_w,n-h}^\times/\mathcal{O}_{D_{F_w,n-h}}^\times$ so that α and $\rho^\vee \otimes \psi$ have the same central character, and where we use associated measures on $GL_{n-h}(F_w)$ and $D_{F_w,n-h}^\times$. (That this second map is well defined follows from corollary I.3.5.) $\mathrm{Red}_\rho^{(h)}$ will denote the composite.

The homomorphism $\mathrm{Red}_\rho^{(h)}$ extends naturally to a homomorphism

$$\mathrm{Red}_\rho^{(h)} : \mathrm{Groth}\,(G(\mathbb{A}^\infty)) \longrightarrow \mathrm{Groth}\,(G^{(h)}(\mathbb{A}^\infty)).$$

We will need a couple of lemmas in local harmonic analysis. It is convenient to fix

- the Haar measure on $F_w^\times$ giving $\mathcal{O}_{F,w}^\times$ volume 1,

- the Haar measure on $D_{F_w,n-h}^\times$ giving $\mathcal{O}_{D_{F_W},n-h}^\times$ volume 1,

- and the Haar measure on $GL_{n-h}(F_w)$ associated to our choice of Haar measure on $D_{F_w,n-h}^\times$.

Lemma V.5.1 *Suppose that $\varphi^0 \in C_c^\infty(D_{F_w,n-h}^\times/\mathcal{O}_{F_w,n-h}^\times)$. Then we can find an element $\mathrm{PC}_\rho(\varphi^0) \in C_c^\infty(GL_{n-h}(F_w))$ with the following properties.*

1. *If π is an irreducible admissible representation of $GL_{n-h}(F_w)$ then*

$$\mathrm{tr}\,\pi(\mathrm{PC}_\rho(\varphi^0)) = \sum_\psi \mathrm{tr}\,\psi(\varphi^0)\mathrm{vol}\,(D_{F_w,n-h}^\times/F_w^\times)^{-1}\mathrm{tr}\,\pi(\varphi_{\mathrm{JL}\,(\rho^\vee \otimes \psi)})$$

   where ψ runs over characters of $D_{F_w,n-h}^\times/\mathcal{O}_{D_{F_w,n-h}}^\times$ so that π and $\rho^\vee \otimes \psi$ have the same central character.

2. *If $g \in GL_{n-h}(F_w)$ is a non-elliptic semi-simple element then*

$$O_g^{GL_{n-h}(F_w)}(\mathrm{PC}_\rho(\varphi^0)) = 0.$$

3. *If $g \in GL_{n-h}(F_w)$ is an elliptic semi-simple element and if $\tilde{g} \in D_{F_w,n-h}^\times$ is an element with the same characteristic polynomial then*

$$O_g^{GL_{n-h}(F_w)}(\mathrm{PC}_\rho(\varphi^0)) =$$
$$(-1)^{(n-h)(1-[F_w(g):F_w]^{-1})}O_{\tilde{g}}^{D_{F_w,n-h}^\times}(\varphi^0)\mathrm{tr}\,\rho(\tilde{g}).$$

4. *If $g \in GL_{n-h}(F_w)$ is in the support of $\mathrm{PC}_\rho(\varphi^0)$ and if λ is an eigenvalue of g then $(n-h)w(\lambda)$ is in the image under $w \circ \det$ of the support of φ^0.*

Proof: Via $w \circ \det : D_{F_w,n-h}^\times/\mathcal{O}_{D_{F_w,n-h}}^\times \xrightarrow{\sim} \mathbb{Z}$ we may and we shall think of $\varphi^0 \in C_c^\infty(\mathbb{Z})$. We will let A^0 denote the set of elements of $GL_{n-h}(F_w)$ all whose eigenvalues have w-valuation in the image under $(n-h)^{-1}w \circ \det$ of the support of φ^0. We will let $A_m \subset GL_{n-h}(F_w)$ denote $(w \circ \det)^{-1}(\{m\})$. We then set

$$\mathrm{PC}_\rho(\varphi^0) = (\varphi^0 \circ w \circ \det)\varphi_{\mathrm{JL}\,(\rho^\vee)}\mathrm{char}\,_{A^0},$$

and

$$\widetilde{\varphi} = (\varphi^0 \circ w \circ \det)\varphi_{\mathrm{JL}\,(\rho^\vee)}.$$

Note that by definition $\mathrm{PC}_\rho(\varphi^0)$ has property 4 of the lemma.

It follows from lemma I.3.1 that for a non-elliptic regular semi-simple element $g \in GL_{n-h}(F_w)$ we have

$$O_g^{GL_{n-h}(F_w)}(\widetilde{\varphi}) = 0,$$

while if $g \in GL_{n-h}(F_w)$ is a elliptic regular semi-simple element and if $\widetilde{g} \in D_{F_w,n-h}^\times$ is an element with the same characteristic polynomial then

$$O_g^{GL_{n-h}(F_w)}(\widetilde{\varphi}) = (-1)^{n-h-1}\mathrm{vol}\,(D_{F_w,n-h}^\times/Z_{D_{F_w,n-h}^\times}(\widetilde{g}))\varphi^0(\widetilde{g})\mathrm{tr}\,\rho(\widetilde{g}).$$

Parts 2 and 3 of the lemma follow (because the valuation of all eigenvalues of an elliptic element of $GL_n(F_w)$ are equal).

It also follows that at all regular semi-simple elements of $GL_{n-h}(F_w)$ the functions $\widetilde{\varphi}$ and $\mathrm{PC}_\rho(\varphi^0)$ have the same orbital integrals. Hence by theorem 2f of appendix 1 of [DKV] we see that for any admissible representation π of $GL_{n-h}(F_w)$ we have

$$\mathrm{tr}\,\pi(\widetilde{\varphi}) = \mathrm{tr}\,\pi(\mathrm{PC}_\rho(\varphi^0)).$$

Thus it suffices to prove the first part of the lemma with $\widetilde{\varphi}$ replacing $\mathrm{PC}_\rho(\varphi^0)$.

Let π be an irreducible admissible representation of $GL_{n-h}(F_w)$ with central character ψ_π. Also let ψ_ρ denote the central character of ρ. If $\psi_\pi|_{\mathcal{O}_{F_w}^\times} \neq \psi_\rho|_{\mathcal{O}_{F_w}^\times}^{-1}$ then $\mathrm{tr}\,\pi(\widetilde{\varphi}) = 0$ as desired. Thus suppose that $\psi_\pi|_{\mathcal{O}_{F_w}^\times} = \psi_\rho|_{\mathcal{O}_{F_w}^\times}$. Then

$$
\begin{aligned}
&\mathrm{tr}\,\pi(\widetilde{\varphi})\\
=\ & \textstyle\sum_{x\in\mathbb{Z}}\varphi^0(x)\mathrm{tr}\,\pi(\varphi_{\mathrm{JL}\,(\rho^\vee)}\mathrm{char}\,_{A_x})\\
=\ & \textstyle\sum_{i=0}^{n-h-1}\sum_{x\equiv i\bmod n-h}\varphi^0(x)(\psi_\pi\psi_\rho)((x-i)/(n-h))\\
& \hspace{6cm} \mathrm{tr}\,\pi(\varphi_{\mathrm{JL}\,(\rho^\vee)}\mathrm{char}\,_{A_i})\\
=\ & \textstyle\sum_{i=0}^{n-h-1}\sum_{x\in\mathbb{Z}}(n-h)^{-1}\sum_{\psi^{n-h}=\psi_\pi\psi_\rho}\varphi^0(x)\psi(x-i)\\
& \hspace{6cm} \mathrm{tr}\,\pi(\varphi_{\mathrm{JL}\,(\rho^\vee)}\mathrm{char}\,_{A_i})\\
=\ & (n-h)^{-1}\textstyle\sum_{\psi^{n-h}=\psi_\pi\psi_\rho}\sum_{x\in\mathbb{Z}}\varphi^0(x)\psi(x)\sum_{i=0}^{n-h-1}\\
& \hspace{6cm} \mathrm{tr}\,\pi(\varphi_{\mathrm{JL}\,(\rho^\vee)\otimes\psi}\mathrm{char}\,_{A_i})\\
=\ & \textstyle\sum_{\psi^{n-h}=\psi_\pi\psi_\rho}\mathrm{tr}\,\psi(\varphi^0)(n-h)^{-1}\mathrm{vol}\,(\mathcal{O}_{D_{F_w,n-h}}^\times)^{-1}\mathrm{tr}\,\pi(\varphi_{\mathrm{JL}\,(\rho^\vee)\otimes\psi})\\
=\ & \textstyle\sum_{\psi^{n-h}=\psi_\pi\psi_\rho}\mathrm{tr}\,\psi(\varphi^0)\mathrm{vol}\,(D_{F_w,n-h}^\times/F_w^\times)^{-1}\mathrm{tr}\,\pi(\varphi_{\mathrm{JL}\,(\rho^\vee)\otimes\psi}).
\end{aligned}
$$

The lemma follows. $\square$

Lemma V.5.2 *Let ρ be an irreducible admissible representation of the group $D^{\times}_{F_w,n-h}$. Suppose that $\varphi^0 \in C^{\infty}_c(D^{\times}_{F_w,n-h}/\mathcal{O}^{\times}_{D_{F_w},n-h})$ and that $\varphi^e \in C^{\infty}_c(GL_h(F_w))$. Suppose moreover that if g^0 (resp. g^e) is in the support of φ^0 (resp. φ^e) then the p-adic valuation of every eigenvalue of g^0 is strictly less than the p-adic valuation of every eigenvalue of g^e. Also fix Haar measures μ_n and μ_h on $GL_n(F_w)$ and $GL_h(F_w)$. Then we can find a function $\mathrm{IPC}_\rho(\varphi^0, \varphi^e; \mu_n, \mu_h) \in C^{\infty}_c(GL_n(F_w))$ with the following properties.*

1. *Let $g \in GL_n(F_w)$ be a semi-simple element. Then*

$$O^{GL_n(F_w)}_g(\mathrm{IPC}_\rho(\varphi^0, \varphi^e; \mu_n, \mu_h)) =$$
$$(-1)^{n-h-1} \sum_{(g^0, g^e)} O^{GL_h(F_w)}_{g^e}(\varphi^e) O^{D^{\times}_{F_w},n-h}_{g^0}(\varphi^0) \mathrm{tr}\, \rho(g^0),$$

   where (g^0, g^e) runs over a set of representatives for conjugacy classes in $D^{\times}_{F_w,n-h} \times GL_h(F_w)$ such that, if $\widetilde{g}^0$ denotes a semi-simple element of $GL_{n-h}(F_w)$ with the same characteristic polynomial as g^0, then $(\widetilde{g}^0, g^e)$ is conjugate to g in $GL_n(F_w)$. Whenever

$$O^{D^{\times}_{F_w},n-h}_{g^0}(\varphi^0) O^{GL_h(F_w)}_{g^e}(\varphi^e) \neq 0$$

   we see that $Z_{GL_{n-h}(F_w) \times GL_h(F_w)}(\widetilde{g}^0 \times g^e) = Z_{GL_n(F_w)}(\widetilde{g}^0, g^e)$ and we take corresponding measures on these two groups in the two sides of this equality of orbital integrals.

2. *If π is an irreducible admissible representation of $GL_n(F_w)$ and if*

$$[J_{N^{\mathrm{op}}_h}(\pi) \otimes \delta^{1/2}_{P_h}] = \sum_{\alpha,\beta} m_{\alpha,\beta}[\alpha \otimes \beta]$$

   with α (resp. β) running over irreducible admissible representations of $GL_{n-h}(F_w)$ (resp. $GL_h(F_w)$) and with $m_{\alpha,\beta} \in \mathbb{Z}$ (and almost all 0), then

$$\mathrm{tr}\, \pi(\mathrm{IPC}_\rho(\varphi^0, \varphi^e; \mu_n, \mu_h)) =$$
$$\sum_{\alpha,\beta,\psi} \mathrm{tr}\, \psi(\varphi^0) m_{\alpha,\beta} \mathrm{vol}\,(D^{\times}_{F_w,n-h}/F^{\times}_w)^{-1} \mathrm{tr}\, \alpha(\varphi_{\mathrm{JL}\,(\rho^{\vee} \otimes \psi)}) \mathrm{tr}\, \beta(\varphi^e)$$

   with α, β, ψ running over irreducible admissible representations of respectively $GL_{n-h}(F_w)$, $GL_h(F_w)$ and $D^{\times}_{F_w,n-h}/\mathcal{O}^{\times}_{D_{F_w},n-h}$ for which α and $\rho^{\vee} \otimes \psi$ have the same central character.

Proof: The choice of measure μ_h on $GL_h(F_w)$ and our fixed choice of Haar measure on $GL_{n-h}(F_w)$ determine a Haar measure on $L_h(F_w)$.

Let $\mathfrak{E}^0$ (resp. $\mathfrak{E}^e$) denote the set of p-adic valuations of elements in the support of φ^0 (resp. φ^e). By our assumptions $\mathfrak{E}^0$ and $\mathfrak{E}^e$ are disjoint finite sets. Let $\mathfrak{S}$ denote the set of elements $g^0 \times g^e \in L_h(F_w)$ such that the p-adic valuations all the eigenvalues of g^0 lie in $\mathfrak{C}^0$ and the p-adic valuations of all the eigenvalues of g^e lie in $\mathfrak{C}^e$. If $g \in \mathfrak{S}$ then $Z_{GL_n(F_w)}(g) = Z_{L_h(F_w)}(g)$. By lemma V.5.1 $\mathfrak{S}$ contains the support of $\mathrm{PC}_\rho(\varphi^0) \times \varphi^e$.

Define a function on $GL_n(F_w)$ by

$$W(g) = \sum_{g'} O_{g'}^{L_h(F_w)}(\mathrm{PC}_\rho(\varphi^0) \times \varphi^e),$$

where the sum is over sets of representatives of $L_h(F_w)$-conjugacy classes contained in the $GL_n(F_w)$ conjugacy class of g. We will show that this function satisfies the hypotheses of theorem B of section 1.n. of [V1]. It will then follow from that theorem that there is a function

$$\mathrm{IPC}_\rho(\varphi^0, \varphi^e; \mu_n, \mu_h) \in C_c^\infty(GL_n(F_w))$$

such that for all $g \in GL_n(F_w)$ we have

$$O_g^{GL_n(F_w)}(\mathrm{IPC}_\rho(\varphi^0, \varphi^e; \mu_n, \mu_h)) = \sum_{g'} O_{g'}^{L_h(F_w)}(\mathrm{PC}_\rho(\varphi^0) \times \varphi^e),$$

where the sum is over sets of representatives of $L_h(F_w)$-conjugacy classes contained in the $GL_n(F_w)$ conjugacy class of g, where we use our fixed measures on $GL_n(F_w)$ and $L_h(F_w)$, and where whenever $O_{g'}^{L_h(F_w)}(\mathrm{PC}_\rho(\varphi^0) \times \varphi^e) \neq 0$ we use conjugate measures on the conjugate groups $Z_{GL_n}(g)(F_w)$ and $Z_{L_h}(g')(F_w)$. The first part of the lemma will then follow from this and lemma V.5.1.

First note that W is clearly invariant under conjugation.

Now suppose that (T, u) is a standard pair (for $GL_n(F_w)$) in the sense of section 1.m of [V1]. If $t \in T$ we will let $\mathfrak{E}(t)$ denote the multiset of p-adic valuations of eigenvalues of t. We will also let $\mathfrak{E}^0(t)$ (resp. $\mathfrak{E}^e(t)$) denote the submultiset of $\mathfrak{E}(t)$ consisting of elements which also lie in $\mathfrak{C}^0$ (resp. $\mathfrak{C}^e$). Note that tu is conjugate to an element of $\mathfrak{S}$ if and only if t is conjugate to an element of $\mathfrak{S}$ if and only if $\#\mathfrak{E}^0 = n - h$ and $\#\mathfrak{E}^e = h$. In particular the support of W in Tu is contained in the set tu such that the p-adic valuation of all the eigenvalues of t lie in the finite set $\mathfrak{C}^0 \cup \mathfrak{C}^e$. Hence W has compact support in Tu.

We will let $\mathfrak{S}(T)$ denote the set of elements $t \in T$ which are conjugate to an element of $\mathfrak{S}$. If $s \in \mathfrak{S}(T)$ and $gsg^{-1} \in \mathfrak{S}$ then

$$gTug^{-1} \subset Z_{GL_n(F_w)}(gtg^{-1}) \subset L_h(F_w).$$

If $t \in T \cap g^{-1}\mathfrak{S}g$ then

$$W(tu) = O_{gtug^{-1}}^{L_h(F_w)}(\mathrm{PC}_\rho(\varphi^0) \times \varphi^e).$$

In particular if s is a regular element of T (in the sense of section 1.1. of [V1]) then we see that W is constant in some neighbourhood of s (by, for instance, theorem A of section 1.n. of [V1]).

Suppose $s \in T$. Let $u_1 = u, u_2, \ldots, u_m$ be unipotent elements of the centraliser $Z_{GL_n(F_w)}(s)$ such that

- for all i the $GL_n(F_w)$-orbit of su_i is open in the union of the $GL_n(F_w)$-orbits of su_j for $j \leq i$;

- and for any unipotent element $v \in Z_{GL_n(F_w)}(s)$, su is in the closure of the $GL_n(F_w)$-orbit of sv if and only if sv is $GL_n(F_w)$ conjugate to some su_j.

(See section 1.j. of [V1].) Fix a Haar measure on $Z_{GL_n}(s)(F_w)$. This determines a canonical Haar measure on each $Z_{GL_n}(su_j)(F_w)$ (see section 1.d. of [V1]). Then we can choose functions $f_1^{GL_n(F_w)}, \ldots, f_m^{GL_n(F_w)} \in C_c^\infty(GL_n(F_w))$ such that

- for $j < i$ the support of $f_i^{GL_n(F_w)}$ does not meet the $GL_n(F_w)$-orbit of su_j,

- if $j \neq i$ then $O_{su_j}^{GL_n(F_w)}(f_i^{GL_n(F_w)}) = 0$,

- and $O_{su_i}^{GL_n(F_w)}(f_i^{GL_n(F_w)}) = 1$.

(See section 1.k. of [V1].) To verify the hypotheses of theorem B of section 1.n. of [V1] it remains to show that there is a neighbourhood V of s in T such that for all $t \in V$ which are regular in T (in the sense of section 1.1. of [V1]) we have the equality

$$W(tu) = \sum_i W(su_i) O_{tu}^{GL_n(F_w)}(f_i^{GL_n(F_w)}).$$

(This equality is independent of the choice of Haar measure on the group $Z_{GL_n}(tu)(F_w)$ as long as in the case $W(tu) \neq 0$ we choose the conjugate Haar measure on $Z_{L_h}(atua^{-1})(F_w) = aZ_{GL_n}(tu)(F_w)a^{-1}$, where $atua^{-1} \in \mathfrak{G}$.)

First suppose that $s \notin \mathfrak{G}(T)$. Then each $W(su_i) = 0$. On the other hand we can find a neighbourhood V of s in T such that if $t \in V$ then $\mathfrak{E}(t) = \mathfrak{E}(s)$. Then $V \cap \mathfrak{G}(T) = \emptyset$ and so for any $t \in V$ we have $W(tu) = 0$ and the desired equality holds.

Now suppose that $s \in \mathfrak{G}(T)$ and that $gsg^{-1} \in \mathfrak{G}$. Replacing T by gTg^{-1}, s by gsg^{-1} and u by gug^{-1} we may suppose that $T \subset L_h$, $s \in \mathfrak{G}$ and $u \in L_h(F_w)$. Because $s \in \mathfrak{G}$ we see

- that for each i, $u_i \in L_h(F_w)$;

- that for all i the $L_h(F_w)$-orbit of su_i is open in the union of the $L_h(F_w)$-orbits of the su_j for $j \leq i$;

- and that for any unipotent element $v \in Z_{GL_n(F_w)}(s)$, su is in the closure of the $L_h(F_w)$-orbit of sv if and only if sv is $L_h(F_w)$-conjugate to some su_j.

It follows from lemma 2.5. of [V1] that we can find a neighbourhood V of s in T such that

- $V \subset \mathfrak{S}$

- and for any compact set $A \subset GL_n(F_w)$ one can find a compact set $C \subset L_h(F_w) \backslash GL_n(F_w)$ such that, if $L_h(F_w)g \notin C$ then for each $i = 1, \ldots, m$

$$g^{-1}Vu_ig \cap A = \emptyset.$$

As in section 2.5. of [V1] it follows that we can find $h \in C_c^\infty(GL_n(F_w))$ such that

$$\int_{L_h(F_w)} h(xg)dx$$

is 1 if $L_h(F_w)g \in C$ and 0 otherwise. If we set

$$f_i^{L_h(F_w)}(g) = \int_{GL_n(F_w)} h(x)f_i^{GL_n(F_w)}(x^{-1}gx)dx$$

then we see that

- $f_i^{L_h(F_w)} \in C_c^\infty(L_h(F_w))$,

- for $j < i$ the support of $f_i^{L_h(F_w)}$ does not meet the $L_h(F_w)$-orbit of su_j,

- for $t \in V$ and for any $i, j = 1, \ldots, m$ we have

$$O_{tu_j}^{GL_n(F_w)}(f_i^{GL_n(F_w)}) = O_{tu_j}^{L_h(F_w)}(f_i^{L_h(F_w)})$$

(argue as page 954 of [V1]),

- if $j \neq i$ then $O_{su_j}^{L_h(F_w)}(f_i^{L_h(F_w)}) = 0$,

- and $O_{su_i}^{L_h(F_w)}(f_i^{L_h(F_w)}) = 1$.

Using theorem A of section 1.n. of [V1] we see that for $t \in V$ which is also regular as an element of T we have

$$
\begin{aligned}
W(tu) &= \sum_i O_{su_i}^{L_h(F_w)}(\mathrm{PC}_\rho(\varphi^0) \times \varphi^e) O_{tu}^{L_h(F_w)}(f_i^{L_h(F_w)}) \\
&= \sum_i W(su_i) O_{tu}^{GL_n(F_w)}(f_i^{GL_n(F_w)}).
\end{aligned}
$$

This completes the proof of the first part of the lemma.

For the second part, the Weyl integration formula tells us that

$$
\begin{aligned}
&\operatorname{tr} \pi(\mathrm{IPC}_\rho(\varphi^0, \varphi^e; \mu_n, \mu_h)) = \sum_T (\#W_G(T))^{-1} \\
&\int_{T^{\mathrm{reg}}} D_G(t) O_t^{GL_n(F_w)}(\mathrm{IPC}_\rho(\varphi^0, \varphi^e; \mu_n, \mu_h)) \chi_\pi(t) dt,
\end{aligned}
$$

where

- T runs over $GL_n(F_w)$-conjugacy classes of maximal tori in $GL_n(F_w)$,

- $W_G(T)$ denotes the normaliser of T in $GL_n(F_w)$ modulo T,

- T^{reg} denotes the subset of regular elements of T,

- $D_G(t) = |\det((\operatorname{ad}(t) - 1)|_{\operatorname{Lie} G / \operatorname{Lie} T})|$, and

- χ_π denotes the character of π.

(See for instance section A.3.f of [DKV] and note that χ_π is locally integrable.) By the first part of the lemma this can be rewritten

$$
\begin{aligned}
&\operatorname{tr} \pi(\mathrm{IPC}_\rho(\varphi^0, \varphi^e; \mu_n, \mu_h)) = \sum_T (\#W_{L_h(F_w)}(T))^{-1} \\
&\int_{T^{\mathrm{reg}}} D_G(t^0 \times t^e) O_{t^0}^{GL_{n-h}(F_w)}(\mathrm{PC}_\rho(\varphi^0)) O_{t^e}^{GL_h(F_w)}(\varphi^e) \chi_\pi(t) dt^0 dt^e,
\end{aligned}
$$

where now

- T runs over $L_h(F_w)$-conjugacy classes of maximal tori in $L_h(F_w)$,

- and $W_{L_h(F_w)}(T)$ denotes the normaliser of T in $L_h(F_w)$ modulo T.

Let $P_{t^0 \times t^e}$ denote the parabolic associated to $t^0 \times t^e$ as in section 2 of [Cas]. If $P_{t^0 \times t^e}$ is not a subset of P_h^{op} then by the assumption on the supports of φ^0 and φ^e and by lemma V.5.1 we see that

$$
O_{t^0}^{GL_{n-h}(F_w)}(\mathrm{PC}_\rho(\varphi^0)) O_{t^e}^{GL_h(F_w)}(\varphi^e) = 0.
$$

If on the other hand $P_{t^0 \times t^e} \subset P_h^{\mathrm{op}}$ then

- $D_G(t^0 \times t^e) = D_{L_h}(t^0 \times t^e) \delta_{P_h}(t^0 \times t^e)$ (from the definitions)

- and $\chi_\pi(t^0 \times t^e) = \chi_{\pi_{N_h^{\mathrm{op}}}}(t^0 \times t^e)$ (by theorem 5.2 of [Cas]).

Thus we obtain

$$\operatorname{tr} \pi(\operatorname{IPC}_\rho(\varphi^0, \varphi^e; \mu_n, \mu_h))$$

$$= \sum_T (\#W_{L_h(F_w)}(T))^{-1} \int_{T^{\mathrm{reg}}} D_{L_h(F_w)}(t^0 \times t^e) O_{t^0}^{GL_{n-h}(F_w)}(\operatorname{PC}_\rho(\varphi^0))$$
$$O_{t^e}^{GL_h(F_w)}(\varphi^e) \chi_{\pi_{N_h^{\mathrm{op}}}}(t) \delta_{P_h}(t) dt$$

$$= \operatorname{tr}(J_{N_h^{\mathrm{op}}}(\pi) \otimes \delta_{P_h}^{1/2})(\operatorname{PC}_\rho(\varphi^0) \times \varphi^e),$$

and the second part of the lemma follows. $\square$

Corollary V.5.3 *If π is an admissible representation of $GL_n(F_w)$ and if φ^0, φ^e, μ_n and μ_h are as in the lemma then*

$$\operatorname{tr} \operatorname{Red}_\rho^{(h)}(\pi)(\varphi^0 \times \varphi^e) = \operatorname{tr} \pi(\operatorname{IPC}_\rho(\varphi^0, \varphi^e; \mu_n, \mu_h)).$$

We can now prove our second main theorem.

Theorem V.5.4 (The second basic identity)

$$\operatorname{Red}_\rho^{(h)}[H(X, \mathcal{L}_\xi)^{\mathbb{Z}_p^\times}] = n[H_c(I^{(h)}, \mathcal{F}_\rho \otimes \mathcal{L}_\xi)]$$

in $\operatorname{Groth}(G^{(h)}(\mathbb{A}^\infty))$.

Proof: By lemma V.1.3, it suffices to check that for any

- $\varphi^p \in C_c^\infty(G(\mathbb{A}^{\infty,p}) \times (\mathbb{Q}_p^\times/\mathbb{Z}_p^\times) \times \prod_{i=2}^r (B_{w_i}^{\mathrm{op}})^\times)$,
- $\varphi^0 \in C_c^\infty(D_{F_w, n-h}^\times/\mathcal{O}_{D_{F_w}, n-h}^\times)$,
- and $\varphi^e \in C_c^\infty(GL_h(F_w))$,

such that $\varphi = \varphi^w \times \varphi^0 \times \varphi^e$ is acceptable, we have

$$\operatorname{tr}(\operatorname{Red}_\rho^{(h)}(H(X, \mathcal{L}_\xi)))(\varphi) = n\operatorname{tr} H_c(I^{(h)}, \mathcal{F}_\rho \otimes \mathcal{L}_\xi)(\varphi).$$

(Note that, if $C \in \mathbb{R}$ and $\psi_0, \ldots, \psi_m$ are characters of $\mathbb{Z}$ we can a function $\varphi' \in C_c^\infty(\mathbb{Z}_{>C})$ such that $\psi_0(\varphi') = 1$ but $\psi_i(\varphi') = 0$ for $i = 1, \ldots, m$.)

But by corollary V.5.3, proposition III.3.1 and lemma V.5.2 we have

$$\operatorname{tr}(\operatorname{Red}_\rho^{(h)}(H(X, \mathcal{L}_\xi)))(\varphi)$$

$$= \operatorname{tr} H(X, \mathcal{L}_\xi)(\varphi^w \times \operatorname{IPC}_\rho(\varphi^0, \varphi^e; \mu_n, \mu_h))$$

$$= (-1)^n \kappa_B n \sum_a [F(a) : F]^{-1} (-1)^{n/[F(a):F]} \operatorname{vol}(Z_G(a)(\mathbb{R})_0^1)^{-1}$$
$$O_a^{G(\mathbb{A}^\infty)}(\varphi^w \times \operatorname{IPC}_\rho(\varphi^0, \varphi^e; \mu_n, \mu_h))$$

$$= (-1)^h \kappa_B n \sum_{[(a,\tilde{w})] \in \operatorname{FP}^{(h)}} [F(a) : F]^{-1} (-1)^{n/[F(a):F] - (n-h)/[F(a)_{\tilde{w}}:F_w]}$$
$$\operatorname{vol}(Z_G(a)(\mathbb{R})_0^1)^{-1} O_{\iota(a)}^{G^{(h)}(\mathbb{A}^\infty)}(\varphi) \operatorname{tr}(\rho \otimes \xi)(\iota(a)),$$

where we drop the term $[F(a) : F]^{-1}$ if $n = 2$ and $F^+ = \mathbb{Q}$. Here $\kappa_B = 2$ if $[B : \mathbb{Q}]/2$ is even and $= 1$ otherwise and the choices of measures are as in proposition V.4.8. Comparing this formula with proposition V.4.8, the trace identity and hence the theorem follows. $\square$

V.6   The main theorem

Combining theorems IV.2.9 and V.5.4 we at once obtain our main result.

To state it simply it is convenient to introduce a little more notation. Let

$$\mathrm{red}_\rho^{(h)} : \mathrm{Groth}\,(GL_n(F_w)) \longrightarrow \mathrm{Groth}\,(GL_h(F_w))$$

to be the composite of the map

$$
\begin{array}{ccc}
\mathrm{Groth}\,(GL_n(F_w)) & \longrightarrow & \mathrm{Groth}\,(GL_{n-h}(F_w) \times GL_h(F_w)) \\
[\pi] & \longmapsto & [J_{N_h^{\mathrm{op}}}(\pi) \otimes \delta_{P_h}^{1/2}]
\end{array}
$$

and the map

$$\mathrm{Groth}\,(GL_{n-h}(F_w) \times GL_h(F_w)) \longrightarrow \mathrm{Groth}\,(GL_h(F_w))$$

which sends $[\alpha \otimes \beta]$ to

$$\mathrm{vol}\,(D_{F_w,n-h}^\times / F_w^\times)^{-1}\,\mathrm{tr}\,\alpha(\varphi_{\mathrm{JL}\,(\rho^\vee)})[\beta]$$

if the product of the central characters of α and ρ is 1 and sends $[\alpha \otimes \beta]$ to 0 otherwise. Then $\mathrm{red}_\rho^{(h)}$ extends to a homomorphism

$$\mathrm{Groth}\,(G(\mathbb{A}^\infty)) \longrightarrow \mathrm{Groth}\,(G(\mathbb{A}^{\infty,p}) \times \mathbb{Q}_p^\times \times GL_h(F_w) \times \prod_{i=2}^r (B_{w_i}^{\mathrm{op}})^\times).$$

Note that

$$\mathrm{Red}_\rho^{(h)}(\pi) = \sum_\psi \psi \otimes \mathrm{red}_{\rho \otimes \psi^{-1}}^{(h)}(\pi),$$

where ψ runs over characters of $D_{F_w,n-h}^\times / \mathcal{O}_{D_{F_w,n-h}}^\times$.

Theorem V.6.1 (Main Theorem) *In* $\mathrm{Groth}_l(G(\mathbb{A}^\infty) \times W_{F_w})$ *we have the equality of* $n[H(X, \mathcal{L}_\xi)^{\mathbb{Z}_p^\times}]$ *and*

$$\sum_{h=0}^{n-1} \sum_\rho \mathrm{Ind}_{P_h(F_w)}^{GL_n(F_w)} \mathrm{red}_\rho^{(h)}[H(X, \mathcal{L}_\xi)^{\mathbb{Z}_p^\times}] *_{|\mathrm{Art}_K^{-1}|} [\Psi_{F_w,l,n-h}(\rho)]$$

where ρ *runs over irreducible admissible representations of* $D_{F_w,n-h}^\times$ *and where* $|\mathrm{Art}_K^{-1}| : W_{F_w} \to p^{\mathbb{Z}} \subset \mathbb{Q}_p^\times$.

As an immediate application let us explain how to recover a result of Kottwitz (see [Ko4]). We will see later how theorem V.6.1 actually allows us to significantly strengthen Kottwitz's result.

Corollary V.6.2 (Kottwitz) *In* $\mathrm{Groth}_l(G(\mathbb{A}^{w,\infty}) \times W_{F_w})$ *we have*

$$n \sum_{\pi} [\pi^w \otimes R_{\xi}(\pi)]$$

$$= \sum_{\pi} [\pi^w] \sum_{i=1}^{n} [|\mathrm{Art}\,{}_K^{-1}|^{(1-n)/2} \pi_{p,0}(|\mathrm{Art}\,{}_K^{-1}|) \chi_{\pi_w,i}^{-1}(\mathrm{Art}\,{}_K^{-1})],$$

where π runs over irreducible representations of $G(\mathbb{A}^{\infty})$ with $\pi_{p,0}$ and π_w unramified, and where

$$\pi_w \cong \chi_{\pi_w,1} \boxplus \cdots \boxplus \chi_{\pi_w,n}.$$

Proof: Theorem V.6.1 tells us that

$$n \sum_{\pi} (\dim \pi_w^{GL_n(\mathcal{O}_{F,w})})[\pi^w \otimes R_{\xi}(\pi)]$$

$$= \sum_{\pi} [\pi_w] \sum_{\psi} \sum_{h=0}^{n-1} (-1)^{n-1-h} \dim(\mathrm{red}_{\psi \circ \det}^{(h)} \pi_w)^{GL_h(\mathcal{O}_{F,w})}$$

$$[\pi_{p,0}(|\mathrm{Art}\,{}_K^{-1}|)\psi(\mathrm{Art}\,{}_K^{-1})],$$

where π runs over irreducible representations of $G(\mathbb{A}^{\infty})$ and where ψ runs over characters of $F_w^{\times}/\mathcal{O}_{F,w}^{\times}$. Thus it suffices to show that if π_w is an irreducible representation of $GL_n(\mathcal{O}_{F,w})$ and if ψ is a character of $F_w^{\times}/\mathcal{O}_{F,w}^{\times}$ then

$$\sum_{h=0}^{n-1} (-1)^{n-1-h} \dim(\mathrm{red}_{\psi \circ \det}^{(h)} \pi_w)^{GL_h(\mathcal{O}_{F,w})}$$

- is zero, if π_w is ramified;

- is the number of indices i for which $\psi = \chi_{\pi_w,i}^{-1} |\ |^{(1-n)/2}$, if $\pi_w \cong \chi_{\pi_w,1} \boxplus \cdots \boxplus \chi_{\pi_w,n}$ is unramified.

Equivalently, it suffices to show that if ψ is a character of $F_w^{\times}/\mathcal{O}_{F,w}^{\times}$, if $n = \sum s_i n_i$ and if π_i is a supercuspidal representation of $GL_{n_i}(F_w)$ then

$$\sum_{h=0}^{n-1} (-1)^{n-1-h} \dim(\mathrm{red}_{\psi \circ \det}^{(h)} n\text{-}\mathrm{Ind}\,(\mathrm{Sp}_{s_1}(\pi_1) \times \cdots \times \mathrm{Sp}_{s_t}(\pi_t)))^{GL_h(\mathcal{O}_{F,w})}$$

- is zero unless $s_i = n_i = 1$ for all i and each π_i is unramified;

- is the number of indices i for which $\psi = \pi_i^{-1} |\ |^{(1-n)/2}$, if $s_i = n_i = 1$ for all i and each π_i is unramified.

However, by lemma I.3.9 and the properties of pseudo-coefficients discussed before lemma I.3.1, we see that

$$\sum_{h=0}^{n-1}(-1)^{n-1-h}$$
$$\dim(\mathrm{red}^{(h)}_{\psi\circ\det}\text{n-Ind}\,(\mathrm{Sp}_{s_1}(\pi_1)\times\cdots\times\mathrm{Sp}_{s_t}(\pi_t)))^{GL_h(\mathcal{O}_{F,w})}$$

$$=\sum_{i\in I}\sum_{\max(1,s_i-1)\leq x\leq s_i}(-1)^{x-1}\mathrm{vol}\,(D^{\times}_{F_w,x}/F^{\times}_w)^{-1}$$
$$\mathrm{tr}\,\mathrm{Sp}_x(\pi_i|\det|^{(n-x)/2})(\varphi_{\mathrm{JL}\,(\psi^{-1}\circ\det)})$$

where I is the set of $i\in\{1,\ldots,t\}$ such that $n_i=1$ and for all $j\neq i$, π_j is non-ramified and $s_j=1$. As

$$\mathrm{JL}\,(\psi^{-1}\circ\det)=\mathrm{Sp}_t(\psi^{-1}|\ |^{(1-x)/2}),$$

we see that in fact

$$\sum_{h=0}^{n-1}(-1)^{n-1-h}$$
$$\dim(\mathrm{red}^{(h)}_{\psi\circ\det}\text{n-Ind}\,(\mathrm{Sp}_{s_1}(\pi_1)\times\cdots\times\mathrm{Sp}_{s_t}(\pi_t)))^{GL_h(\mathcal{O}_{F,w})}$$
$$=\sum_{i\in J}\sum_{\max(1,s_i-1)\leq x\leq s_i}(-1)^{x-1},$$

where now J is the set of $i\in\{1,\ldots,t\}$ such that $n_i=1$, $\pi_i=\psi^{-1}|\ |^{(1-n)/2}$ and for all $j\neq i$, π_j is non-ramified and $s_j=1$. This is easily seen to be equivalent to the desired formula. $\square$

Corollary V.6.3 (Kottwitz) *Let π be an irreducible representation of the group $G(\mathbb{A}^{\infty})$. For all but finitely many rational primes q which split in E and for all places x of F above q the representations $R^i_{\xi}(\pi)$ are unramified at x, π_q is unramified and*

$$n[R_{\xi}(\pi)|_{W_{F_x}}]=\dim[R_{\xi}(\pi)]\sum_{i=1}^{n}[\pi_{p,0}(|\mathrm{Art}\,_K^{-1}|)|\mathrm{Art}\,_K^{-1}|^{(1-n)/2}\chi_{x,i}(\mathrm{Art}\,_K^{-1})],$$

where $\pi_x\cong\chi_{q,1}\boxplus\cdots\boxplus\chi_{q,n}$.

Proof: We may suppose that $[R_{\xi}(\pi)]\neq 0$. For all but finitely many primes q which split in E, if π'_q is an unramified irreducible representation of $G(\mathbb{Q}_q)$ with $[R_{\xi}(\pi^q\times\pi'_q)]\neq 0$ then we must have $\pi'_q\cong\pi_q$. Thus this corollary follows from the previous one. $\square$

It follows from this corollary and from proposition III.2.1 that for all i, $n\dim R^i_{\xi}(\pi)/|\dim[R_{\xi}(\pi)]|$ is an integer and that for all but finitely many places x of F lying above a rational prime q which splits in E,

$$n\dim R^i_{\xi}(\pi)/|\dim[R_{\xi}(\pi)]|$$

is the number of $j \in \{1, \ldots, n\}$ such that

$$|\imath \chi_{x,j}(\varpi_x)| = \pi_{q,0}(\#k(x))(\#k(x))^{(1+i+w(\xi)-n)/2},$$

where $\pi_x \cong \chi_{q,1} \boxplus \cdots \boxplus \chi_{q,n}$. The sum of these numbers must be n, so that

$$\sum_i \dim R^i_\xi(\pi) = |\dim[R_\xi(\pi)]| = \pm \sum_i (-1)^{n-1-i} \dim R^i_\xi(\pi).$$

Thus if $\dim R^i_\xi(\pi) \neq 0 \neq \dim R^{i'}_\xi(\pi)$, then $i \equiv i'$ mod 2. We also see that if π' is another irreducible representation of $G(\mathbb{A}^\infty)$ with $\pi^S \cong (\pi')^S$ for some finite set of primes S, then

$$n \dim R^i_\xi(\pi)/|\dim[R_\xi(\pi)]| = n \dim R^i_\xi(\pi')/|\dim[R_\xi(\pi')]|.$$

Thus we obtain the following corollary.

Corollary V.6.4 (Kottwitz) *Let π and π' be irreducible representations of $G(\mathbb{A}^\infty)$ and suppose that for some finite set of primes S we have $\pi^S \cong (\pi')^S$. Suppose also that $R^i_\xi(\pi) \neq (0)$ and $R^{i'}_\xi(\pi') \neq 0$. Then $i \equiv i'$ mod 2. In particular $\dim[R_\xi(\pi)] \neq 0$.*

Chapter VI

Automorphic forms

So far our arguments have been essentially geometric and algebraic. (Although we have occasionally appealed to an analytic result, this was simply because it provided a convenient reference.) We now need to make serious use of some analytic results. Nothing in this chapter is new in any significant way. Rather we collect results of Clozel, Labesse and Vigneras in a form that will be convenient for the next chapter.

VI.1 The Jacquet-Langlands correspondence

Let $S(B)$ denote the set of places of F at which B ramifies. Recall that we are assuming that at such a place B_x is a division algebra. The following theorem was proved by Vigneras in her unpublished manuscript [V2], which relied on a seminar of Langlands which to the best of our knowledge was never written up. We will explain how it follows easily from an important theorem of Arthur and Clozel [AC].

Theorem VI.1.1   *1. If ρ is an irreducible automorphic representation of*
$(B^{\mathrm{op}} \otimes \mathbb{A})^{\times}$ then there is a unique irreducible automorphic representation $\mathrm{JL}\,(\rho)$ of $GL_n(\mathbb{A}_F)$, which occurs in the discrete spectrum and for which

$$\mathrm{JL}\,(\rho)^{S(B)} \cong \rho^{S(B)}.$$

2. If $x \in S(B)$ and $\mathrm{JL}\,(\rho_x) = \mathrm{Sp}_{s_x}(\pi_x)$ then

- *either $\mathrm{JL}\,(\rho)_x \cong \mathrm{Sp}_{s_x}(\pi_x)$,*
- *or $\mathrm{JL}\,(\rho)_x \cong \pi_x \boxplus \cdots \boxplus (\pi_x \otimes |\det|^{s_x - 1})$.*

3. *The image of* JL *is the set of irreducible automorphic representations* π *of* $GL_n(\mathbb{A}_F)$ *such that*

- π *occurs in the discrete spectrum,*

- *and for every* $x \in S(B)$ *there is a positive integer* $s_x | n$ *and an irreducible supercuspidal representation* π'_x *of* $GL_{n/s_x}(F_x)$ *so that either* $\mathrm{JL}\,(\rho)_x \cong \mathrm{Sp}_{s_x}(\pi'_x)$ *or* $\mathrm{JL}\,(\rho)_x \cong \pi'_x \boxplus \cdots \boxplus (\pi'_x \otimes |\det|^{s_x-1})$.

4. *If* ρ_1 *and* ρ_2 *are two irreducible automorphic representations of* $(B^{\mathrm{op}} \otimes \mathbb{A})^\times$ *such that for all but finitely many places* x *of* F *we have* $\rho_{1x} \cong \rho_{2x}$, *then* $\rho_1 = \rho_2$ *(i.e.* $\rho_1 \cong \rho_2$ *and this representation occurs with multiplicity 1 in the space of automorphic forms).*

Proof: Let H denote the algebraic group over $\mathbb{Q}$ such that $H(R) = (B^{\mathrm{op}} \otimes_{\mathbb{Q}} R)^\times$ for any $\mathbb{Q}$-algebra R. We will confuse irreducible admissible representations occurring discretely in the space of automorphic forms (with fixed central character restricted to $\mathbb{R}^\times_{>0}$) with their completions in L^2. By twisting, we need only consider representations which are trivial on $\mathbb{R}^\times_{>0}$.

Then by theorem B of [AC] we have, in the notation of [AC],

$$a^H_{\mathrm{disc}}(\rho) = \sum_\pi a^{GL_n \times F}_{\mathrm{disc}}(\pi)\delta(\pi,\rho),$$

where the sum is over (pre)unitary representations of $GL_n(\mathbb{A}_F)^1$. Let us write $\mathrm{JL}\,(\rho_x) = \mathrm{Sp}_{s_x(\rho)}(\pi'_x(\rho))$ for $x \in S(B)$. Moreover for $T \subset S(B)$ let $\pi(T, \rho)$ denote

$$\rho^{S(B)} \times \prod_{x \in T} \mathrm{Sp}_{s_x(\rho)}(\pi'_x(\rho)) \times$$

$$\times \prod_{x \in S(B)-T} (\pi'_x(\rho) \boxplus \cdots \boxplus (\pi'_x(\rho) \otimes |\det|^{s_x(\rho)-1})).$$

Then, using section 8 of [AC], lemma I.3.4 and corollary I.3.7 we see that $\delta(\pi, \rho) = 0$ unless $\pi \cong \pi(T, \rho)$ for some $T \subset S(B)$, in which case $\delta(\pi, \rho) = \pm 1$. Thus the equality of theorem B of [AC] becomes

$$a^H_{\mathrm{disc}}(\rho) = \sum_{T \subset S(B)} \pm a^{GL_n \times F}_{\mathrm{disc}}(\pi(T, \rho)).$$

The coefficients $a^H_{\mathrm{disc}}(\rho)$ are just the multiplicity of ρ in the space of automorphic forms trivial on $\mathbb{R}^\times_{>0}$. The coefficients $a^{GL_n \times F}_{\mathrm{disc}}(\pi(T, \rho))$ are defined by the equalities

$$\sum a_{\text{disc}}^{GL_N \times F}(\pi) \text{tr}\, \pi(f) =$$
$$\sum_{L \in \mathcal{L}} |W_0^L| |W_0^{GL_n \times F}|^{-1} \sum_{s \in W(a_L)_{\text{reg}}} |\det(s-1)_{a_L^{GL_n \times F}}|^{-1}$$
$$\text{tr}\,(M(s,0)\rho_{Q,t}(0,f))$$

in the notation of section 9 of chapter 2 of [AC]. Choose $x \in S(B)$. Then for any z in the Bernstein centre for $GL_n(F_x)$ we see that

$$\sum a_{\text{disc}}^{GL_N \times F}(\pi) \text{tr}\, \pi(f)\pi_x(z) = \sum_{L \in \mathcal{L}} |W_0^L| |W_0^{GL_n \times F}|^{-1}$$
$$\sum_{s \in W(a_L)_{\text{reg}}} |\det(s-1)_{a_L^{GL_n \times F}}|^{-1} \text{tr}\,(M(s,0)\rho_{Q,t}(0,f))\rho_{Q,t}(0)_x(z).$$

Let D denote the variety of unramified twists of the representation

$$\pi_x'(\rho)^{s_x(\rho)}$$

of $GL_{n/s_x(\rho)}(F_x)^{s_x(\rho)}$ and suppose that z corresponds as in [Bern] to a regular function on $D/W(GL_{n/s_x(\rho)}(F_x)^{s_x(\rho)}, D)$ (in the notation of [Bern]). If $Q \neq GL_n \times F$ then either $\rho_{Q,t}(0)_x(z) = 0$ for all z which correspond to regular functions on $D/W(GL_{n/s_x(\rho)}(F_x)^{s_x(\rho)}, D)$, or $\rho_{Q,t}(0)_x$ maps to a 0-cycle on $D/W(GL_{n/s_x(\rho)}(F_x)^{s_x(\rho)}, D)$ supported away from $\pi_x'(\rho) \times \cdots \times (\pi_x'(\rho) \otimes |\det|^{s_x(\rho)})$. Choose z in the Bernstein centre corresponding to a regular function on the space $D/W(GL_{n/s_x(\rho)}(F_x)^{s_x(\rho)}, D)$ which is 1 at $\pi_x'(\rho) \times \cdots \times (\pi_x'(\rho) \otimes |\det|^{s_x(\rho)})$ and zero at all other terms occurring in our sum (which is finite for any given f). Then only the term corresponding to $L = GL_n \times F$ persists on the right hand side of the above formula, and we see that

$$\sum_\pi a_{\text{disc}}^{GL_N \times F}(\pi) \text{tr}\, \pi(f) = \sum_\pi m(\pi) \text{tr}\, \pi(f),$$

where

- $m(\pi)$ denotes the multiplicity of π in the discrete part of the space of automorphic forms invariant by $\mathbb{R}_{>0}^\times$;

- and where both sums run over irreducible representations π such that π_w maps to $\pi_x'(\rho) \times \cdots \times (\pi_x'(\rho) \otimes |\det|^{s_x(\rho)})$ in in the quotient $D/W(GL_{n/s_x(\rho)}(F_x)^{s_x(\rho)}, D)$.

We deduce that $a_{\text{disc}}^{GL_N \times F}(\pi)$ is just the multiplicity of π in the discrete part of the space of automorphic forms invariant by $\mathbb{R}_{>0}^\times$.

Using the strong multiplicity one theorem for $GL_n(\mathbb{A}_F)$, the theorem follows. $\square$

Combining this with the main result of [MW] we obtain the following corollary.

Corollary VI.1.2 *Suppose that ρ is an irreducible automorphic representation of $(B^{\mathrm{op}} \otimes \mathbb{A})^{\times}$. Then the following are equivalent.*

1. *JL (ρ) is cuspidal.*

2. *For one place $x \notin S(B)$ the component ρ_x is generic.*

3. *For all places $x \notin S(B)$ the component ρ_x is generic.*

VI.2   Clozel's base change

For this section fix an embedding $\tau : F \hookrightarrow \mathbb{C}$. We will give a description of automorphic representations of $G_\tau(\mathbb{A})$ in terms of automorphic representations on $(B^{\mathrm{op}} \otimes_{\mathbb{Q}} \mathbb{A})^{\times}$ together with some applications. This description is basically due to Clozel (see [Cl1] and [Cl2]), but there were a number of gaps in his argument which were repaired with the help of Labesse (see [Lab] and [CL]).

Recall that we have fixed $\imath : \mathbb{Q}_l^{\mathrm{ac}} \xrightarrow{\sim} \mathbb{C}$. We will write $\xi' = \imath(\xi)$, an irreducible algebraic representation of G_τ over $\mathbb{C}$. Note that

$$\mathrm{RS}^E_{\mathbb{Q}}(G_\tau \times E) \times \mathbb{C} \cong (G_\tau \times \mathbb{C}) \times_{\mathbb{C}} (G_\tau \times \mathbb{C}),$$

where the first factor corresponds to $\tau : E \hookrightarrow \mathbb{C}$ and the second factor to $\tau \circ c$. We will let ξ'_E denote the representation $\xi' \otimes \xi'$ of $\mathrm{RS}^E_{\mathbb{Q}}(G_\tau \times E)$ over $\mathbb{C}$. It restricts to the representation $\xi' \otimes (\xi' \circ c)$ of $G_\tau(E_\infty) = \mathrm{RS}^E_{\mathbb{Q}}(G_\tau \times E)(\mathbb{R})$. We will also use ξ'_E for the restriction of this representation to $GL_n(F_\infty) \subset E_\infty^{\times} \times GL_n(F_\infty) \cong G_\tau(E_\infty)$ (see section I.7).

We will call an irreducible admissible representation π_∞ of $G_\tau(\mathbb{R})$ (resp. Π_∞ of $GL_n(F_\infty)$) *cohomological for ξ'* (resp. *cohomological for ξ'_E*) if for some i,

$$H^i((\mathrm{Lie}\, G_\tau(\mathbb{R})) \otimes_{\mathbb{R}} \mathbb{C}, U_\tau, \pi_\infty \otimes \xi') \neq (0),$$

(resp.

$$H^i(M_n(F_\infty) \otimes_{\mathbb{R}} \mathbb{C}, U(0,n)^{[F^+ : \mathbb{Q}]}, \Pi_\infty \otimes \xi'_E) \neq (0).)$$

Suppose that x is a place of $\mathbb{Q}$ which splits as $x = yy^c$ in E. Recall (from chapter I) that the choice of a place $y|x$ allows us to consider $\mathbb{Q}_x \xrightarrow{\sim} E_y$ as an E-algebra and hence to identify

$$G(\mathbb{Q}_x) \cong (B_y^{\mathrm{op}})^{\times} \times \mathbb{Q}_x^{\times}.$$

If π is an irreducible admissible representation of $G(\mathbb{Q}_x)$ we can then decompose

$$\pi \cong \pi_y \otimes \psi_{\pi,y^c}.$$

If we vary our choice of y we find that $\pi_{y^c} = \pi_y^{\#}$ and that $\psi_{\pi,y} = \psi_{\pi_y}\psi_{\pi,y^c}$. (Here we set $\pi_y^{\#}(g) = \pi_y(g^{-\#})$.) We define BC (π) to be the representation

$$\pi_y \otimes \pi_{y^c} \otimes (\psi_{\pi,y^c} \circ c) \otimes (\psi_{\pi,y} \circ c)$$

of

$$G(E_x) \cong (B_x^{\mathrm{op}})^{\times} \times E_x^{\times} \cong (B_y^{\mathrm{op}})^{\times} \times (B_{y^c}^{\mathrm{op}})^{\times} \times E_y^{\times} \times E_{y^c}^{\times}.$$

Now suppose that x is a finite place of $\mathbb{Q}$ which is inert in E but such that

- x is unramified in F;

- $(B_x^{\mathrm{op}}, \#) \cong (M_n(F_x), \dagger)$, where $g^{\dagger} = w(g^c)^t w^{-1}$ with w the antidiagonal matrix with ones on the antidiagonal.

These latter two conditions exclude only finitely many places of $\mathbb{Q}$. If x is such a place then $G(\mathbb{Q}_x)$ is quasi-split and split over an unramified extension. We will fix a maximal torus T_x in a Borel subgroup B_x in $G \times F_x$ so that $B_x(F_x)$ consists of elements of $G(F_x)$ which correspond to upper triangular elements of $M_n(F_x)$ and $T_x(F_x)$ consists of elements of G which correspond to diagonal elements of $M_n(F_x)$. Thus $T_x(\mathbb{Q}_x)$ can be identified with the set of elements $(d_0; d_1, \ldots, d_n) \in \mathbb{Q}_x^{\times} \times (F_x^{\times})^n$ such that $d_0 = d_i d_{n+1-i}^c$ for $i = 1, \ldots, n$. If ψ is a character of $T(\mathbb{Q}_x)$ we define a character BC (ψ) of $E_x^{\times} \times (F_x^{\times})^n$ by

$$\mathrm{BC}\,(\psi)(d_0; d_1, \ldots, d_n) = \psi(d_0 d_0^c; d_0 d_1/d_n^c, \ldots, d_0 d_n/d_1^c).$$

Let B denote the Borel subgroup of upper triangular elements of GL_n. If π is an unramified representation of $G(\mathbb{Q}_x)$ which is a subquotient of the induced representation n-$\mathrm{Ind}_{B_x(\mathbb{Q}_x)}^{G(\mathbb{Q}_x)}(\psi)$ then we will denote by BC (π) the unique unramified representation of $E_x^{\times} \times GL_n(F_x)$ which is a subquotient of the normalised induction from $E_x^{\times} \times B(F_x)$ of BC (ψ).

If Π is an irreducible automorphic representation of $(B^{\mathrm{op}} \otimes_{\mathbb{Q}} \mathbb{A})^{\times}$ then define $\Pi^{\#}$ by

$$\Pi^{\#}(g) = \Pi(g^{-\#}).$$

Using the strong multiplicity one theorem we see that

$$\mathrm{JL}\,(\Pi^{\#}) = \mathrm{JL}\,(\Pi)^{\vee} \circ c.$$

Theorem VI.2.1 *Suppose that π is an irreducible automorphic representation of $G_\tau(\mathbb{A})$ such that π_∞ is cohomological for ξ'. Then there is a unique irreducible automorphic representation $\mathrm{BC}(\pi) = (\psi, \Pi)$ of $\mathbb{A}_E^\times \times (B^{op} \otimes_\mathbb{Q} \mathbb{A})^\times$ such that*

1. *$\psi = \psi_\pi|_{\mathbb{A}_E^\times}^c$;*

2. *if x is a place of $\mathbb{Q}$ which splits in E then $\mathrm{BC}(\pi)_x = \mathrm{BC}(\pi_x)$;*

3. *for all but finitely many places x of $\mathbb{Q}$ which are inert in E we have $\mathrm{BC}(\pi)_x = \mathrm{BC}(\pi_x)$;*

4. *Π_∞ is cohomological for ξ'_E;*

5. *$\psi_\infty^c = \xi'|_{E_\infty^\times}^{-1}$ (where $E_\infty^\times \subset G_\tau(\mathbb{R})$);*

6. *$\psi_\Pi|_{\mathbb{A}_E^\times} = \psi^c/\psi$;*

7. *$\Pi^\# \cong \Pi$.*

Proof: We will deduce this from theorem A.5.2 of [CL].

Let $T/\mathbb{Q}$ denote the torus $\mathrm{RS}_\mathbb{Q}^E(\mathbb{G}_m)$ and let $T^1 \subset T$ denote the kernel of the norm homomorphism $T \to \mathbb{G}_m$. We have a natural morphism $T \times G_{\tau,1} \to G_\tau$ which is surjective on geometric points. It has kernel T^1 embedded by $t \mapsto (t, t^{-1})$. If π is an admissible representation of $G_\tau(\mathbb{A})$ we obtain an admissible representation $\pi|_{(T \times G_{\tau,1})(\mathbb{A})}$ of $(T \times G_{\tau,1})(\mathbb{A})$ by composing π with the homomorphism $T \times G_{\tau,1} \to G_\tau$. (Use the fact that $(T \times G_{\tau,1})(\mathbb{A}) \to G_\tau(\mathbb{A})$ is continuous and open.) If π is automorphic then $\pi|_{(T \times G_{\tau,1})(\mathbb{A})}$ is a direct sum of irreducible subrepresentations. As far as we can see not all these direct summands are automorphic for $T \times G_{\tau,1}$. Rather suppose that $g_i \in G_\tau(\mathbb{A})$ form a set of representatives for $\nu(G_\tau(\mathbb{A}))/\nu(G_\tau(\mathbb{Q}))\mathbf{N}(\mathbb{A}_E^\times)$. Then we get a bijection of spaces of automorphic forms

$$
\begin{array}{ccc}
\mathcal{A}(G_\tau(\mathbb{Q})\backslash G_\tau(\mathbb{A})) & \overset{\sim}{\to} & \bigoplus_i \mathcal{A}((T \times G_{\tau,1})(\mathbb{Q})\backslash(T \times G_{\tau,1})(\mathbb{A}))^{g_i T^1(\mathbb{A})g_i^{-1}} \\
f & \mapsto & (g_i(f)|_{(T \times G_{\tau,1})(\mathbb{A})})_i.
\end{array}
$$

If π' is an irreducible subquotient of $\pi|_{(T \times G_{\tau,1})(\mathbb{A})}$ then π' may not be automorphic but π' composed with conjugation by one of the g_i will be. Note that (because $G_\tau(\mathbb{Q})$ is dense in $G_\tau(\mathbb{R})$) one may assume that $g_{i,\infty} = 1$ for all i.

If x is a place of $\mathbb{Q}$ which splits in E then we get an exact sequence

$$
(0) \longrightarrow T^1(\mathbb{Q}_x) \longrightarrow (T \times G_1)(\mathbb{Q}_x) \longrightarrow G(\mathbb{Q}_x) \longrightarrow (0).
$$

If π_x is an irreducible admissible representation of $G(\mathbb{Q}_x)$ then the restriction $\pi_x|_{(T \times G_1)(\mathbb{Q}_x)}$ is also irreducible.

If x is a finite place of $\mathbb{Q}$ which is inert in E, but such that

- x is unramified in F,

- $(B_x^{op}, \#) \cong (M_n(F_x), \dagger)$, where $g^\dagger = w(g^c)^t w^{-1}$ with w the antidiagonal matrix with ones on the antidiagonal;

then we get an exact sequence

$$(0) \longrightarrow T^1(\mathbb{Q}_x) \longrightarrow (T \times G_1)(\mathbb{Q}_x) \longrightarrow G(\mathbb{Q}_x) \longrightarrow (0),$$

if n is odd, while if n is even we get an exact sequence

$$(0) \longrightarrow T^1(\mathbb{Q}_x) \longrightarrow (T \times G_1)(\mathbb{Q}_x) \longrightarrow G(\mathbb{Q}_x) \xrightarrow{x \circ \nu} (\mathbb{Z}/2\mathbb{Z}) \longrightarrow (0).$$

In either case if π_x is an unramified irreducible representation of $G(\mathbb{Q}_x)$ then $\pi_x|_{(T \times G_1)(\mathbb{Q}_x)}$ contains a unique unramified subquotient which we will denote $\pi_x|^0_{(T \times G_1)(\mathbb{Q}_x)}$.

If $[F^+ : \mathbb{Q}] > 1$ or $n > 2$ then we have an exact sequence

$$(0) \longrightarrow T^1(\mathbb{R}) \longrightarrow (T \times G_{\tau,1})(\mathbb{R}) \longrightarrow G_\tau(\mathbb{R}) \longrightarrow (0),$$

while if $F^+ = \mathbb{Q}$ and $n = 2$ then we get an exact sequence

$$(0) \longrightarrow T^1(\mathbb{R}) \longrightarrow (T \times G_{\tau,1})(\mathbb{R}) \longrightarrow G_\tau(\mathbb{R}) \longrightarrow \{\pm 1\} \longrightarrow (0).$$

If π_∞ is a irreducible admissible representation of $G_\tau(\mathbb{R})$ then it is cohomological for ξ' if and only if

1. $\pi_\infty|_{E_\infty^\times} = \xi'|_{E_\infty^\times}^{-1}$,

2. and there is an irreducible constituent π'_∞ of $\pi_\infty|_{G_{\tau,1}(\mathbb{R})}$ and an $i \in \mathbb{Z}_{\geq 0}$ such that

$$H^i(\operatorname{Lie} G_{\tau,1}(\mathbb{R}) \otimes_\mathbb{R} \mathbb{C}, U_\infty, \pi'_\infty \otimes \xi'|_{G_{\tau,1}}) \neq (0).$$

Thus if π is an irreducible automorphic representation of $G_\tau(\mathbb{A})$ such that π_∞ is cohomological for ξ' we can find an irreducible automorphic representation π' for $(T \times G_{\tau,1})(\mathbb{A})$ such that

- $\pi'|_{T^1(\mathbb{A})} = 1$,

- if x is a place of $\mathbb{Q}$ which splits in E then $\pi'_x = \pi_x|_{(T \times G_1)(\mathbb{Q}_x)}$;

- for all but finitely many places x of $\mathbb{Q}$ which are inert in E we have $\pi'_x = \pi_x|^0_{(T \times G_1)(\mathbb{Q}_x)}$;

- $\pi'_\infty|_{E_\infty^\times} = \xi'|_{E_\infty^\times}^{-1}$;

- and for some i we have

$$H^i(\text{Lie } G_{\tau,1}(\mathbb{R}) \otimes_{\mathbb{R}} \mathbb{C}, U_\infty, (\pi'_\infty \otimes \xi')|_{G_{\tau,1}(\mathbb{R})}) \neq (0).$$

Thus π' is of the form $\psi^c \otimes \pi'_1$, where ψ is a character of $E^\times \backslash \mathbb{A}_E^\times$ and π'_1 is an automorphic representation of $G_{\tau,1}(\mathbb{A})$. Moreover $\psi|_{T^1(\mathbb{A})} = \psi_{\pi'_1}|_{T^1(\mathbb{A})}^{-1}$.

Now we apply theorem A.5.2 of [CL] to π'_1 and we conclude that there exists an automorphic representation Π of $(B^{\text{op}} \otimes \mathbb{A})^\times$ such that

- if x is a place of $\mathbb{Q}$ which splits in E then $\Pi_x = \text{BC}(\pi_x)|_{(B_x^{\text{op}})^\times}$;

- for all but finitely many places x of $\mathbb{Q}$ which are inert in E we have $\Pi_x = \text{BC}(\pi_x)|_{(B_x^{\text{op}})^\times}$;

- for some i we have

$$H^i(M_n(F_\infty) \otimes_{\mathbb{R}} \mathbb{C}, U(0,n)^{[F^+:\mathbb{Q}]}, \Pi_\infty \otimes \xi'_E) \neq (0).$$

(The first of these properties is not explicitly stated in theorem A.5.2 of [CL]. However it follows easily from a slight modification of the proof. In the notation of the proof of theorem A.5.2 in [CL] divide the set S as a disjoint union $S = S_1 \cup S_2$ where S_1 consists of places which split in E and S_2 of places which are inert. Now take $f_S = (\prod_{x \in S_1} f_x) \times f_{S_2}$ where for $x \in S_1$, f_x is any element of $C_c^\infty(G_0(\mathbb{Q}_x))$ and f_{S_2} is the characteristic function of a sufficiently small open compact subgroup, K_{S_2}. Then we can still find a function φ_S associated to f_S. We conclude as in [CL] that if we fix an unramified representation ρ of $G_0(\mathbb{A}^{S \cup \{\infty\}})$

$$\sum_\pi \text{ep}(\pi_\infty) \dim \pi_{S_2}^{K_{S_2}} \text{tr BC}(\pi_{S_1})(\varphi_{S_1}) =$$
$$\sum_\Pi \text{tr}(\Pi_\infty(\varphi_{ep}^{G,\theta})\Pi_{S_2}(\varphi_{S_2})I_\theta)\text{tr}\,\Pi_{S_1}(\varphi_{S_1}),$$

where π runs over automorphic representations of $G_0(\mathbb{A})$ with $\pi^{S \cup \{\infty\}} \cong \rho$ and Π runs over automorphic representations of $G_0(\mathbb{A}_E)$ with $\Pi^{S \cup \{\infty\}} \cong \text{BC}(\rho)$. We deduce that if we fix an irreducible, unramified outside S_1, representation ρ of $G_0(\mathbb{A}^{S_2 \cup \{\infty\}})$, then

$$\sum_\pi \text{ep}(\pi_\infty) \dim \pi_{S_2}^{K_{S_2}} = \sum_\Pi \text{tr}(\Pi_\infty(\varphi_\infty)\Pi_{S_2}(\varphi_{S_2})I_\theta),$$

where π runs over automorphic representations of $G_0(\mathbb{A})$ with $\pi^{S_2 \cup \{\infty\}} \cong \rho$ and Π runs over automorphic representations of $G_0(\mathbb{A}_E)$ with $\Pi^{S_2 \cup \{\infty\}} \cong \text{BC}(\rho)$. The rest of the argument is as in [CL].)

It is now easy to check that $(\psi_\pi|_{\mathbb{A}_E^\times}, \Pi)$ satisfies the first six properties of the theorem. Uniqueness follows from theorem VI.1.1. The final property also follows from VI.1.1, because for all but finitely many places x of $\mathbb{Q}$ we have $\Pi_x^\# \cong \Pi_x$. $\square$

We remark that theorem A.5.2 of [CL] relies on the work of Kottwitz [Ko4] via theorem A.4.2 of [CL]. However one can replace the appeal to theorem A.4.2 of [CL], and hence to Kottwitz's work, by an appeal to corollary V.6.4 of this book.

Corollary VI.2.2 *If π and π' are irreducible automorphic representations of $G_\tau(\mathbb{A})$ such that π_∞ and π'_∞ are cohomological for ξ' and such that $\pi_x \cong \pi'_x$ for all but finitely many places x of $\mathbb{Q}$, then $\pi_x \cong \pi'_x$ for all places x of $\mathbb{Q}$ which split in E.*

Proof: This follows from theorems VI.1.1 and VI.2.1. $\square$

Combining this with proposition III.2.1 we get the next corollary.

Corollary VI.2.3 *Suppose that π and π' are irreducible admissible representations of $G(\mathbb{A}^\infty)$ such that*

- $\pi^p \cong (\pi')^p$,

- $[R_\xi(\pi)] \neq 0$,

- *and $[R_\xi(\pi')] \neq 0$.*

Then $\pi_p \cong \pi'_p$.

Corollary VI.2.4 *Let π be an irreducible representation of $G(\mathbb{A}^\infty)$ and let $\tau : F \hookrightarrow \mathbb{C}$. The following are equivalent.*

1. *There exists an irreducible representation π_∞ of $G_\tau(\mathbb{R})$ such that*

   - *π_∞ is cohomological for $\imath\xi$,*

   - *$\imath(\pi) \otimes \pi_\infty$ occurs in the space of automorphic forms on $G_\tau(\mathbb{A})$ and*

   - *$\mathrm{BC}\,(\imath(\pi) \otimes \pi_\infty) = (\psi, \Pi)$ with $\mathrm{JL}\,(\Pi)$ cuspidal.*

2. *$[R_\xi(\pi)] \neq 0$ and there exists a place x of F such that*

   - *$x|_\mathbb{Q}$ splits in E,*

   - *B_x^{op} is split,*

   - *and π_x is generic.*

3. *$[R_\xi(\pi)] \neq 0$ and for any place x of F such that*

   - *$x|_\mathbb{Q}$ splits in E and*

- B_x^{op} is split,

π_x is generic.

Proof: This follows from proposition III.2.1, theorem VI.2.1, and corollary VI.1.2. □

Corollary VI.2.5 *Suppose that S is a finite set of finite places of F such that*

- *if $x \in S$ then B_x is split and $x|_{\mathbb{Q}}$ splits in E,*

- *and if x and y are two elements of S with the same restriction to F^+ then $x = y$.*

For $x \in S$ let π_x^0 be a square integrable representation of $GL_n(F_x)$. Also let ξ be an irreducible representation of G over $\overline{\mathbb{Q}}_l^{\mathrm{ac}}$. Then we can find an irreducible admissible representation π of $G(\mathbb{A}^\infty)$ such that

- $\dim[R_\xi(\pi)] \neq (0)$,

- *for $x \in S$ we have $\pi_x \cong \pi_x^0 \otimes (\psi_x \circ \det)$ for some character ψ_x of $F_x^\times / \mathcal{O}_{F,x}^\times$,*

- *and for $x \in S$ we have $\psi_\pi|_{\mathcal{O}_{E,x|_E^c}^\times} = 1$.*

Proof: Choose an embedding $\tau : F \hookrightarrow \mathbb{C}$. We may suppose that for some $x \in S$ the representation π_x^0 is supercuspidal. Let $S(\mathbb{Q})$ denote the set of places of $\mathbb{Q}$ below places in S. If $y \in S(\mathbb{Q})$ then it gives rise to a distinguished place $\widetilde{y}$ of E above y, i.e. $x|_E$ for any place $x \in S$ above y. Decompose

$$G(\mathbb{A}^\infty) \cong G(\mathbb{A}^{S(\mathbb{Q}) \cup \{\infty\}}) \times \prod_{y \in S(\mathbb{Q})} (\mathbb{Q}_y^\times \times \prod_{x|\widetilde{y}} (B_x^{\mathrm{op}})^\times).$$

If $U \subset G(\mathbb{A}^{S(\mathbb{Q}) \cup \{\infty\}})$ is an open compact subgroup, then set

$$\varphi_U = \mathrm{char}_U \times \prod_{y \in S(\mathbb{Q})} \mathrm{char}_{\mathbb{Z}_y^\times} \times \prod_{x|\widetilde{y}} \varphi_x,$$

where

- if $x \notin S$ then φ_x is the characteristic function of some open compact subgroup,

- while if $x \in S$ then then φ_x is the product of a pseudo-coefficient $\varphi_{\pi_x^0}$ for π_x^0 with the characteristic function of $\det^{-1}(\mathcal{O}_{F,x}^\times)$.

We may and will choose U sufficiently small so that the only element of finite order in the intersection of $G_\tau(\mathbb{Q})$ with the support of φ_U is 1.

Then proposition III.3.1 and lemma I.3.1 tell us that

$$\mathrm{vol}\,(G_\tau(\mathbb{R})_0^1)\mathrm{tr}\,(\varphi_U|H(X,\mathcal{L}_\xi)) = n\kappa_B(\dim\xi)\prod_{x\in S}(\dim\,\mathrm{JL}^{-1}(\pi_x^0)),$$

where we use Tamagawa measure on $G_\tau(\mathbb{A}) = G(\mathbb{A}^\infty) \times G_\tau(\mathbb{R})$ and a measure on $G_\tau(\mathbb{R})_0^1$ compatible with this, the association of measures on $G_\tau(\mathbb{R})$ and $G_\tau(\mathbb{R})_0$, the exact sequence

$$(0) \longrightarrow G_\tau(\mathbb{R})_0^1 \longrightarrow G_\tau(\mathbb{R})_0 \xrightarrow{\,|\nu|\,} \mathbb{R}_{>0}^\times \longrightarrow (0),$$

and the measure dt/t on $\mathbb{R}_{>0}^\times$. In particular we see that

$$\sum_\pi (\mathrm{tr}\,\pi(\varphi_U))(\dim[R_\xi(\pi)]) \neq 0,$$

where π runs over irreducible admissible representations of $G(\mathbb{A}^\infty)$.

We may therefore choose an irreducible admissible representation π of $G(\mathbb{A}^\infty)$ such that both

- $\dim[R_\xi(\pi)] \neq 0$

- and $\mathrm{tr}\,\pi(\varphi_U) \neq 0$.

The second condition implies that

- for $y \in S(\mathbb{Q})$ we have $\psi_\pi|_{\mathcal{O}_{E,\bar{y}^c}^\times} = 1$,

- and, for $x \in S$, we have

$$\mathrm{tr}\,\pi_x(\varphi_{\pi_x^0\otimes(\psi_x\circ\det)}) \neq 0$$

for some character ψ_x of $F_x^\times/\mathcal{O}_{F,x}^\times$ (see the argument for lemma V.5.1).

Thus for any x in S with π_x^0 supercuspidal we see that $\pi_x \cong \pi_x^0 \otimes (\psi_x \circ \det)$ (use lemma I.3.4). As we are assuming that some such place x exists it follows from corollary VI.2.4 that for all $x \in S$ the representation π_x is generic and hence it follows from corollary I.3.6 that $\pi_x \cong \pi_x^0 \otimes (\psi_x \circ \det)$ for all $x \in S$. Thus π is our desired representation of $G(\mathbb{A}^\infty)$. $\square$

Corollary VI.2.6 *Suppose that L is a CM field which is the composite of a totally real field L^+ and an imaginary quadratic field M. Suppose that S is a finite set of places of L such that*

- *if $x \in S$ then $x|_{\mathbb{Q}}$ splits in E,*

- *and if x and y are elements of S with the same restriction to L^+ then $x = y$.*

For $x \in S$ suppose that Π_x^0 is a square integrable representation of $GL_g(L_x)$. Suppose also that Ξ is an algebraic representation of $\mathrm{RS}_{\mathbb{Q}}^L(GL_g)$ over $\mathbb{C}$ such that $\Xi^c \cong \Xi^\vee$ (where c acts on $\mathrm{RS}_{\mathbb{Q}}^L(GL_g)$ via its action on L). Then we can find a cuspidal automorphic representation Π of $GL_g(\mathbb{A}_L)$ such that

1. *$\Pi^c \cong \Pi^\vee$,*

2. *Π_∞ has the same infinitesimal character as $\Xi^\vee|_{GL_g(L_\infty)}$,*

3. *and for all $x \in S$ there is a character ψ_x of $L_x^\times / \mathcal{O}_{L,x}^\times$ such that*

$$\Pi_x \cong \Pi_x^0 \otimes (\psi_x \circ \det).$$

Proof: We may assume that for some $x \in S$ the representation Π_x^0 is supercuspidal. Choose $(E, F, B, *, (\ ,\ ))$ as in section I.7 with $E = M$, $F = L$, $[B : F] = g^2$ and B_x split for all $x \in S$. We can choose an algebraic representation ξ of G over $\mathbb{Q}_l^{\mathrm{ac}}$ such that $\imath(\xi)_E|_{GL_n(L_\infty)} = \Xi|_{GL_n(L_\infty)}$. The corollary now follows from corollary VI.2.5 and theorem VI.2.1. $\square$

Corollary VI.2.7 (Clozel) *Let π be an irreducible admissible representation of $G(\mathbb{A}^\infty)$ over $\mathbb{Q}_l^{\mathrm{ac}}$ and let $\tau : F \hookrightarrow \mathbb{C}$. Suppose that one of the following two conditions hold.*

1. *There exists an irreducible representation π_∞ of $G_\tau(\mathbb{R})$, which is cohomological for $\imath(\xi)$, and such that $\imath(\pi) \otimes \pi_\infty$ is an automorphic representation of $G_\tau(\mathbb{A})$ with $\mathrm{BC}\,(\imath(\pi) \otimes \pi_\infty) = (\psi, \Pi)$ where $\mathrm{JL}\,(\Pi)$ is cuspidal.*

2. *For some rational prime q which splits in E and for some place x of F above q at which B splits, the representation π_x is generic.*

Then we have the following assertions.

1. *$R_\xi^i(\pi^\infty) = (0)$ unless $i = n - 1$.*

2. *If $\imath(\pi) \otimes \pi'_\infty$ is an automorphic representation of $G_\tau(\mathbb{A})$ with π'_∞ cohomological for $\imath(\xi)$, then π'_∞ is a discrete series representation of $G_\tau(\mathbb{R})$.*

3. *If $n = 2$ and $F^+ = \mathbb{Q}$ then there is one discrete series representation $\pi_\infty(\imath\xi)$ of $G_\tau(\mathbb{R})$ which is cohomological for $\imath(\xi)$. This representation $\pi_\infty(\imath\xi)$ is cohomological only in degree 1 and*

$$\dim H^1(\operatorname{Lie} G_\tau(\mathbb{R}), U_\tau, \pi_\infty(\xi) \otimes \imath(\xi)) = 2.$$

Otherwise there are n discrete series representations, which we will denote $\pi_\infty^0(\imath\xi), \ldots, \pi_\infty^{n-1}(\imath\xi)$, of $G_\tau(\mathbb{R})$ which are cohomological for $\imath(\xi)$. In each case $\pi_\infty^j(\imath\xi)$ is cohomological only in degree $n-1$ and

$$\dim H^{n-1}(\operatorname{Lie} G_\tau(\mathbb{R}), U_\tau, \pi_\infty^j(\imath\xi) \otimes \imath(\xi)) = 1.$$

Each of the representations $\imath(\pi) \otimes \pi_\infty^j(\imath\xi)$ is automorphic with the same multiplicity.

Proof: First note that by corollary VI.2.4, the second assumption implies that either $R_\xi^i(\pi) = (0)$ for all i or the first assumption holds. Thus we will restrict attention to the first assumption.

Because $\Pi_\infty^c \cong \Pi_\infty^\vee$ we see that $\psi_{\Pi_\infty}|_{(F_\infty^+)^\times} = 1$. Hence if for a place x of F, $\Pi_x \cong \chi_{x,1} \boxplus \cdots \boxplus \chi_{x,n}$ is unramified we have $|\psi_{\Pi_x}| = 1$ and

$$(\#k(x))^{-1/2} < |\chi_{x,i}(\varpi_x)| < (\#k(x))^{1/2}$$

(see corollary 2.5 of [JS1]). Thus for all but finitely many primes q which split $q = yy^c$ in E we have $\pi_q = \pi_{q,0} \times \prod_{x|y} \pi_x$ as x runs over primes of F above y, with $|\imath\psi_{\pi_x}| = 1$ and

$$(\#k(x))^{-1/2} < |\imath\chi_{x,i}(\varpi_x)| < (\#k(x))^{1/2}$$

if $\pi_x \cong \chi_{x,1} \boxplus \cdots \boxplus \chi_{x,n}$ is unramified. On the other hand ξ restricted to the central $\mathbb{G}_m$ sends $t \mapsto t^{-w(\xi)}$ and so $\psi_{\pi_\infty}(t) = t^{w(\xi)}$ for $t \in \mathbb{R}^\times$. Hence for $a \in \mathbb{Q}_p^\times$ we have

$$|\imath\pi_{p,0}(a)|^2 = |\imath\psi_{\pi,p}(a)| = |a|_p^{w(\xi)}.$$

Combining this with corollary V.6.3 we see that if α is an eigenvalue of Frob_x on $R_\xi^i(\pi)$ then

$$(\#k(x))^{(w(\xi)+n)/2-1} < |\imath\alpha| < (\#k(x))^{(w(\xi)+n)/2}.$$

This combined with the purity assertion in proposition III.2.1 gives the first part of the corollary.

The second part follows from the first as the only unitary (up to twist) representations π_∞ of $G_\tau(\mathbb{R})$ with

$$H^{n-1}(\operatorname{Lie} G_\tau(\mathbb{R}), U_\tau, \pi_\infty \otimes \imath(\xi)) \neq (0)$$

but

$$H^i(\operatorname{Lie} G_\tau(\mathbb{R}), U_\tau, \pi_\infty \otimes \imath(\xi)) = (0)$$

for all $i \neq n - 1$ are discrete series representations. (This follows from theorems 5.5, 5.6 and 6.16 of [VZ]. In the notation of that paper, note that $\wedge^i(\mathfrak{l} \cap \mathfrak{p}) \cong \bigoplus_{p+q=i} \wedge^p(\mathfrak{l} \cap \mathfrak{p}^+) \otimes \wedge^q(\mathfrak{l} \cap \mathfrak{p}^-)$ and that $\mathfrak{l} \cap \mathfrak{p}^+$ and $\mathfrak{l} \cap \mathfrak{p}^-$ are dual as $\mathfrak{l} \cap \mathfrak{k}$-modules. Thus $A_\mathfrak{q}(\lambda)$ has cohomology in dimension $\dim(\mathfrak{u} \cap \mathfrak{p}) + p$ for $p = 0, \ldots, \dim(\mathfrak{l} \cap \mathfrak{p}^+)$. If it has cohomology in only one dimension we deduce that $\mathfrak{l} \cap \mathfrak{p}^+ = (0)$. Hence in this case we also have $\mathfrak{l} \cap \mathfrak{p} = (0)$ so that $\mathfrak{l} \subset \mathfrak{k}$ and $P^d = G$. This implies that $A_\mathfrak{q}(\lambda)$ is discrete series.)

Everything except the last sentence of the third part is standard. To prove the last sentence let $f_{j,\infty}$ denote $(-1)^{n-1}$ times a pseudo-coefficient for $\pi_\infty^j(\imath\xi)$. If $f^\infty \in C_c^\infty(G(\mathbb{A}^\infty))$ then equation (4.1), lemma 4.1 and the remark at the start of the second paragraph after lemma 3.1 in [Ko4] show that

$$\sum_\pi m_\tau(\imath(\pi) \otimes \pi_\infty^j(\imath\xi)) \operatorname{tr}(\imath(\pi) \otimes \pi_\infty^j(\imath\xi))(f^\infty \times f_{j,\infty})$$

is independent of j, where $m_\tau(\pi')$ denotes the multiplicity of π' in the space of automorphic forms on $G_\tau(\mathbb{R})$. Hence

$$m_\tau(\imath(\pi) \otimes \pi_\infty^j(\imath\xi))$$

is also independent of j. $\square$

Corollary VI.2.8 *Let $\sigma : F \hookrightarrow \mathbb{Q}_l^{\mathrm{ac}}$ and let $\tau = \imath\sigma$. Let π be an irreducible representation of $G(\mathbb{A}^\infty)$ over $\mathbb{Q}_l^{\mathrm{ac}}$. Suppose that one of the following two conditions hold.*

1. *There exists an irreducible representation π_∞ of $G_\tau(\mathbb{R})$, which is cohomological for $\imath(\xi)$, and such that $\imath(\pi) \otimes \pi_\infty$ is an automorphic representation of $G_\tau(\mathbb{A})$ with $\mathrm{BC}(\imath(\pi) \otimes \pi_\infty) = (\psi, \Pi)$ where $\mathrm{JL}(\Pi)$ is cuspidal.*

2. *For some rational prime q which splits in E and for some place x of F above q at which B splits, the representation π_x is generic.*

Then

$$n \dim \operatorname{gr}^j D_{\mathrm{DR},\sigma}(R_\xi^{n-1}(\pi)) = \dim R_\xi^{n-1}(\pi)$$

if $j = p - \vec{a}(\xi, \sigma|_E)_{\sigma,n-p} - a_0(\xi, \sigma|_E)$ for a necessarily unique $p \in \{0, \ldots, n-1\}$, and

$$\dim \operatorname{gr}^j D_{\mathrm{DR},\sigma}(R_\xi^{n-1}(\pi)) = 0$$

otherwise. In particular

$$\mathrm{gr}^{\,j} D_{\mathrm{DR},\sigma}(R_\xi^{n-1}(\pi)) \neq (0)$$

for exactly n values of j.

Proof: We will treat the case $n > 2$ or $F^+ \neq 2$. The case $n = 2$ and $F^+ = \mathbb{Q}$ is very similar, but the book-keeping is slightly different. We leave the necessary changes to the reader. We will use without comment the notation established in section III.2.

Recall that we are identifying the dominant weights of a maximal torus in Q_τ / N_τ with

$$\mathbb{Z} \times (X(\mathbb{G}_m^{n-1})^+ \times \mathbb{Z}) \times \prod_{\tau'} X(\mathbb{G}_m^n)^+$$

where τ' runs over embeddings $F \hookrightarrow \mathbb{C}$ with $\tau'|_E = \tau|_E$ but $\tau' \neq \tau$. For $\lambda = (\lambda_0, (\vec{\lambda}_\tau, \lambda_1), \vec{\lambda}_{\tau'})$ in the real cone spanned by this set with $(\vec{\lambda}_\tau, \lambda_1) + \rho$ and $\vec{\lambda}_{\tau'} + \rho \in X(\mathbb{G}_m^n)$ we will let $\pi(\lambda)$ denote the discrete series representation of $G_\tau(\mathbb{R})$ with Harish-Chandra parameter λ.

Then we may take (combining the notation of section III.2 and of the last corollary)

$$\pi_\infty^p(\xi) = \pi(a_0(\xi, \sigma|_E), w_p(\vec{a}(\xi, \sigma|_E)_\sigma + \rho), \vec{a}(\xi, \sigma|_E)_{\iota^{-1}\tau'} + \rho)^\vee,$$

for $p = 0, \ldots, n-1$. According to theorem 3.4 of [Har1] we have

$$\dim H^{n-1-p}(\mathrm{Lie}\,Q_\tau, U_\tau, \pi_\infty^p(\xi) \otimes \mu^p(\xi)) = 1.$$

For such p set $j_p = p - \vec{a}(\xi, \sigma|_E)_{\sigma, n-p} - a_0(\xi, \sigma|_E)$. Then, by proposition III.2.1, we have

$$\dim \mathrm{gr}^{\,j_p} D_{\mathrm{DR},\sigma}(R_\xi^{n-1}(\pi)) \geq \# \ker^1(\mathbb{Q}, G_\tau) m_\tau(\iota(\pi) \otimes \pi_\infty^p(\xi)).$$

Combining this with proposition III.2.1 and the last lemma we get

$$
\begin{aligned}
&\# \ker^1(\mathbb{Q}, G_\tau) \sum_{p=0}^{n-1} m_\tau(\iota(\pi) \otimes \pi_\infty^p(\xi)) \\
= \;& \dim R_\xi^{n-1}(\pi) \\
= \;& \sum_j \dim \mathrm{gr}^{\,j} D_{\mathrm{DR},\sigma}(R_\xi^{n-1}(\pi)) \\
\geq \;& \sum_{p=0}^{n-1} \dim \mathrm{gr}^{\,j_p} D_{\mathrm{DR},\sigma}(R_\xi^{n-1}(\pi)) \\
\geq \;& \# \ker^1(\mathbb{Q}, G_\tau) \sum_{p=0}^{n-1} m_\tau(\iota(\pi) \otimes \pi_\infty^p(\xi))
\end{aligned}
$$

and conclude that all the inequalities are in fact equalities. In particular

$$
\begin{aligned}
\dim \mathrm{gr}^{\,j_p} D_{\mathrm{DR},\sigma}(R_\xi^{n-1}(\pi)) &= \# \ker^1(\mathbb{Q}, G_\tau) m_\tau(\iota(\pi) \otimes \pi_\infty^p(\xi)) \\
&= (\dim R_\xi^{n-1}(\pi))/n
\end{aligned}
$$

and, if $j \neq j_p$ for any p, then

$$\mathrm{gr}^j D_{\mathrm{DR},\sigma}(R_\xi^{n-1}(\pi)) = (0).$$

$\square$

We now turn our attention to descending representations from $(B^{\mathrm{op}} \otimes \mathbb{A})^\times$ to G_τ.

Theorem VI.2.9 *Suppose that Π is an irreducible automorphic representation of $(B^{\mathrm{op}} \otimes \mathbb{A})^\times$ and that ψ is a character of $\mathbb{A}_E^\times / E^\times$ such that*

1. $\Pi \cong \Pi^\#$,

2. $\psi_\Pi|_{\mathbb{A}_E^\times} = \psi^c/\psi$,

3. Π_∞ *is cohomological for* ξ'_E,

4. *and* $\xi'|_{E_\infty^\times}^{-1} = \psi|_{E_\infty^\times}^c$ *(where* $E_\infty^\times \subset G_\tau(\mathbb{R})$*).*

Then there is an irreducible automorphic representation π of $G_\tau(\mathbb{A})$ such that

1. $\mathrm{BC}(\pi) = (\psi, \Pi)$,

2. *and π_∞ is cohomological for ξ'.*

Moreover

$$\dim[R_{\iota^{-1}(\xi')}(\iota^{-1}\pi^\infty)] \neq 0.$$

Proof: The last assertion follows from the others and corollary V.6.4. The rest of this theorem will follow from proposition 2.3 of [Cl2]. (We caution the reader that the proof of proposition 2.3 in [Cl2] seems to us rather sketchy (see in particular the first paragraph of section 2.5 of [Cl2]). However this is remedied in theorem A.3.1 of [CL] where a complete derivation of the key trace identity is given.)

We will use the notation of the proof of theorem VI.2.1. Let π_1 be the automorphic representation of $G_{\tau,1}(\mathbb{A})$ whose existence is guaranteed by proposition 2.3 of [Cl2]. Then $\psi^c \times \pi_1$ is an irreducible automorphic representation of $(T \times G_{\tau,1})(\mathbb{A})$ which vanishes on $T^1(\mathbb{A})$ (because if $b \in \mathbb{A}_E^\times$ then $\psi^c(b/b^c) = \psi_\Pi(b) = \psi_{\pi_1}(b/b^c)$, the latter equality following from Clozel's definition of "base change lift"). Thus $\psi^c \times \pi_1$ is a subrepresentation of the restriction of some automorphic representation π of $G_\tau(\mathbb{A})$ to $(T \times G_{\tau,1})(\mathbb{A})$. Again from Clozel's definition of "base change lift" we see that $\mathrm{BC}(\pi) = (\psi_\pi|_{\mathbb{A}_E^\times}, \Pi)$. Finally because π_1 is cohomological for

$\xi'|_{G_{\tau,1}(\mathbb{R})}$ and because $\psi^c|_{E_\infty^\times}^{-1} = \xi'|_{E_\infty^\times}^{-1}$ we see that $\psi_\infty^c \times \pi_{1,\infty}$ and hence π_∞ is cohomological for ξ'. $\square$

It may sometimes be useful to combine theorem VI.2.9 with the following lemma.

Lemma VI.2.10 *Suppose that Π is an automorphic representation of $(B^{\mathrm{op}} \otimes \mathbb{A})^\times$ such that*

1. *$\Pi \cong \Pi^\#$,*

2. *and Π_∞ has the same infinitesimal character as some algebraic representation of $\mathrm{RS}_{\mathbb{Q}}^F(GL_n)$ over $\mathbb{C}$.*

Then we can find a character ψ of $\mathbb{A}_E^\times / E^\times$ and an algebraic representation ξ' of G over $\mathbb{C}$ such that

1. *$\psi_\Pi|_{\mathbb{A}_E^\times} = \psi^c/\psi$,*

2. *Π_∞ is cohomological for ξ'_E,*

3. *$\xi'|_{E_\infty^\times}^{-1} = \psi_\infty^c$ (where $E_\infty^\times \subset G_\tau(\mathbb{R})$),*

4. *and $\psi|_{\mathcal{O}_{E,u}^\times} = 1$.*

Proof: Because Π_∞ has the same infinitesimal character as an algebraic representation of $\mathrm{RS}_{\mathbb{Q}}^F(GL_n)$ over $\mathbb{C}$ we see that

$$\psi_{\Pi_\infty} : F_\infty^\times \longrightarrow \mathbb{C}^\times$$

is of the form

$$(x_{\tau'}) \longmapsto \prod_{\tau'} \tau'(x_{\tau'})^{a_{\tau'}} \tau'(x_{\tau'}^c)^{b_{\tau'}},$$

where $a_{\tau'}, b_{\tau'} \in \mathbb{Z}$ and where τ' runs over

$$\mathrm{Hom}_{\mathbb{Q}}(F^+, \mathbb{R}) \cong \mathrm{Hom}_{E,\tau|_E}(F, \mathbb{C}).$$

As $\Pi_\infty^\vee \cong \Pi_\infty^c$ we see that for all τ' we have $a_{\tau'} + b_{\tau'} = 0$. Thus if $x \in E_\infty^\times$ we have

$$\psi_{\Pi_\infty}(x) = (\tau(x)/\tau(x^c))^a,$$

where $a = \sum a_{\tau'} \in \mathbb{Z}$. Define

$$\psi_\infty : E_\infty^\times \longrightarrow \mathbb{C}^\times$$

by

$$\psi_\infty(x) = \tau(x)^{-a}.$$

Then $\psi_\infty \times \Pi_\infty$ defines a representation of $G_\tau(E_\infty) \cong E_\infty^\times \times GL_n(F_\infty)$ such that

$$(\psi_\infty \times \Pi_\infty) \circ c = \psi_\infty^c \psi_{\Pi_\infty}^c \times \Pi_\infty^{c\vee} \cong \psi_\infty \times \Pi_\infty.$$

(Note that the first c refers to the G_τ-structure, the second and third to the GL_1-structure and the fourth to the GL_n-structure.)

We may choose an algebraic representation Ξ of $\mathrm{RS}_\mathbb{Q}^E G_\tau$ over $\mathbb{C}$ such that $\psi_\infty \times \Pi_\infty$ and $\Xi^\vee$ have the same infinitesimal character. As $(\psi_\infty \times \Pi_\infty) \circ c \cong \psi_\infty \times \Pi_\infty$ we see that

$$\Xi \circ c \cong \Xi.$$

(Here c acts on E and so gives a $\mathbb{C}$-morphism from $\mathrm{RS}_\mathbb{Q}^E(G_\tau) \times \mathbb{C}$ to itself.) Under the isomorphism

$$(\mathrm{RS}_\mathbb{Q}^E G_\tau) \times \mathbb{C} \cong (G_\tau \times \mathbb{C}) \times_\mathbb{C} (G_\tau \times \mathbb{C})$$

we see that Ξ corresponds to $\xi' \otimes \xi'$ for some algebraic representation ξ' of G_τ over $\mathbb{C}$, i.e. $\Xi = \xi'_E$. Using τ to view $\mathbb{C}$ as an E-algebra we get an identification

$$G_\tau(\mathbb{C}) \cong \mathbb{C}^\times \times GL_n(F_\infty)$$

and hence an embedding

$$i : (E_\infty^\times)^2 \hookrightarrow G_\tau(\mathbb{C})$$

which sends (x_1, x_2) to $(\tau(x_1), x_2)$. Embedding $G_\tau(\mathbb{C})$ as the first factor of $\mathrm{RS}_\mathbb{Q}^E(G_\tau)(\mathbb{C}) \cong G_\tau(\mathbb{C}) \times G_\tau(\mathbb{C})$ we can extend i to a homomorphism

$$i : (E_\infty^\times)^2 \hookrightarrow \mathrm{RS}_\mathbb{Q}^E(G_\tau)(\mathbb{C}).$$

If instead we identify

$$\begin{aligned} \mathrm{RS}_\mathbb{Q}^E(G_\tau)(\mathbb{C}) &\cong (E \otimes_\mathbb{Q} \mathbb{C})^\times \times GL_n(F \otimes_\mathbb{Q} \mathbb{C}) \\ &\cong (\mathbb{C} \oplus \mathbb{C})^\times \times GL_n(F_\infty \oplus F_\infty) \end{aligned}$$

(using the identifications

$$((\tau \otimes 1) \oplus ((\tau \circ c) \otimes 1)) : E \otimes_\mathbb{Q} \mathbb{C} \xrightarrow{\sim} \mathbb{C}^2$$

and

$$((1 \otimes \tau^{-1}) \oplus (1 \otimes (c \circ \tau^{-1}))) : F \otimes_\mathbb{Q} \mathbb{C} \xrightarrow{\sim} F_\infty^2)$$

then

$$i(x_1, x_2) = (\tau(x_1), 1; x_2, 1).$$

Thus $(\xi' \circ i)(x_1, x_2) = \tau(x_1/x_2)^a$. If $x \in E_\infty^\times \subset G_\tau(\mathbb{R}) \subset G_\tau(\mathbb{C})$ then we can identify $x \in G_\tau(\mathbb{C})$ with $i(xx^c, x)$. Thus $\xi'(x) = \tau(x^c)^a$, i.e.

$$\xi'|_{E_\infty^\times}^{-1} = \psi_\infty^c.$$

Recall the exact sequence.

$$(0) \to \mathbb{Q}^\times \backslash \mathbb{A}^\times \to E^\times \backslash \mathbb{A}_E^\times \xrightarrow{1-c} E^\times \backslash \mathbb{A}_E^\times \xrightarrow{1+c} \mathbb{Q}^\times \backslash \mathbb{A}^\times \to \mathrm{Gal}\,(E/\mathbb{Q}) \to (0).$$

Note that $\psi_\Pi^{-1} = \psi_\Pi^c$ and so ψ_Π is trivial on $(\mathbb{A}_E^\times)^{1+c}$. From the formulae at the start of this proof we see that $\psi_{\Pi_\infty}(-1) = 1$. But from the exact sequence we see that $\mathbb{A}^\times = \mathbb{Q}^\times (\mathbb{A}_E^\times)^{1+c} \mathbb{R}^\times$, and so we deduce that

$$\psi_\Pi|_{\mathbb{A}^\times} = 1.$$

Again from the above exact sequence we see that

$$\psi_\Pi = \psi' \circ (c - 1)$$

for some character ψ' of $(\mathbb{A}_E^\times)^{c-1}/\{\pm 1\}$. Again from the explicit formulae at the start of this proof we see that

$$\psi'|_{(E_\infty^\times)^{c-1}} = \psi_\infty|_{(E_\infty^\times)^{c-1}}.$$

Thus we can choose a character ψ of $\mathbb{A}_E^\times/E^\times$ such that

- $\psi|_{(\mathbb{A}_E^\times)^{c-1}} = \psi'$,

- $\psi|_{E_\infty^\times} = \psi_\infty$,

- and $\psi|_{\mathcal{O}_{E,u}^\times} = 1$.

Then $\psi_\Pi|_{\mathbb{A}_E^\times} = \psi \circ (c - 1)$.

Finally because $(\xi_E')^\vee$ and Π_∞ have the same infinitesimal character we see by theorems 6.1 and 7.1 of [En] that Π_∞ is cohomological for ξ_E'. $\square$

Lemma VI.2.11 *Suppose that $s|n$ is a positive integer and that π_w^0 is an irreducible supercuspidal representation of $GL_{n/s}(F_w)$. Then we can find an irreducible representation ξ of G over $\mathbb{Q}_l^{ac}$, an irreducible admissible representation π^∞ of $G(\mathbb{A}^\infty)$ and a character ψ_w^0 of $F_w^\times/\mathcal{O}_{F,w}^\times$ such that*

   1. $\dim[R_\xi(\pi^\infty)] \neq 0$,

2. $\pi_w \cong (\pi_w^0 \boxplus \cdots \boxplus (\pi_w^0 \otimes |\det|^{s-1})) \otimes (\psi_w^0 \circ \det)$,

3. $\pi_{p,0}|_{\mathbb{Z}_p^\times} = 1$.

Proof: Choose $\tau : F \hookrightarrow \mathbb{C}$. Let $S(B)$ denote the set of places of F at which B is ramified. Recall that if $x \in S(B)$ then $x|_{\mathbb{Q}}$ splits in E.

As

$$\mathrm{RS}_{\mathbb{Q}}^F(GL_n) \times \mathbb{C} \cong (GL_n \times \mathbb{C})^{\mathrm{Hom}\,(F,\mathbb{C})},$$

we may choose a maximal torus $T \cong (GL_1^n \times \mathbb{C})^{\mathrm{Hom}\,(F,\mathbb{C})}$ and a Borel subgroup $B \supset T$ consisting of upper triangular matrices in $(GL_n \times \mathbb{C})^{\mathrm{Hom}\,(F,\mathbb{C})}$. Then we can identify

$$X^*(T) \cong (\mathbb{Z}^n)^{\mathrm{Hom}\,(F,\mathbb{C})}$$

in such a way that the set of B-positive weights, $X^*(T)_+$, consists of vectors $(x_{\tau',i})$ with $x_{\tau',i} \geq x_{\tau',j}$ whenever $i \leq j$. We will let $\rho \in X^*(T)_+$ denote half the sum of the positive roots, i.e.

$$\rho_{\tau',i} = (n+1)/2 - i.$$

(Note that perversely we are using a different notation in this lemma from the one we used in the rest of this book, particularly section III.2.) If Ξ is an irreducible algebraic representation of $\mathrm{RS}_{\mathbb{Q}}^F(GL_n)$ over $\mathbb{C}$ we will let $x(\Xi) \in X^*(T)_+$ denote its heighest weight. Note that $\Xi^c \cong \Xi^\vee$ (where c acts on $\mathrm{RS}_{\mathbb{Q}}^F(GL_n)$ via its action on F) if and only if

$$x(\Xi)_{\tau',i} + x(\Xi)_{c\tau',n+1-i} = 0$$

for all $\tau' \in \mathrm{Hom}\,(F,\mathbb{C})$ and all $i = 1, \ldots, n$. We will use exactly similar notation for $\mathrm{RS}_{\mathbb{Q}}^F(GL_{n/s})$.

Choose an irreducible algebraic representation Ξ' of $\mathrm{RS}_{\mathbb{Q}}^F(GL_{n/s})$ over $\mathbb{C}$ such that

- $(\Xi')^\vee \cong (\Xi')^c$

- and $x(\Xi')_{\tau',i+1} \geq x(\Xi')_{\tau',i} + (s-1)$ for all $\tau' \in \mathrm{Hom}\,(F,\mathbb{C})$ and all $i = 1, \ldots, n-1$.

By corollary VI.2.6 we may choose a cuspidal automorphic representation Π' of $GL_{n/s}(\mathbb{A}_F)$ such that

- $(\Pi')^\vee \cong (\Pi')^c$,

- Π'_∞ has the same infinitesimal character as $(\Xi')^\vee$,

- Π'_x is supercuspidal for all $x \in S(B)$,

- $\Pi'_w \cong \pi^0_w \otimes (\psi^1_w \circ \det)$ for some character ψ^1_w of $F^\times_w/\mathcal{O}^\times_{F,w}$.

Also choose a character ϕ of $\mathbb{A}^\times_F/F^\times$ such that

- $\phi^{-1} = \phi^c$;

- for every $\tau' : F \hookrightarrow \mathbb{C}$ defining a place y of F we have

$$\phi_y : z \longmapsto (\tau'(z)/|\tau'(z)|)^{\delta_{\tau'}},$$

where $\delta_{\tau'} = 0$ if either s or n/s is odd, while $\delta_{\tau'} = \pm 1$ if both both s and n/s are even;

- and ϕ_w is unramified.

(The existence of such a character ϕ is proved exactly as in the proof of corollary VII.2.8.) Note that $\Pi'_\infty \otimes \phi_\infty$ has infinitesimal character parametrised by $x' \in X^*(T)$, where

$$x'_{\tau',i} = (n/s+1)/2 - i - x(\Xi')_{\tau',n/s+1-i} + \delta_{\tau'}/2.$$

According to the main theorem of [MW] there is an irreducible automorphic representation Π' of $GL_n(\mathbb{A}_F)$ which occurs in the discrete spectrum and which is a subquotient of

$$\text{n-Ind}^{GL_n(\mathbb{A}_F)}_{Q(\mathbb{A}_F)}(\Pi' \otimes \phi \otimes |\det|^{(1-s)/2}) \times \cdots \times (\Pi' \otimes \phi \otimes |\det|^{(s-1)/2}),$$

where $Q \subset GL_n$ is a parabolic subgroup with Levi component $GL^s_{n/s}$. Moreover we have the following properties.

1. $\Pi^\vee = \Pi^c$ (by the strong multiplicity one theorem).

2. Π_∞ has the same infinitesimal character as the algebraic representation Ξ of $\text{RS}^F_{\mathbb{Q}}(GL_n)$ over $\mathbb{C}$ with

$$x(\Xi)_{\tau',(I-1)s+J} = 1/2(s-1)(2I-1-n/s) + x(\Xi')_{\tau',I} - \delta_{\tau'}/2,$$

for $I = 1, \ldots, n/s$ and $J = 1, \ldots, s$. (Note that by our assumptions on $\delta_{\tau'}$ and $x(\Xi')$ we do have $x(\Xi) \in X^*(T)_+$ and so $\Xi^\vee$ has infinitesimal character with parameter $x \in X^*(T)$ given by

$$x_{\tau',(I-1)s+J} =$$

$$(n+1)/2 - (I-1)S - J - 1/2(s-1)(2(n/s+1-I)-1-n/s)$$
$$-x(\Xi')_{\tau',n/s+1-I} + \delta_{\tau'}/2 =$$

$$(n/s+1)/2 - I + (s+1)/2 - J - x(\Xi')_{\tau',n/s+1-I} + \delta_{\tau'}/2,$$

for $I = 1, \ldots, n/s$ and $J = 1, \ldots, s$. This element $x \in X^*(T)$ also parametrises the infinitesimal character of any subquotient of

$$\text{n-Ind}_{Q(F_\infty)}^{GL_n(F_\infty)}(\Pi'_\infty \otimes \phi_\infty \otimes |\det|^{(1-s)/2}) \times \ldots$$
$$\times (\Pi'_\infty \otimes \phi_\infty \otimes |\det|^{(s-1)/2}).)$$

3. For each $x \in S(B)$, $\Pi_x \cong (\pi_x^0 \boxplus \cdots \boxplus (\pi_x^0 \otimes |\det|^{s-1}))$, for some super-cuspidal representation π_x^0 of $GL_{n/s}(F_x)$ (see [MW] and proposition 2.10 of [Ze]).

4. $\Pi_w \cong (\pi_w^0 \boxplus \cdots \boxplus (\pi_w^0 \otimes |\det|^{s-1})) \otimes (\psi_w^0 \circ \det)$, for some character ψ_w^0 of $F_w^\times / \mathcal{O}_{F,w}^\times$ (see [MW] and proposition 2.10 of [Ze]).

Applying theorem VI.1.1, lemma VI.2.10 and theorem VI.2.9 we see that there is an irreducible automorphic representation π of $G_\tau(\mathbb{A})$ such that

- BC$(\pi) = (\psi, \text{JL}^{-1}(\Pi))$ (for some character ψ)

- and $\dim[R_\xi(\imath^{-1}(\pi^\infty))] \neq 0$.

Then $\imath^{-1}(\pi^\infty)$ is our desired representation of $G(\mathbb{A}^\infty)$. $\square$

Chapter VII

Applications

VII.1   Galois representations

Let us start by rephrasing our main theorem V.6.1. Define

$$\text{n-red}_\rho^{(h)} : \text{Groth}\,(G(\mathbb{A}^\infty))$$

$$\to \text{Groth}\,(G(\mathbb{A}^{\infty,p}) \times \mathbb{Q}_p^\times \times GL_h(F_w) \times \prod_{i=2}^r (B_{w_i}^{\text{op}})^\times)$$

in the same manner we defined $\text{red}_\rho^{(h)}$ (see section V.6) except that we replace $J_{N_h^{\text{op}}}(\pi) \otimes \delta_{P_h}^{1/2}$ by simply $J_{N_h^{\text{op}}}(\pi)$. Using corollary VI.2.3 we get the following reformulation of theorem V.6.1.

Theorem VII.1.1 *Suppose that $\pi = \pi^p \times \pi_{p,0} \times \pi_{w_1} \times \cdots \times \pi_{w_r}$ is an irreducible admissible representation of $G(\mathbb{A}^\infty)$ such that $\pi_{p,0}|_{\mathbb{Z}_p^\times} = 1$. Then in $\text{Groth}_l(GL_n(F_w) \times W_{F_w})$ we have*

$$n[\pi_w][R_\xi(\pi)|_{W_{F_w}}] \;=\; (\dim[R_\xi(\pi)]) \sum_{h=0}^{n-1} \sum_\rho \text{n-Ind}_{P_h(F_w)}^{GL_n(F_w)}$$
$$((\text{n-red}_\rho^{(h)}[\pi_w])[\Psi_{F_w,l,n-h}(\rho) \otimes ((\pi_{p,0}^{-1} \otimes |\ |_p^{-h/2}) \circ \text{Art}_{\mathbb{Q}_p}^{-1})|_{W_{F_w}}])$$

where ρ runs over irreducible admissible representations of $D_{F_w,n-h}^\times$.

As a special case we get the following consequence.

Corollary VII.1.2 *Suppose that $\pi = \pi^p \times \pi_{p,0} \times \pi_{w_1} \times \cdots \times \pi_{w_r}$ is an irreducible admissible representation of $G(\mathbb{A}^\infty)$ such that $\pi_{p,0}|_{\mathbb{Z}_p^\times} = 1$ and π_w is supercuspidal. Then in $\text{Groth}_l(GL_n(F_w) \times W_{F_w})$ we have an equality between*

$$n[\pi_w][R_\xi(\pi)|_{W_{F_w}}]$$

217

and

$$(\dim[R_\xi(\pi)])[\Psi_{F_w,l,n}(\text{JL}^{-1}(\pi_w)^\vee) \otimes (\pi_{p,0}^{-1} \circ \text{Art}_{\mathbb{Q}_p}^{-1})|_{W_{F_w}}].$$

Proof: Use the following facts.

- If π_w is supercuspidal and if N is the unipotent radical of a proper parabolic subgroup of $GL_n(F_w)$ then $J_N(\pi_w) = (0)$.

- If π_w is a supercuspidal representation of $GL_n(F_w)$ and if π'_w is a square integrable representation of $GL_n(F_w)$ with the same central character, then it follows from the results listed in section I.3 that $\text{tr } \pi_w(\varphi_{\pi'_w}) = 0$ unless $\pi'_w \cong \pi_w$ in which

$$\text{tr } \pi_w(\varphi_{\pi'_w}) = \text{vol}\,(D_{F_w,n}^\times/F_w^\times).$$

□

Theorem VII.1.3 *Suppose that K is a p-adic field and $l \neq p$ is a prime. Suppose that π is a supercuspidal representation of $GL_g(K)$. Then there is a continuous semi-simple representation*

$$r_l(\pi) : W_K \longrightarrow GL_g(\mathbb{Q}_l^{\text{ac}})$$

such that

$$[\Psi_{K,l,g}(\text{JL}^{-1}(\pi^\vee))] = [\pi \otimes r_l(\pi)].$$

Proof: Choose $(E, F^+, w, B, *, (\,,\,), \Lambda_i, \xi)$ as in section I.7 and such that $F_w^+ \cong K$ and $[B : F] = g^2$. By corollaries VI.2.5 and VI.2.7 we may choose an irreducible admissible representation $\widetilde{\pi}$ of $G(\mathbb{A}^\infty)$ such that

- $[R_\xi(\widetilde{\pi})] = [R_\xi^{g-1}(\widetilde{\pi})] \neq 0$,

- $\widetilde{\pi}_w \cong \pi \otimes (\psi \circ \det)$ for some unramified character ψ of $K^\times/\mathcal{O}_K^\times$,

- and $\widetilde{\pi}_{p,0}|_{\mathbb{Z}_p^\times} = 1$.

By the previous corollary we see that in $\text{Groth}_l(GL_g(K) \times W_K)$ we have

$$g[\widetilde{\pi}_w \otimes R_\xi(\widetilde{\pi})|_{W_K}]$$
$$= (\dim[R_\xi(\widetilde{\pi})])[\Psi_{K,l,g}(\text{JL}^{-1}(\widetilde{\pi}_w^\vee)) \otimes (\widetilde{\pi}_{p,0}^{-1} \circ \text{Art}_{\mathbb{Q}_p}^{-1})|_{W_K}].$$

Thus, if we set

$$[r_l(\pi)] = g(\dim[R_\xi(\widetilde{\pi})])^{-1}[R_\xi(\widetilde{\pi})|_{W_K} \otimes (\widetilde{\pi}_{p,0} \circ \text{Art}_{\mathbb{Q}_p}^{-1})|_{W_K} \otimes (\psi \circ \text{Art}_K^{-1})],$$

we see that

$$[r_l(\pi)] \in \text{Groth}_l(W_K),$$

that

$$[\Psi_{K,l,g}(\text{JL}^{-1}(\pi^\vee))] = [\pi][r_l(\pi)],$$

that

$$\dim[r_l(\pi)] = g$$

and that $[r_l(\pi)]$ can be represented by a true representation $r_l(\pi)$. □

The following lemma follows from the definitions, from lemma II.2.9 and from lemma II.2.11.

Lemma VII.1.4 *Suppose that K and K' are p-adic fields and $l \neq p$ is a prime.*

1. *If $\sigma : K \xrightarrow{\sim} K'$ is an isomorphism of $\mathbb{Q}_p$-algebras then for any irreducible supercuspidal representation π' of $GL_g(K')$ we have*

$$r_l(\pi' \circ \sigma) = r_l(\pi')^\sigma,$$

where, if we fix an extension $\tilde{\sigma}$ of σ to an isomorphism $\tilde{\sigma} : K^{\text{ac}} \xrightarrow{\sim} (K')^{\text{ac}}$, then $r_l(\pi')^\sigma(x) = r_l(\pi')(\tilde{\sigma}x\tilde{\sigma}^{-1}).$

2. *If $\sigma \in \text{Aut}(\mathbb{Q}_l^{\text{ac}})$ and if π is an irreducible supercuspidal representation of $GL_g(K)$ then*

$$r_l(\sigma(\pi)) = \sigma(r_l(\pi)).$$

3. *If π is an irreducible supercuspidal representation of $GL_g(K)$ and if ψ is a character of $K^\times / \mathcal{O}_K^\times$ then*

$$r_l(\pi \otimes (\psi \circ \det)) = r_l(\pi) \otimes (\psi^{-1} \circ \text{Art}_K^{-1}).$$

4. *If π is a character of $K^\times$ then $r_l(\pi) = \pi^{-1} \circ \text{Art}_K^{-1}$.*

For the rest of this section we will use without comment the notation established in section I.3.

Theorem VII.1.5 *Suppose that K is a p-adic field and $l \neq p$ is a prime. Suppose also that s and g are positive integers and that π is an irreducible supercuspidal representation of $GL_g(K)$. Then*

$$[\Psi_{K,l,gs}(\text{JL}^{-1}(\text{Sp}_s(\pi)^\vee))] =$$
$$\sum_{j=1}^{s}(-1)^{s-j}[\text{Sp}_j(\pi) \boxplus (\pi \otimes |\det|^j) \boxplus \cdots \boxplus (\pi \otimes |\det|^{s-1})]$$
$$[r_l(\pi \otimes |\det|^{j-1}) \otimes |\text{Art}_K^{-1}|^{g(1-s)/2}].$$

Proof: We will argue by induction on s. The case $s = 1$ is just the previous theorem. Thus suppose the theorem is proved for all $s' < s$.

Choose $(E, F^+, w, B, *, (\ ,\ ), \Lambda_i, \xi)$ as in section I.7 and such that $F_w^+ \cong K$ and $[B : F] = g^2 s^2$. By corollary VI.2.5 we may choose an irreducible admissible representation $\widetilde{\pi}$ of $G(\mathbb{A}^\infty)$ such that

- $\dim[R_\xi(\widetilde{\pi})] \neq 0$,

- $\widetilde{\pi}_w \cong \mathrm{Sp}_s(\pi) \otimes (\psi \otimes \det)$ for some character ψ of $K^\times / \mathcal{O}_K^\times$,

- and $\widetilde{\pi}_{p,0}|_{\mathbb{Z}_p^\times} = 1$.

If we write π' for $\pi \otimes (\psi \circ \det)$, then by theorem VII.1.1 and lemma I.3.3 we see that

$$gs[\mathrm{Sp}_s(\pi')][R_\xi(\widetilde{\pi})] =$$

$$(\dim[R_\xi(\widetilde{\pi})]) \sum_{h'=0}^{s-1} \text{n-Ind}_{P_{gh'}(K)}^{GL_{sg}(K)} [\mathrm{Sp}_{h'}(\pi' \otimes |\det|^{s-h'}) \times$$
$$\Psi_{K,l,g(s-h')}(\mathrm{JL}^{-1}(\mathrm{Sp}_{s-h'}(\pi')^\vee)) \otimes ((\widetilde{\pi}_{p,0}^{-1} \otimes |\ |_p^{-gh'/2}) \circ \mathrm{Art}_{\mathbb{Q}_p}^{-1})|_{W_K}].$$

By the inductive hypothesis we may rewrite this

$$gs[\mathrm{Sp}_s(\pi')][R_\xi(\widetilde{\pi}) \otimes ((\widetilde{\pi}_{p,0} \otimes |\ |_p^{g(s-1)/2}) \circ \mathrm{Art}_{\mathbb{Q}_p}^{-1})|_{W_K}] =$$

$$(\dim[R_\xi(\widetilde{\pi})])[\Psi_{K,l,gs}(\mathrm{JL}^{-1}(\mathrm{Sp}_s(\pi')^\vee)) \otimes |\mathrm{Art}_K^{-1}|^{g(s-1)/2}] +$$
$$(\dim[R_\xi(\widetilde{\pi})]) \sum_{h'=1}^{s-1} \sum_{j=1}^{s-h'} (-1)^{s-h'-j} \text{n-Ind}_{P_{h'g}(K)}^{GL_{gs}(K)}[(\mathrm{Sp}_j(\pi') \boxplus$$
$$(\pi' \otimes |\det|^j) \boxplus \cdots \boxplus (\pi' \otimes |\det|^{s-h'-1})) \times \mathrm{Sp}_{h'}(\pi' \otimes |\det|^{s-h'})]$$
$$[r_l(\pi' \otimes |\det|^{j-1})].$$

By lemma I.3.2 this can be rewritten

$$[\Psi_{K,l,gs}(\mathrm{JL}^{-1}(\mathrm{Sp}_s(\pi')^\vee)) \otimes |\mathrm{Art}_K^{-1}|^{g(s-1)/2}] =$$

$$gs(\dim[R_\xi(\widetilde{\pi})])^{-1}[\mathrm{Sp}_s(\pi')][R_\xi(\widetilde{\pi}) \otimes ((\widetilde{\pi}_{p,0} \otimes |\ |_p^{g(s-1)/2}) \circ \mathrm{Art}_{\mathbb{Q}_p}^{-1})|_{W_K}] -$$
$$\sum_{h'=1}^{s-1}[\mathrm{Sp}_s(\pi')][r_l(\pi' \otimes |\det|^{s-1-h'})] -$$
$$\sum_{h'=1}^{s-1} \sum_{j=1}^{s-1-h'} (-1)^{s-h'-j}[\omega(\vec{\Gamma}_{j,s-1-h'-j,h'+1}(\pi'))][r_l(\pi' \otimes |\det|^{j-1})] -$$
$$\sum_{h'=1}^{s-1} \sum_{j=1}^{s-h'} (-1)^{s-h'-j}[\omega(\vec{\Gamma}_{j,s-h'-j,h'}(\pi'))][r_l(\pi' \otimes |\det|^{j-1})] =$$

$$gs(\dim[R_\xi(\widetilde{\pi})])^{-1}[\mathrm{Sp}_s(\pi')][R_\xi(\widetilde{\pi}) \otimes ((\widetilde{\pi}_{p,0} \otimes |\ |_p^{g(s-1)/2}) \circ \mathrm{Art}_{\mathbb{Q}_p}^{-1})|_{W_K}] -$$
$$\sum_{h'=1}^{s-1}[\mathrm{Sp}_s(\pi')][r_l(\pi' \otimes |\det|^{s-1-h'})] +$$
$$\sum_{j=1}^{s-2} \sum_{h''=2}^{s-j} (-1)^{s-h''-j}[\omega(\vec{\Gamma}_{j,s-h''-j,h''}(\pi'))][r_l(\pi' \otimes |\det|^{j-1})] -$$
$$\sum_{j=1}^{s-1} \sum_{h'=1}^{s-j} (-1)^{s-h'-j}[\omega(\vec{\Gamma}_{j,s-h'-j,h'}(\pi'))][r_l(\pi' \otimes |\det|^{j-1})] =$$

$$gs(\dim[R_\xi(\widetilde{\pi})])^{-1}[\operatorname{Sp}_s(\pi')][R_\xi(\widetilde{\pi}) \otimes ((\widetilde{\pi}_{p,0} \otimes | \ |_p^{g(s-1)/2}) \circ \operatorname{Art}_{\mathbb{Q}_p}^{-1})_{W_K}] -$$
$$\sum_{h'=1}^{s-1}[\operatorname{Sp}_s(\pi')][r_l(\pi' \otimes | \det |^{s-1-h'})]+$$
$$\sum_{j=1}^{s-1}(-1)^{s-j}[\omega(\widetilde{\Gamma}_{j,s-1-j,1}(\pi'))][r_l(\pi' \otimes | \det |^{j-1})] =$$

$$\sum_{j=1}^{s}[\operatorname{Sp}_j(\pi') \boxplus (\pi' \otimes | \det |^j) \boxplus \cdots \boxplus (\pi' \otimes | \det |^{s-1})][r_l(\pi' \otimes | \det |^{j-1})]$$
$$+[\operatorname{Sp}_s(\pi')][A'],$$

for some $[A'] \in \operatorname{Groth}(W_K)$. Then by lemma II.2.9 we see that

$$[\Psi_{K,l,gs}(\operatorname{JL}^{-1}(\operatorname{Sp}_s(\pi)^\vee))] =$$

$$\sum_{j=1}^{s}[\operatorname{Sp}_j(\pi) \boxplus (\pi \otimes | \det |^j) \boxplus \cdots \boxplus (\pi \otimes | \det |^{s-1})]$$
$$[r_l(\pi' \otimes | \det |^{j-1}) \otimes |\operatorname{Art}_K^{-1}|^{g(1-s)/2}] + [\operatorname{Sp}_s(\pi)][A],$$

for some $[A] \in \operatorname{Groth}(W_K)$.

It remains to show that $[A] = 0$. By lemma VI.2.11 we may choose an irreducible representation $\widetilde{\xi}$ of G over $\mathbb{Q}_l^{ac}$ and an irreducible admissible representation $\widetilde{\pi}'$ of $G(\mathbb{A}^\infty)$ such that

- $\dim[R_{\widetilde{\xi}}(\widetilde{\pi}')] \neq 0$,

- $\widetilde{\pi}'_w \cong (\pi \boxplus \cdots \boxplus (\pi \otimes | \det |^{s-1})) \otimes (\psi' \circ \det)^{-1}$ for some character ψ' of $K^\times / \mathcal{O}_K^\times$,

- and $\widetilde{\pi}'_{p,0}|_{\mathbb{Z}_p^\times} = 1$.

Now let π' denote $\pi \otimes (\psi' \circ \det)^{-1}$. By theorem VII.1.1 and lemma I.3.3 we also see that

$$gs[\pi' \boxplus \cdots \boxplus (\pi' \otimes | \det |^{s-1})][R_{\widetilde{\xi}}(\widetilde{\pi}')] =$$

$$(\dim[R_{\widetilde{\xi}}(\widetilde{\pi}')]) \sum_{h'=0}^{s-1}(-1)^{s-1-h'}\text{n-Ind}_{P_{gh'}(K)}^{GL_{gs}(K)}[(\pi' \boxplus \ldots$$
$$\boxplus(\pi' \otimes | \det |^{h'-1})) \times \Psi_{K,l,g(s-h')}(\operatorname{JL}^{-1}(\operatorname{Sp}_{s-h'}(\pi' \otimes | \det |^{h'})^\vee)) \otimes$$
$$(((\widetilde{\pi}'_{p,0})^{-1} \otimes | \ |_p^{-gh'/2}) \circ \operatorname{Art}_{\mathbb{Q}_p}^{-1})|_{W_K}].$$

By the inductive hypothesis and what we have already proved, this can be rewritten

$$gs[\pi' \boxplus \cdots \boxplus (\pi' \otimes | \det |^{s-1})][R_{\widetilde{\xi}}(\widetilde{\pi}')|_{W_K} \otimes (\widetilde{\pi}'_{p,o} \circ \operatorname{Art}_{\mathbb{Q}_p}^{-1})|_{W_K}] =$$

$$(-1)^{s-1}(\dim[R_{\widetilde{\xi}}(\widetilde{\pi}')])[\operatorname{Sp}_s(\pi')][A \otimes (\psi' \circ \det)]+$$
$$(\dim[R_{\widetilde{\xi}}(\widetilde{\pi}')]) \sum_{h'=0}^{s-1} \sum_{j=1}^{s-h'}(-1)^{j-1}\text{n-Ind}_{P_{gh'}(K)}^{GL_{gs}(K)}[(\pi' \boxplus \ldots$$
$$\boxplus(\pi' \otimes | \det |^{h'-1}))$$
$$\times(\operatorname{Sp}_j(\pi' \otimes | \det |^{h'}) \boxplus (\pi' \otimes | \det |^{h'+j}) \boxplus \cdots \boxplus (\pi' \otimes | \det |^{s-1}))]$$
$$[r_l(\pi' \otimes | \det |^{h'+j-1}) \otimes |\operatorname{Art}_K^{-1}|^{g(1-s)/2}].$$

Again using lemma I.3.2 this becomes

$$gs[\pi' \boxplus \cdots \boxplus (\pi' \otimes |\det|^{s-1})][R_{\tilde{\xi}}(\tilde{\pi}')|_{W_K} \otimes (\tilde{\pi}'_{p,o} \circ \operatorname{Art}_{\mathbb{Q}_p}^{-1})|_{W_K}] =$$

$$(-1)^{s-1}(\dim[R_{\tilde{\xi}}(\tilde{\pi}')])[\operatorname{Sp}_s(\pi')][A \otimes (\psi' \circ \det)]+$$
$$(\dim[R_{\tilde{\xi}}(\tilde{\pi}')]) \sum_{h'=0}^{s-1} \sum_{j=1}^{s-h'} (-1)^{j-1}[\omega(\vec{\Gamma}'_{h',j,s-1-h'-j}(\pi'))]$$
$$[r_l(\pi' \otimes |\det|^{h'+j-1}) \otimes |\operatorname{Art}_K^{-1}|^{g(1-s)/2}]-$$
$$\sum_{h''=0}^{s-2} \sum_{j'=2}^{s-h''} (-1)^{j'-1}[\omega(\vec{\Gamma}'_{h'',j',s-1-h''-j'}(\pi'))]$$
$$[r_l(\pi' \otimes |\det|^{h''+j-1}) \otimes |\operatorname{Art}_K^{-1}|^{g(1-s)/2}] =$$

$$(-1)^{s-1}(\dim[R_{\tilde{\xi}}(\tilde{\pi}')])[\operatorname{Sp}_s(\pi')][A \otimes (\psi' \circ \det)]+$$
$$(\dim[R_{\tilde{\xi}}(\tilde{\pi}')]) \sum_{h'=0}^{s-1}[\pi' \boxplus \cdots \boxplus (\pi' \otimes |\det|^{s-1})]$$
$$[r_l(\pi' \otimes |\det|^{h'}) \otimes |\operatorname{Art}_K^{-1}|^{g(1-s)/2}].$$

In particular we see that $[A] = 0$, as desired. $\square$

Suppose that π is an irreducible admissible representation of $GL_g(K)$. Then we can find

- a parabolic subgroup $P \subset GL_n$ with a Levi component isomorphic to $GL_{g_1} \times \cdots \times GL_{g_t}$,

- and an irreducible supercuspidal representation π_i of $GL_{g_i}(K)$;

such that π is a subquotient of

$$\operatorname{n-Ind}_{P(K)}^{GL_g(K)}(\pi_1 \times \cdots \times \pi_t).$$

Moreover the multiset $\{\pi_i\}$ is independent of all choices. Thus we may define a semi-simple representation $r_l(\pi)$ of W_K by

$$[r_l(\pi)] = \sum_{i=1}^{t}[r_l(\pi_i) \otimes |\operatorname{Art}_K^{-1}|^{(g_i-g)/2}].$$

The following lemma follows at once from lemma VII.1.4.

Lemma VII.1.6 *Suppose that K and K' are p-adic fields and $l \neq p$ is a prime.*

1. *If $\sigma : K \xrightarrow{\sim} K'$ is an isomorphism of $\mathbb{Q}_p$-algebras then for any irreducible admissible representation π' of $GL_g(K')$ we have*

$$r_l(\pi' \circ \sigma) = r_l(\pi')^{\sigma},$$

where, if we fix an extension $\tilde{\sigma}$ of σ to an isomorphism $\tilde{\sigma} : K^{\mathrm{ac}} \xrightarrow{\sim} (K')^{\mathrm{ac}}$, then $r_l(\pi')^{\sigma}(x) = r_l(\pi')(\tilde{\sigma}x\tilde{\sigma}^{-1})$.

2. *If $\sigma \in \mathrm{Aut}\,(\mathbb{Q}_l^{ac})$ and if π is an irreducible admissible representation of $GL_g(K)$ then*

$$r_l(\sigma(\pi)) = \sigma(r_l(\pi)).$$

3. *If π is an irreducible admissible representation of $GL_g(K)$ and if ψ is a character of $K^\times / \mathcal{O}_K^\times$ then*

$$r_l(\pi \otimes (\psi \circ \det)) = r_l(\pi) \otimes (\psi^{-1} \circ \mathrm{Art}\,_K^{-1}).$$

We now return to the analysis of $[R_\xi(\pi)|_{W_{F_w}}]$.

Theorem VII.1.7 *Suppose that π is an irreducible admissible representation of $G(\mathbb{A}^\infty)$. Then*

$$n[R_\xi(\pi)|_{W_{F_w}}] = (\dim[R_\xi(\pi)])[r_l(\pi_w) \otimes (\pi_{p,0} \circ |\mathrm{Art}\,_{F_w}^{-1}|)].$$

Proof: Consider the two homomorphisms

$$\Theta_1, \Theta_2 : \mathrm{Groth}\,(GL_n(F_w)) \longrightarrow \mathrm{Groth}\,_l(GL_n(F_w) \times W_{F_w})$$

defined by

$$\Theta_1([\pi]) = [\pi \otimes r_l(\pi)]$$

for any irreducible π and

$$\Theta_2([\pi]) = \sum_{h=0}^{n-1} \sum_\rho \text{n-Ind}\,_{P_h(F_w)}^{GL_n(F_w)} \text{n-red}_\rho^{(h)}[\pi][\Psi_{F_w,l,n-h}(\rho) \otimes |\mathrm{Art}\,_K^{-1}|^{-h/2}]$$

for any admissible π (where ρ runs over irreducible admissible representations of $D_{F_w,n-h}^\times$). By theorem VII.1.1 we only need to show that $\Theta_1 = \Theta_2$. Moreover it follows from lemma A.4.f of [DKV] that we only need check that $\Theta_1([\pi]) = \Theta_2([\pi])$ when π is a full induced from square integrable.

Thus suppose that we have positive integers $s_1, \ldots, s_t$ and $n_1, \ldots, n_t$ such that $n = s_1 n_1 + \cdots + s_t n_t$. Suppose also that for $i = 1, \ldots, t$ we have an irreducible supercuspidal representation π_i of $GL_{n_i}(F_w)$. Let $P \subset GL_n$ be a parabolic subgroup with Levi component $GL_{s_1 n_1} \times \cdots \times GL_{s_t n_t}$ and write π for

$$\text{n-Ind}\,_{P(F_w)}^{GL_n(F_w)} (\mathrm{Sp}\,_{s_1}(\pi_1) \times \cdots \times \mathrm{Sp}\,_{s_t}(\pi_t)).$$

We must check that $\Theta_1([\pi]) = \Theta_2([\pi])$.

If $h_1, \ldots, h_t$ are positive integers such that $h_i \leq s_i$ we will let

- $P'_{h_i} \subset GL_{s_i n_i}$ denote a parabolic subgroup with Levi component $GL_{h_i n_i} \times GL_{(s_i - h_i) n_i}$,

- $h = h_1 n_1 + \cdots + h_t n_t$,

- $P_{\{h_i\}} \subset GL_h$ denote a parabolic subgroup with Levi component $GL_{h_1 n_1} \times \cdots \times GL_{h_t n_t}$,

- $P'_{\{h_i\}} \subset GL_{n-h}$ denote a parabolic subgroup with Levi component $GL_{(s_1 - h_1) n_1} \times \cdots \times GL_{(s_t - h_t) n_t}$,

By lemma I.3.9 we see that

$$\Theta_2([\pi]) =$$

$$\sum_{h_i} \sum_\rho \mathrm{vol}\,(D^\times_{F_w, n-h} / F_w^\times)^{-1}$$
$$\mathrm{tr}\,(\text{n-Ind}_{P'_{\{h_i\}}(F_w)}^{GL_{n-h}(F_w)} (\mathrm{Sp}_{s_1 - h_1}(\pi_1) \times \cdots \times \mathrm{Sp}_{s_t - h_t}(\pi_t)))(\varphi_{\mathrm{JL}\,(\rho^\vee)})$$
$$\text{n-Ind}_{P_h(F_w)}^{GL_n(F_w)} [(\text{n-Ind}_{P_{\{h_i\}}(F_w)}^{GL_h(F_w)} \mathrm{Sp}_{h_1}(\pi_1 \otimes |\det|^{s_1 - h_1}) \times \cdots$$
$$\times \mathrm{Sp}_{h_t}(\pi_t \otimes |\det|^{s_t - h_t})) \times (\Psi_{F_w, l, n-h}(\rho) \otimes |\mathrm{Art}_{F_w}^{-1}|^{-h/2})],$$

where $h_1, \ldots, h_t$ run over positive integers with $h_i \leq s_i$, where ρ runs over irreducible admissible representations of $D^\times_{F_w, n-h}$. Using the second part of lemma I.3.4 we see that this becomes

$$\Theta_2([\pi]) =$$

$$\sum_{i=1}^t \sum_{h_i=0}^{s_i - 1} \text{n-Ind}_{P(F_w)}^{GL_n(F_w)} [\mathrm{Sp}_{s_1}(\pi_1) \times \cdots \times$$
$$(\text{n-Ind}_{P'_{h_i}(F_w)}^{GL_{s_i n_i}(F_w)} \mathrm{Sp}_{h_i}(\pi_i \otimes |\det|^{s_i - h_i}) \times$$
$$(\Psi_{F_w, l, (s_i - h_i) n_i}(\mathrm{JL}^{-1}(\mathrm{Sp}_{s_i - h_i}(\pi_i)^\vee)) \otimes |\mathrm{Art}_{F_w}^{-1}|^{(n_i(s_i - h_i) - n)/2}))$$
$$\times \ldots \mathrm{Sp}_{s_t}(\pi_t)].$$

Now by theorem VII.1.5 this becomes

$$\Theta_2([\pi]) =$$

$$\sum_{i=1}^t \sum_{h_i=0}^{s_i - 1} \sum_{j=1}^{s_i - h_i} (-1)^{s_i - h_i - j} \text{n-Ind}_{P(F_w)}^{GL_n(F_w)} [\mathrm{Sp}_{s_1}(\pi_1) \times \cdots \times$$
$$(\text{n-Ind}_{P'_{h_i}(F_w)}^{GL_{s_i n_i}(F_w)} (\mathrm{Sp}_j(\pi_i) \boxplus (\pi_i \otimes |\det|^j) \boxplus \cdots \boxplus (\pi_i \otimes |\det|^{s_i - 1 - h_i})$$
$$\times \mathrm{Sp}_{h_i}(\pi_i \otimes |\det|^{s_i - h_i}))$$
$$\times \cdots \times \mathrm{Sp}_{s_t}(\pi_t)][r_l(\pi_i \otimes |\det|^{j-1}) \otimes |\mathrm{Art}_K^{-1}|^{(n_i - n)/2}].$$

Next by lemma I.3.2 we rewrite this

$$\Theta_2([\pi]) =$$

$$\sum_{i=1}^{t} \sum_{h_i=1}^{s_i-1} \sum_{j=1}^{s_i-h_i} (-1)^{s_i-h_i-j}$$
$$\text{n-Ind}_{P(F_w)}^{GL_n(F_w)} [\text{Sp}_{s_1}(\pi_1) \times \cdots \times \omega(\vec{\Gamma}_{j,s_i-h_i-j,h_i}) \times \cdots \times \text{Sp}_{s_t}(\pi_t)]$$
$$[r_l(\pi_i \otimes |\det|^{j-1}) \otimes |\text{Art}_K^{-1}|^{(n_i-n)/2}]+$$
$$\sum_{i=1}^{t} \sum_{h_i=0}^{s_i-1} \sum_{j=1}^{s_i-h_i} (-1)^{s_i-h_i-j}$$
$$\text{n-Ind}_{P(F_w)}^{GL_n(F_w)} [\text{Sp}_{s_1}(\pi_1) \times \cdots \times \omega(\vec{\Gamma}_{j,s_i-1-h_i-j,h_i+1}) \times \cdots \times \text{Sp}_{s_t}(\pi_t)]$$
$$[r_l(\pi_i \otimes |\det|^{j-1}) \otimes |\text{Art}_K^{-1}|^{(n_i-n)/2}] =$$

$$\sum_{i=1}^{t} \sum_{j=1}^{s_i} \text{n-Ind}_{P(F_w)}^{GL_n(F_w)} [\text{Sp}_{s_1}(\pi_1) \times \cdots \times \omega(\vec{\Gamma}_{j,-1,s_i+1-j}) \times$$
$$\cdots \times \text{Sp}_{s_t}(\pi_t)][r_l(\pi_i \otimes |\det|^{j-1}) \otimes |\text{Art}_K^{-1}|^{(n_i-n)/2}] =$$

$$[\pi] \sum_{i=1}^{t} \sum_{j=0}^{s_i-1} [r_l(\pi_i \otimes |\det|^j) \otimes |\text{Art}_K^{-1}|^{(n_i-n)/2}] =$$

$$\Theta_1([\pi]).$$

The theorem follows. $\square$

The proof of the next proposition was explained by one of us (R.T.) in a letter to Clozel in 1991 [Tay1], assuming the truth of corollary VI.2.8. The proof of this corollary was probably considered standard, but to the best of our knowledge it is written down for the first time in this book.

Proposition VII.1.8 *Let π be an irreducible admissible representation of $G(\mathbb{A}^\infty)$ over $\mathbb{Q}_l^{ac}$. Suppose that one of the following two conditions hold.*

1. *There exists $\tau : F \hookrightarrow \mathbb{C}$ and an irreducible representation π_∞ of $G_\tau(\mathbb{R})$, which is cohomological for $\iota(\xi)$, and such that $\iota(\pi) \otimes \pi_\infty$ is an automorphic representation of $G_\tau(\mathbb{A})$ with $BC(\iota(\pi) \otimes \pi_\infty) = (\psi, \Pi)$ where $JL(\Pi)$ is cuspidal.*

2. *For some rational prime q which splits in E and for some place x of F above q at which B splits, the representation π_x is generic.*

Then there is a true continuous representation

$$\widetilde{R}_\xi(\pi) : \text{Gal}(F^{ac}/F) \longrightarrow GL_n(\mathbb{Q}_l^{ac})$$

such that

$$n[R_\xi(\pi)] = (\dim[R_\xi(\pi)])[\widetilde{R}_\xi(\pi)].$$

Proof: Note that by corollary VI.2.7 we have $[R_\xi(\pi)] = [R_\xi^{n-1}(\pi)]$. The proposition will follow from lemma I.2.2 if we can check the two hypotheses of that lemma. The first is satisfied because of the Cebotarev density theorem and corollary V.6.3.

For the second hypothesis note that $R_\xi^{n-1}(\pi)$ is defined over some finite extension $L/\mathbb{Q}_l$ inside $\mathbb{Q}_l^{ac}$ (by the Baire category theorem). We will write $W'_{R_\xi^{n-1}(\pi)}$ for the L-vector space on which it can be realised. Let $\mathfrak{g}$ denote the Lie algebra of the image of $R_\xi^{n-1}(\pi)$ and let $\mathfrak{g}_L$ denote its L-span. Let $\sigma = \imath^{-1}\tau$ and let λ denote the place of F above l induced by σ. By corollary VI.2.8 (and corollary VI.2.4) we have a decomposition

$$W'_{R_\xi^{n-1}(\pi)} \otimes_{\mathbb{Q}_l} \widehat{F}_\lambda^{ac}$$

$$\cong \bigoplus_{\mu:L\hookrightarrow F_\lambda^{ac}} \bigoplus_{p=0}^{n-1} \widehat{F}_\lambda^{ac}(a_0(\xi,\mu^{-1}|_E) + \vec{a}(\xi,\mu^{-1}|_E)_{\mu,n-p} - p)^d$$

as $L \otimes_{\mathbb{Q}_l} \widehat{F}_\lambda^{ac}$-modules with continuous L-linear but $\widehat{F}_\lambda^{ac}$-semilinear actions of $\mathrm{Gal}\,(F_\lambda^{ac}/F_\lambda)$. Here L acts on the factor indexed by μ via μ, and $dn = \dim R_\xi^{n-1}(\pi)$. By theorem 1 of [Sen] there is an element $\phi \in \mathfrak{g}\otimes_{\mathbb{Q}_l} \widehat{F}_\lambda^{ac}$ which acts as a on each factor $\widehat{F}_\lambda^{ac}(a)$. Thus there are elements $\phi_\mu \in \mathfrak{g}_L \otimes_{L,\mu} \widehat{F}_\lambda^{ac}$ with eigenvalues a on each $\widehat{F}_\lambda^{ac}(a)$-factor of

$$W'_{R_\xi^{n-1}(\pi)} \otimes_{L,\mu} \widehat{F}_\lambda^{ac} \cong \bigoplus_{p=0}^{n-1} \widehat{F}_\lambda^{ac}(a_0(\xi,\mu^{-1}|_E) + \vec{a}(\xi,\mu^{-1}|_E)_{\mu,n-p} - p)^d.$$

Thus the second hypothesis is also verified. $\square$

We now use our results to improve a theorem of Clozel [Cl1].

Theorem VII.1.9 *Suppose that L is a CM field and that Π is a cuspidal automorphic representation of $GL_g(\mathbb{A}_L)$ satisfying the following conditions:*

- $\Pi^\vee \cong \Pi^c$,

- *Π_∞ has the same infinitesimal character as some algebraic representation $\Xi^\vee$ over $\mathbb{C}$ of the restriction of scalars from L to $\mathbb{Q}$ of GL_g,*

- *and for some finite place x of L the representation Π_x is square integrable.*

Thinking of Ξ as a representation of $GL_g^{\mathrm{Hom}\,(L,\mathbb{C})}$ it is parametrised by some element

$$(\vec{a}_\tau) \in (X(\mathbb{G}_m^n)^+)^{\mathrm{Hom}\,(L,\mathbb{C})}.$$

(See section III.2. Also note the dual.)

Then there is a continuous representation

$$R_l(\Pi) = R_{l,\imath}(\Pi) : \mathrm{Gal}\,(L^{\mathrm{ac}}/L) \longrightarrow GL_n(\mathbb{Q}_l^{\mathrm{ac}})$$

with the following properties.

1. *For any finite place y of L not dividing l we have*

$$[R_l(\Pi)|_{W_{F_y}}] = [r_l(\imath^{-1}\Pi_y)].$$

   Moreover if $\sigma \in W_{F_y}$ and if α is an eigenvalue of $R_l(\Pi)(\sigma)$ then $\alpha \in \mathbb{Q}^{\mathrm{ac}}$ and for every embedding $\mathbb{Q}^{\mathrm{ac}} \hookrightarrow \mathbb{C}$ we have $|\alpha|^2 \in (\#k(y))^{\mathbb{Z}}$.

2. *If y is a finite place of L not dividing l with Π_y unramified then $R_l(\Pi)$ is unramified at y and for every eigenvalue α of $R_l(\Pi)(\mathrm{Frob}_y)$ and for every embedding of α into $\mathbb{C}$ we have $|\alpha|^2 = (\#k(y))^{(g-1)/2}$.*

3. *If y is a place of L dividing l then $R_l(\Pi)$ is potentially semi-stable at y. If, moreover, Π_y is unramified then $R_l(\Pi)$ is crystalline at y.*

4. *If $\sigma : L \hookrightarrow \mathbb{Q}_l^{\mathrm{ac}}$ then*

$$\mathrm{gr}^j D_{\mathrm{DR},\sigma} R_l(\Pi) = (0)$$

   unless $j = p - \vec{a}_{\imath\sigma,n-p}$ for some $p \in \{0, 1, \ldots, p-1\}$, in which case

$$\dim \mathrm{gr}^j D_{\mathrm{DR},\sigma} R_l(\Pi) = 1$$

Proof: Suppose first that $L = EF^+$ where E is an imaginary quadratic field such that l and $x|_{\mathbb{Q}}$ split in E and where F^+ is a totally real field with $[F^+ : \mathbb{Q}]$ even. In this case we will write F for L.

Choose $(B, *, (\ ,\ ), \Lambda_i)$ as in section I.7 such that B is a division algebra with centre F and

- $[B : F] = g^2$,

- B splits at all places of F other than x and x^c

- and B_x and B_{x^c} are division algebras.

(Here we are using the assumption that $[F^+ : \mathbb{Q}]$ is even.) By theorem VI.2.9, lemma VI.2.10 and corollary VI.2.7, for each $\tau : F \hookrightarrow \mathbb{C}$ we can find

- an algebraic representation ξ of G over $\mathbb{Q}_l^{\mathrm{ac}}$,

- and an automorphic representation π of $G_\tau(\mathbb{A})$

such that, if we set $\mathrm{BC}\,(\pi) = (\psi, \widetilde{\Pi})$, then

- $\mathrm{JL}\,(\widetilde{\Pi}) = \Pi$

- and $[R_\xi(\imath^{-1}\pi^\infty)] = [R_\xi^{n-1}(\imath^{-1}\pi^\infty)] \neq 0$.

We may also arrange that ψ_l is unramified. Set

$$R_l(\Pi, \tau) = \widetilde{R}_\xi(\imath^{-1}\pi^\infty) \otimes \mathrm{rec}(\psi^c)|_{\mathrm{Gal}\,(F^{\mathrm{ac}}/F)}.$$

By proposition III.2.1 we see that for any finite place y of F not dividing l, for any $\sigma \in W_{F_w}$ and any eigenvalue α of $R_l(\Pi, \tau)(\sigma)$ we have $\alpha \in \mathbb{Q}^{\mathrm{ac}}$ and for any embedding $\mathbb{Q}^{\mathrm{ac}} \hookrightarrow \mathbb{C}$,

$$|\alpha|^2 \in (\#k(y))^{\mathbb{Z}}.$$

Also $R_l(\Pi, \tau)$ is potentially semi-stable.

Moreover from theorem VII.1.7 it follows that for all finite places y of F such that

- $y \nmid x x^c$,

- $y|_{\mathbb{Q}}$ splits in E

- and $y \nmid l$,

we have

$$[R_l(\Pi, \tau)|_{W_{F_y}}] = [r_l(\imath^{-1}\Pi_y)].$$

If Π_y is unramified then $R_l(\Pi, \tau)$ is unramified at y and every eigenvalue α of $R_l(\Pi, \tau)(\mathrm{Frob}_y)$ satisfies $|\alpha|^2 = (\#k(y))^{(g-1)/2}$ for every embedding of α into $\mathbb{C}$ (see lemma III.4.2). If moreover $y|l$ and Π_y is unramified then $R_l(\Pi, \tau)$ is crystalline at y (see lemma III.4.2). Finally by corollary VI.2.8

$$\mathrm{gr}^j D_{\mathrm{DR}, \imath^{-1} \circ \tau} R_l(\Pi, \tau) = (0)$$

unless $j = p - \vec{a}_{\tau, n-p}$ for some $p \in \{0, 1, \ldots, p-1\}$, in which case

$$\dim \mathrm{gr}^j D_{\mathrm{DR}, \imath^{-1} \circ \tau} R_l(\Pi, \tau) = 1$$

The Cebotarev density theorem implies that $R_l(\Pi, \tau)$ is independent of τ. Thus we will write simply $R_l(\Pi)$.

Next suppose that $y|_{\mathbb{Q}}$ is inert in E or that $y|x x^c$. Let p denote the rational prime under y. We can find a real quadratic field A such that

- $A_p \cong E_p$

- and $x|_{\mathbb{Q}}$ splits in A.

Let E' denote the third quadratic subfield of AE, let $(F^+)' = AF^+$ and let $F' = E'(F^+)' = AF$. Note that E' is an imaginary quadratic field in which p, l and $x|_{\mathbb{Q}}$ split. Let x' denote a prime of F' above x and let y' denote a prime of F' above y which does not divide $x'(x')^c$. Note also that $F_x \overset{\sim}{\to} F'_{x'}$ and $F_y \overset{\sim}{\to} F'_{y'}$. The x' component of the automorphic restriction $\mathrm{Res}^F_{F'}(\Pi)$ is square integrable and hence $\mathrm{Res}^F_{F'}(\Pi)$ is cuspidal. Moreover by strong multiplicity one $\mathrm{Res}^F_{F'}(\Pi)^c \cong \mathrm{Res}^F_{F'}(\Pi)^\vee$, and $\mathrm{Res}^F_{F'}(\Pi)$ has the same infinitesimal character as an algebraic representation of $GL_g(F' \otimes_{\mathbb{Q}} \mathbb{C})$ over $\mathbb{C}$. Thus we can associate to $\mathrm{Res}^F_{F'}(\Pi)$ a continuous representation $R_l(\mathrm{Res}^F_{F'}(\Pi))$ of $\mathrm{Gal}\,(F^{\mathrm{ac}}/F')$ such that for all places z of F' for which

- $z \nmid x'(x')^c$,

- $z|_{\mathbb{Q}}$ splits in E'

- and $z \nmid l$,

we have

$$[R_l(\mathrm{Res}^F_{F'}(\Pi))|_{W_{F'_z}}] = [r_l(\imath^{-1}\mathrm{Res}^F_{F'}(\Pi)_z)].$$

It follows from the Cebotarev density theorem that

$$[R_l(\mathrm{Res}^F_{F'}(\Pi))] = [R_l(\Pi)|_{\mathrm{Gal}\,(F^{\mathrm{ac}}/F')}].$$

Thus

$$[R_l(\Pi)|_{W_{F_y}}] = [R_l(\mathrm{Res}^F_{F'}(\Pi))|_{W_{F'_{y'}}}] = [r_l(\imath^{-1}\mathrm{Res}^F_{F'}(\Pi)_{y'})] = [r_l(\imath^{-1}\Pi_y)],$$

and we have established the theorem in the special case.

Now we will turn to the proof in the general case.

If A is an imaginary quadratic field such that l and $x|_{\mathbb{Q}}$ split in A then set

- F_A^+ to be the maximal totally real subfield of AL,

- $F_A = AF_A^+ = AL$,

- x_A a prime of F_A above x,

- σ_A to be the non-trivial element of $\mathrm{Gal}\,(F_A/L)$

- and ϵ_A to be the non-trivial character of $\mathrm{Gal}\,(A/\mathbb{Q})$.

Then $[F_A^+ : \mathbb{Q}]$ is even and $x_A|_{\mathbb{Q}}$ splits in A. Moreover $L_x \overset{\sim}{\to} F_A|_{x_A}$ and so as before we see that $\mathrm{Res}^F_{F_A}(\Pi)$ continues to satisfy the conditions of the theorem, but for F_A. Thus, from what we have already proved, there is a continuous representation $R_l(\mathrm{Res}^L_{F_A}(\Pi))$ of $\mathrm{Gal}\,(L^{\mathrm{ac}}/F_A)$ such that for all places $z \nmid l$ of F_A we have

$$[R_l(\mathrm{Res}^L_{F_A}(\Pi))|_{W_{F_{A,z}}}] = [r_l(\imath\mathrm{Res}^L_{F_A}(\Pi)_z)].$$

It follows from the Cebotarev density theorem that

$$[R_l(\mathrm{Res}^L_{F_A}(\Pi))^{\sigma_A}] = [R_l(\mathrm{Res}^L_{F_A}(\Pi))]$$

and that if A' is a second such field then

$$[R_l(\mathrm{Res}^L_{F_A}(\Pi))|_{\mathrm{Gal}\,(L^{\mathrm{ac}}/F_A F_{A'})}] = [R_l(\mathrm{Res}^L_{F_{A'}}(\Pi))|_{\mathrm{Gal}\,(L^{\mathrm{ac}}/F_A F_{A'})}].$$

Fix one such quadratic extension A_0. Let $\{\rho_i\}$ be a set of representatives of the equivalence classes of irreducible continuous representations of $\mathrm{Gal}\,(L^{\mathrm{ac}}/F_{A_0})$ on finite dimensional $\overline{\mathbb{Q}}_l^{\mathrm{ac}}$-vector spaces. Let I be the set of indices such that $\rho_i^{\sigma_{A_0}} \cong \rho_i$ and ρ_i is a constituent of $[R_l(\mathrm{Res}^L_{F_{A_0}}(\Pi))]$. Let J be the set of indices such that $\rho_i^{\sigma_{A_0}} \not\cong \rho_i$ and ρ_i is a constituent of $[R_l(\mathrm{Res}^L_{F_{A_0}}(\Pi))]$. For $i \in I$ choose an extension $\widetilde{\rho}_i$ of ρ_i to $\mathrm{Gal}\,(L^{\mathrm{ac}}/L)$. Also write

$$[R_l(\mathrm{Res}^L_{F_{A_0}}(\Pi))] = \sum_{i \in I \cup J} b_i[\rho_i].$$

Let

$$\rho = \bigoplus_{i \in I \cup J} \rho_i.$$

Let H denote the Zariski closure of the image of ρ and let H^0 denote the connected component of the identity in H. Also let M/F_{A_0} denote the fixed field of $\rho^{-1}H^0$. If N/L is a finite Galois extension disjoint from M then $\rho\mathrm{Gal}\,(F^{\mathrm{ac}}/NA_0)$ is Zariski dense in H and so we have the following results.

1. If $i \in I \cup J$ then $\rho_i|_{\mathrm{Gal}\,(L^{\mathrm{ac}}/NA_0)}$ is irreducible.

2. If $i \in I$ then $\widetilde{\rho}_i|_{\mathrm{Gal}\,(L^{\mathrm{ac}}/N)}$ is irreducible.

3. If $i,j \in I \cup J$ and $\rho_i|_{\mathrm{Gal}\,(L^{\mathrm{ac}}/NA_0)} \cong \rho_j|_{\mathrm{Gal}\,(L^{\mathrm{ac}}/NA_0)}$ then $i = j$.

4. If $i,j \in I$, if $\delta = 0$ or 1 and if $\widetilde{\rho}_i|_{\mathrm{Gal}\,(L^{\mathrm{ac}}/N)} \cong \widetilde{\rho}_j|_{\mathrm{Gal}\,(L^{\mathrm{ac}}/N)} \otimes \epsilon^\delta_{A_0}$ then $i = j$ and $\delta = 0$.

In particular if $A/\mathbb{Q}$ is a quadratic extension as above such that F_A is linearly disjoint from M over L then

- $\rho_i|_{\mathrm{Gal}\,(L^{\mathrm{ac}}/F_{A_0}A)}$ is irreducible for all $i \in I \cup J$

- and if, for some $i, j \in I \cup J$, we have

$$\rho_i^{\sigma_{A_0}}|_{\mathrm{Gal}\,(L^{\mathrm{ac}}/F_{A_0}A)} \cong \rho_j|_{\mathrm{Gal}\,(L^{\mathrm{ac}}/F_{A_0}A)}$$

then $\rho_i^{\sigma_{A_0}} = \rho_j$.

We have also seen that

$$[R_l(\mathrm{Res}_{F_A}^L(\Pi))|_{\mathrm{Gal}\,(L^{\mathrm{ac}}/F_{A_0}A)}] = \sum_{i \in I \cup J} b_i [\rho_i|_{\mathrm{Gal}\,(L^{\mathrm{ac}}/F_{A_0}A)}].$$

Thus we must have

$$[R_l(\mathrm{Res}_{F_A}^L(\Pi))] =$$
$$\sum_{i \in J} (b_i/2)[(\mathrm{Ind}_{\mathrm{Gal}\,(L^{\mathrm{ac}}/F_{A_0})}^{\mathrm{Gal}\,(L^{\mathrm{ac}}/L)} \rho_i)|_{\mathrm{Gal}\,(L^{\mathrm{ac}}/F_A)}] +$$
$$\sum_{i \in I} b_i [(\widetilde{\rho}_i \otimes \epsilon_{A_0}^{\delta_{Ai}})|_{\mathrm{Gal}\,(L^{\mathrm{ac}}/F_A)}],$$

where $\delta_{Ai} = 0$ or 1.

Choose such an extension A_1 so that F_{A_1} is linearly disjoint from M over L. Set

$$[R_l(\Pi)] = \sum_{i \in J} (b_i/2)[\mathrm{Ind}_{\mathrm{Gal}\,(L^{\mathrm{ac}}/F_{A_0})}^{\mathrm{Gal}\,(L^{\mathrm{ac}}/L)} \rho_i] + \sum_{i \in I} b_i [(\widetilde{\rho}_i \otimes \epsilon_{A_0}^{\delta_{A_1 i}})].$$

Suppose now that A is such a quadratic extension of $\mathbb{Q}$ such that F_A is linearly disjoint from MA_1 over L. Then

$$\sum_{i \in I} b_i [(\widetilde{\rho}_i \otimes \epsilon_{A_0}^{\delta_{A_1 i}})|_{\mathrm{Gal}\,(L^{\mathrm{ac}}/F_{A_1}A)}]) = \sum_{i \in I} b_i [(\widetilde{\rho}_i \otimes \epsilon_{A_0}^{\delta_{Ai}})|_{\mathrm{Gal}\,(L^{\mathrm{ac}}/F_A A_1)}])$$

and so $\delta_{Ai} = \delta_{A_1 i}$ for all $i \in I$. Hence

$$[R_l(\Pi)|_{\mathrm{Gal}\,(L^{\mathrm{ac}}/F_A)}] = [R_l(\mathrm{Res}_{F_A}^L(\Pi))].$$

Given any finite place y of L we can choose an imaginary quadratic extension $A/\mathbb{Q}$ such that

- LA is disjoint from MA_1 over L,

- y splits as $y'y''$ in LA

- and l and $x|_{\mathbb{Q}}$ split in A.

If $y \nmid l$ then

$$[R_l(\Pi)|_{W_{F_y}}] = [R_l(\mathrm{Res}_{F_A}^L(\Pi))|_{W_{F_{y'}}}] = [r_l(\imath^{-1}\Pi_y)],$$

and so

$$[R_l(\Pi)|_{W_{F_y}}] = [r_l(\imath^{-1}\Pi_y)].$$

If $y|l$ and Π_y is unramified then

$$[R_l(\Pi)|_{\mathrm{Gal}\,(F_y^{\mathrm{ac}}/F_y)}] = [R_l(\mathrm{Res}_{F_A}^L(\Pi))|_{\mathrm{Gal}\,(F_{y'}^{\mathrm{ac}}/F_{y'})}]$$

is crystalline. The theorem follows because it suffices to check all other asserted properties of $R_l(\Pi)$ after restriction to any quadratic extension. $\square$

(Constructions of Galois representations by piecing together representations over many quadratic extensions are not new (see for instance [BR]).)

Corollary VII.1.10 *Let π be an automorphic representation of $G_\tau(\mathbb{A})$ such that π_∞ is cohomological for ξ. Let $\mathrm{BC}\,(\pi) = (\psi, \Pi)$ and suppose that $\mathrm{JL}\,(\Pi)$ is cuspidal. Then $\mathrm{JL}\,(\Pi)$ satisfies the hypotheses of theorem VII.1.9 and*

$$n[R_\xi(\imath^{-1}\pi^\infty)] = (\dim[R_\xi(\imath^{-1}\pi^\infty)])[R_{l,\imath}(\mathrm{JL}\,(\Pi)) \otimes \mathrm{rec}(\psi)|_{\mathrm{Gal}\,(F^{\mathrm{ac}}/F)}^{-1}].$$

In particular if $y \nmid l$ is a place of F then

$$n[R_\xi(\imath^{-1}\pi^\infty)] = (\dim[R_\xi(\imath^{-1}\pi^\infty)])[r_l(\imath^{-1}\mathrm{JL}\,(\Pi)_y) \otimes (\psi_{y|E}^{-1} \circ \mathrm{Art}_{E_y}^{-1})|_{\mathrm{Gal}\,(F^{\mathrm{ac}}/F)}].$$

Proof: Using the Cebotarev density theorem, this follows easily from theorems VII.1.7 and VII.1.9. $\square$

The next corollary results from theorem VII.1.9 and corollary VII.2.18 below. (Note that because ψ_Π is algebraic and satisfies $\psi_\Pi\psi_\Pi^c = 1$, we must have $|\psi_{\Pi,\infty}| \equiv 1$ and hence $|\psi_\Pi| \equiv 1$.) We choose to state this result here (rather than after corollary VII.2.18) as it fits better with the discussion of this section.

Corollary VII.1.11 *Suppose that L is a CM field and that Π is a cuspidal automorphic representation of $GL_g(\mathbb{A}_L)$ satisfying the following conditions:*

- *$\Pi^\vee \cong \Pi^c$,*

- *Π_∞ has the same infinitesimal character as some algebraic representation over $\mathbb{C}$ of the restriction of scalars from L to $\mathbb{Q}$ of GL_g,*

- *and for some finite place x of L the representation Π_x is square integrable.*

If y is a finite place of L then Π_y is tempered.

VII.2 The local Langlands conjecture

We will start this section by checking various basic functoriality properties of our map r_l. *Throughout this section K will denote a finite extension of $\mathbb{Q}_p$ and l will denote a prime other than p. Recall that we have fixed an isomorphism $\imath : \overline{\mathbb{Q}}_l^{\mathrm{ac}} \xrightarrow{\sim} \mathbb{C}$, which we will often suppress in our notation.*

Lemma VII.2.1 *Suppose that π is an irreducible admissible representation of $GL_g(K)$ and that χ is a smooth character of $K^\times$. Then*

$$[r_l(\pi \otimes (\chi \circ \det))] = [r_l(\pi) \otimes (\chi^{-1} \circ \mathrm{Art}_K^{-1})].$$

Proof: It is easy to reduce to the case that π is supercuspidal, so suppose that π is supercuspidal. Choose an imaginary quadratic field M in which p splits and a totally real field L^+ with a place $x(+)$ such that $L^+_{x(+)} \cong K$. Set $L = ML^+$ and choose a place x of L above $x(+)$. Thus $L_x \cong K$. Let $\widetilde{\chi}$ be a continuous character of $\mathbb{A}_L^\times / L^\times L_\infty^\times$ such that $\chi \widetilde{\chi}_x^{-1}$ is unramified. By corollary VI.2.6 we may choose a cuspidal automorphic representation Π of $GL_g(\mathbb{A}_L)$ such that

- $\Pi^c \cong \Pi^\vee$,

- Π_∞ has the same infinitesimal character as some algebraic representation of $\mathrm{RS}_{\mathbb{Q}}^L(GL_g)$,

- and $\Pi_x \cong \pi \otimes (\psi_x \circ \det)$ for some character ψ_x of $K^\times / \mathcal{O}_K^\times$.

From theorem VII.1.9 (applied at good places of L) and from the Cebotarev density theorem we see that

$$[R_l(\Pi \otimes (\widetilde{\chi} \circ \det))] = [R_l(\Pi) \otimes \mathrm{rec}_{l,\imath}(\widetilde{\chi})^{-1}].$$

Applying theorem VII.1.9 at x we conclude that

$$[r_l(\Pi_x \otimes \widetilde{\chi}_x)] = [r_l(\Pi_x) \otimes (\widetilde{\chi}_x^{-1} \circ \mathrm{Art}_K^{-1})].$$

The lemma now follows from lemma VII.1.4. $\square$

Lemma VII.2.2 *Suppose that π is an irreducible admissible representation of $GL_g(K)$ with central character ψ_π. Then*

$$\det r_l(\pi) = (\psi_\pi \otimes |\ |^{g(g-1)/2})^{-1} \circ \mathrm{Art}_K^{-1}.$$

Proof: Again it is easy to reduce to the case that π is supercuspidal, so suppose that π is supercuspidal. Choose an imaginary quadratic field M in which p splits and a totally real field L^+ with a place $x(+)$ such that $L^+_{x(+)} \cong K$. Set $L = ML^+$ and choose a place x of L above $x(+)$. Thus $L_x \cong K$. By corollary VI.2.6 we may choose a cuspidal automorphic representation Π of $GL_g(\mathbb{A}_L)$ such that

- $\Pi^c \cong \Pi^\vee$,

- Π_∞ has the same infinitesimal character as some algebraic representation of $\mathrm{RS}^L_{\mathbb{Q}}(GL_g)$,

- and $\Pi_x \cong \pi \otimes (\psi_x \circ \det)$ for some character ψ_x of $K^\times / \mathcal{O}_K^\times$.

From theorem VII.1.9 (applied at good places of L) and from the Cebotarev density theorem we see that for all $\sigma \in \mathrm{Gal}(\mathbb{Q}^{\mathrm{ac}}/\mathbb{Q})$ there exist elements $\alpha_1(\sigma), \ldots, \alpha_n(\sigma) \in \mathbb{Q}_l^{\mathrm{ac}}$ such that

- $R_l(\Pi)(\sigma)$ has eigenvalues $\alpha_1(\sigma), \ldots, \alpha_n(\sigma)$,

- and $\alpha_1(\sigma) \ldots \alpha_n(\sigma) = \mathrm{rec}_{l,\imath}(\psi_\pi \otimes | \ |^{g(g-1)/2})^{-1}(\sigma)$.

(See for instance the first paragraph of the proof of proposition 1 of [Tay2] for more details of this sort of argument.) Applying theorem VII.1.9 at x we conclude that

$$\det r_l(\Pi_x) = (\psi_{\Pi,x} \otimes | \ |^{g(g-1)/2})^{-1} \circ \mathrm{Art}_K^{-1}.$$

The lemma now follows from lemma VII.1.4. $\square$

Lemma VII.2.3 *Suppose that π is an irreducible admissible representation of $GL_g(K)$. Then*

$$[r_l(\pi^\vee)] = [r_l(\pi \otimes | \det |^{1-g})^\vee].$$

Proof: Again it is easy to reduce to the case that π is supercuspidal, so suppose that π is supercuspidal. Choose an imaginary quadratic field M in which p splits and a totally real field L^+ with a place $x(+)$ such that $L^+_{x(+)} \cong K$. Set $L = ML^+$ and choose a place x of L above $x(+)$. Thus $L_x \cong K$. By corollary VI.2.6 we may choose a cuspidal automorphic representation Π of $GL_g(\mathbb{A}_L)$ such that

- $\Pi^c \cong \Pi^\vee$,

- Π_∞ has the same infinitesimal character as some algebraic representation of $\mathrm{RS}^L_{\mathbb{Q}}(GL_g)$,

- and $\Pi_x \cong \pi \otimes (\psi_x \circ \det)$ for some character ψ_x of $K^\times / \mathcal{O}_K^\times$.

From theorem VII.1.9 (applied at good places of L) and from the Cebotarev density theorem we see that

$$[R_l(\Pi^\vee)] = [R_l(\Pi \otimes | \det |^{1-g})^\vee].$$

Applying theorem VII.1.9 at x we conclude that

$$[r_l(\Pi_x^\vee)] = [r_l(\Pi_x \otimes | \det |^{1-g})^\vee].$$

The lemma now follows from lemma VII.1.4. $\square$

Lemma VII.2.4 *Suppose that π is an irreducible admissible representation of $GL_g(K)$. Then*

$$[r_l(\mathrm{Res}_{K'}^K(\pi))] = [r_l(\pi)|_{W_{K'}}].$$

Proof: Again one may reduce to the case that π is square integrable, so suppose that π is square integrable. (See section 6.2 of chapter 1 and pages 59 and 60 of [AC].) Choose an imaginary quadratic field M in which p splits. Also choose a cyclic Galois extension $(L')^+/L^+$ of totally real fields and a place $x(+)$ of L^+ such that $x(+)$ is inert in $(L')^+$ and the extension $(L')_{x(+)}^+/L_{x(+)}^+$ is isomorphic to the extension K'/K. Set $L = ML^+$ (resp. $L' = M(L')^+$) and choose a place x of L above $x(+)$. Thus the extension L'_x/L_x is also isomorphic to K'/K. Choose a place y of L which splits completely in L' and which lies above a rational prime other than p which splits in M. By corollary VI.2.6 we may choose a cuspidal automorphic representation Π of $GL_g(\mathbb{A}_L)$ such that

- $\Pi^c \cong \Pi^\vee$,

- Π_∞ has the same infinitesimal character as some algebraic representation of $\mathrm{RS}_{\mathbb{Q}}^L(GL_g)$,

- $\Pi_x \cong \pi \otimes (\psi_x \circ \det)$ for some character ψ_x of $K^\times/\mathcal{O}_K^\times$,

- and Π_y is supercuspidal.

Then there is an "induced from cuspidal" representation $\mathrm{Res}_{L'}^L(\Pi)$ of the group $GL_g(\mathbb{A}_{L'})$ such that for all places w of L we have

- $\mathrm{Res}_{L'}^L(\Pi)_w = \mathrm{Res}_{L'_w^w}^{L_w}(\Pi_w)$ if w is inert in L',

- and $\mathrm{Res}_{L'}^L(\Pi)_w = \Pi_w^{\otimes q}$ if w splits in L'.

From the second of these conditions we see that $\mathrm{Res}_{L'}^L(\Pi)$ is supercuspidal at every place above y and hence is cuspidal automorphic. From the second of these conditions we also see that $\mathrm{Res}_{L'}^L(\Pi)_\infty$ has the same infinitesimal character as some algebraic representation of $\mathrm{RS}_{\mathbb{Q}}^{L'}(GL_g)$.

If w is a finite place of L at which Π_w is unramified and if w' is a prime of L' above w then, from the compatibility of base change with parabolic induction and from the explicit description of base change when $g = 1$ (see part (d) of theorem 6.2 of chapter 1 of [AC]), we see that if w' is a prime of L' above w then

$$[R_l(\mathrm{Res}_{L'}^L(\Pi))|_{W_{L'_{w'}}}] = [r_l(\Pi_w)|_{W_{L'_{w'}}}].$$

Thus, from theorem VII.1.9 (applied at good places of L') and from the Cebotarev density theorem we see that

$$[R_l(\text{Res}_{L'}^L(\Pi)] = [R_l(\Pi)|_{\text{Gal}\,((L')^{ac}/L')}].$$

Applying theorem VII.1.9 at x we conclude that

$$[r_l(\text{Res}_{K'}^K(\Pi_x))] = [r_l(\Pi_x)|_{W_{L'_{x'}}}].$$

The lemma now follows from lemma VII.1.4. $\square$

Lemma VII.2.5 *Suppose that π' is a generic irreducible admissible representation of $GL_g(K')$ which is $\text{Gal}\,(K'/K)$-regular. Then*

$$[r_l(\text{Ind}_{K'}^K(\pi') \otimes |\det|^{g(1-q)/2})] = [(\text{Ind}_{W_{K'}}^{W_K} r_l(\pi'))].$$

Proof: One may reduce to the case that $\pi' = \text{Sp}_2(\pi^0)$, where π^0 is supercuspidal, so suppose that π' has this form. (See corollary 5.5 and theorem 5.6 of [HH]. One could also try reducing to the case where π' is supercuspidal, but this case seems to be harder to treat directly.) Note in particular that in this case g is even. Choose an imaginary quadratic field M in which p splits. Also choose a cyclic Galois extension $(L')^+/L^+$ of totally real fields and a place $x(+)$ of L^+ such that $x(+)$ is inert in $(L')^+$ and the extension $(L')^+_{x(+)}/L^+_{x(+)}$ is isomorphic to the extension K'/K. Set $L = ML^+$ (resp. $L' = M(L')^+$) and choose a place x of L above $x(+)$. Thus the extension L'_x/L_x is also isomorphic to K'/K. Choose a generator σ of $\text{Gal}\,(L'/L)$. Choose a place y of L which is inert in L' and which lies above a rational prime other than p which splits in M. Choose a supercuspidal representation π_y of $GL_g(L'_y)$ such that $\pi_y \not\cong \pi_y^\sigma \otimes \psi$ for any character ψ of $(L'_y)^\times/\mathcal{O}_{L',y}^\times$. (To see that this is possible one may argue as follows. First choose any irreducible supercuspidal representation π_y^0 of $GL_g(L'_y)$. Then take $\pi_y = \pi_y^0 \otimes \chi$, where χ is a character of $(L'_y)^\times$ such that

$$(\psi_{\pi_y}\chi^g)|_{\mathcal{O}_{L',y}^\times} \neq (\psi_{\pi_y}\chi^g)|_{\mathcal{O}_{L',y}^\times} \circ \sigma.)$$

By corollary VI.2.6 we may choose a cuspidal automorphic representation Π of $GL_g(\mathbb{A}_L)$ such that

- $\Pi^c \cong \Pi^\vee$,

- Π_∞ has infinitesimal character parametrised by

$$(\vec{a}_\tau) \in (\rho + X(\mathbb{G}_m^g)^+)^{\text{Hom}\,(L',\mathbb{C})}$$

with $\vec{a}_{\tau,i} \neq \vec{a}_{\tau',i'}$ whenever $(\tau,i) \neq (\tau',i')$ but $\tau|_L = \tau'|_L$ (we are using the notation of section III.2, in particular $\rho = ((1-g)/2, (3-g)/2, \ldots, (g-1)/2))$,

- $\Pi_x \cong \pi \otimes (\psi_x \circ \det)$ for some character ψ_x of $(K')^\times / \mathcal{O}_{K'}^\times$,

- and $\Pi_y \cong \pi_y \otimes (\psi_y \circ \det)$ for some character ψ_y of $(L'_y)^\times / \mathcal{O}_{L',y}^\times$.

Note in particular that Π_∞ has the same infinitesimal character as some algebraic representation of $\mathrm{RS}_{\mathbb{Q}}^{L'}(GL_g)$ and that $\Pi^\sigma \not\cong \Pi$.

Consider the cuspidal automorphic representation $\mathrm{Ind}\,_{L'}^{L}\Pi$ of $GL_{qg}(\mathbb{A}_L)$. Using the the strong multiplicity one theorem we see that

$$(\mathrm{Ind}\,_{L'}^{L}\Pi)^\vee \cong (\mathrm{Ind}\,_{L'}^{L}\Pi)^c.$$

If w is an infinite place of L which splits as $w_1 \dots w_q$ in L', then $(\mathrm{Ind}\,_{L'}^{L}\Pi)_w$ is a subquotient of n-$\mathrm{Ind}_{Q(L_w)}^{GL_{qg}(L_w)}(\Pi_{w_1} \times \cdots \times \Pi_{w_q})$, where Q is a parabolic subgroup of GL_{qg} with Levi component GL_g^q. In particular this allows one to check that $\mathrm{Ind}\,_{L'}^{L}(\Pi)_\infty$ has the same infinitesimal character as some algebraic representation of $\mathrm{RS}_{\mathbb{Q}}^{L}(GL_{qg})$ (use the fact that g is even). Moreover by proposition 5.5 of [HH] we see that $\mathrm{Ind}\,_{L'}^{L}(\Pi)_y$ is supercuspidal.

From theorem VII.1.9 (applied at good places of L) and from the Cebotarev density theorem we see that

$$[R_l(\mathrm{Ind}\,_{L'}^{L}(\Pi) \otimes |\det|^{g(1-q)/2})] = [\mathrm{Ind}\,_{\mathrm{Gal}\,((L')^{\mathrm{ac}}/L')}^{\mathrm{Gal}\,((L')^{\mathrm{ac}}/L)} R_l(\Pi)].$$

Applying theorem VII.1.9 at x we conclude that

$$[r_l(\mathrm{Ind}\,_{K'}^{K}(\Pi_x) \otimes |\det|^{g(1-q)/2})] = [(\mathrm{Ind}\,_{W_{K'}}^{W_K} r_l(\Pi_x))].$$

The lemma now follows from lemma VII.1.4. $\square$

If π is an irreducible admissible representation of $GL_g(K)$ we will set

$$\mathrm{rec}_l(\pi) = r_l(\pi^\vee \otimes |\det|^{(1-g)/2}).$$

With this new normalisation we have the following restatement of lemma VII.1.6 and of the preceding lemmas.

Lemma VII.2.6 *Let K'/K be a cyclic Galois extension of prime degree q and let π be an irreducible admissible representation of $GL_g(K)$. Then we have the following results.*

1. *If $\tau \in \mathrm{Gal}\,(K^{\mathrm{ac}}/\mathbb{Q}_p)$ then $\mathrm{rec}_l(\pi \circ \tau) = \mathrm{rec}_l(\pi)^\tau$.*

2. *If $g = 1$ then $\mathrm{rec}_l(\pi) = \pi \circ \mathrm{Art}\,_K^{-1}$.*

3. *If ψ_π is the central character of π then $\det \mathrm{rec}_l(\pi) = \mathrm{rec}_l(\psi_\pi)$.*

4. *If χ is a character of $K^\times$ then $\mathrm{rec}_l(\pi \otimes (\chi \circ \det)) = \mathrm{rec}_l(\pi) \otimes \mathrm{rec}_l(\chi)$.*

5. $\text{rec}_l(\text{Res}^K_{K'}(\pi)) = \text{rec}_l(\pi)|_{W_{K'}}$.

6. *If π' is an irreducible admissible representation of $GL_g(K')$ then*
$$\text{rec}_l(\text{Ind}^K_{K'}(\pi')) = \text{Ind}^{W_K}_{W_{K'}} \text{rec}_l(\pi').$$

We next turn to some cases of non-Galois global automorphic induction, which were established by one of us (M.H.) in [Har3]. Indeed in a sense the rest of this section is superfluous, as we could simply refer to section 4 of [Har3]. However we will repeat the arguments here in somewhat greater detail, as we can now be slightly more direct. We will repeat not only arguments of [Har3], but also arguments of Henniart from [BHK] and [He6].

Proposition VII.2.7 *Suppose that $L_3 \supset L_2 \supset L_1$ are CM-fields with L_3/L_1 soluble and Galois. Suppose that χ is a character of $\mathbb{A}^\times_{L_2}/L^\times_2$ such that*

1. $\chi^c = \chi^{-1}$;

2. *for every embedding $\tau : L_2 \hookrightarrow \mathbb{C}$ giving rise to an infinite place x we have*
$$\chi_x : z \twoheadrightarrow (\tau z/c\tau z)^{p_\tau}$$
where $p_\tau \in \mathbb{Z}$ and if $\tau \neq \tau'$ then $p_\tau \neq p'_\tau$;

3. *there is a finite place y of L_1 which is inert in L_3, which does not divide l and and for which the stabiliser of the character $\chi_y \circ N_{L_3/L_2}$ of $(L_3)^\times_y$ in $\text{Gal}(L_3/L_1)$ is $\text{Gal}(L_3/L_2)$.*

Let ϕ be a character of $\mathbb{A}^\times_{L_1}/L^\times_1$ such that

1. $\phi^c = \phi^{-1}$;

2. *if $[L_2 : L_1]$ is odd then $\phi_\infty = 1$;*

3. *if $[L_2 : L_1]$ is even, then for every embedding $\tau : L_1 \hookrightarrow \mathbb{C}$ giving rise to an infinite place x we have*
$$\phi_x : z \twoheadrightarrow (\tau z/|\tau z|)^{\pm 1};$$

4. ϕ_y *is unramified.*

Then there is a cuspidal automorphic representation $I^{L_1}_{L_2}(\chi)$ of the group $GL_{[L_2:L_1]}(\mathbb{A}_{L_1})$ such that

- $I^{L_1}_{L_2}(\chi)^c \cong I^{L_1}_{L_2}(\chi)^\vee$;

- $(I_{L_2}^{L_1}(\chi) \otimes (\phi \circ \det))_\infty$ has the same infinitesimal character as an algebraic representation of $\mathrm{RS}_{\mathbb{Q}}^{L_1}(GL_{[L_2:L_1]})$;

- $I_{L_2}^{L_1}(\chi)_y$ is supercuspidal;

- and

$$[R_l(I_{L_2}^{L_1}(\chi) \otimes (\phi \circ \det))] =$$
$$[\mathrm{Ind}_{\mathrm{Gal}(L_3^{ac}/L_2)}^{\mathrm{Gal}(L_3^{ac}/L_1)} \mathrm{rec}_{l,i}(\chi^{-1}(\phi^{-1} \circ N_{L_2/L_1})| \ |^{(1-[L_2:L_1])/2})].$$

Proof: The proof will be by induction on $[L_3 : L_1]$, there being nothing to prove in the case $[L_3 : L_1] = 1$.

Now consider the inductive step. Because L_3/L_1 is soluble we may choose a subextension $L_3 \supset L_4 \supset L_1$ with L_4/L_1 cyclic Galois with prime degree q. Let σ be a generator of $\mathrm{Gal}(L_4/L_1)$ and let $\tilde{\sigma}$ be a lift of σ to $\mathrm{Gal}(L_3/L_1)$. We will consider separately the cases $L_4 \subset L_2$ and $L_4 \cap L_2 = L_1$.

Suppose first that $L_4 \subset L_2$. Set $\phi' = \phi \circ N_{L_4/L_1}$ unless $[L_2 : L_4]$ is odd in which case set $\phi' = 1$. Then from the inductive hypothesis we see that there is a cuspidal automorphic representation $I_{L_2}^{L_4}(\chi)$ of $GL_{[L_2:L_4]}(\mathbb{A}_{L_4})$ such that

- $I_{L_2}^{L_4}(\chi)^c \cong I_{L_2}^{L_4}(\chi)^\vee$;

- $(I_{L_2}^{L_4}(\chi) \otimes (\phi' \circ \det))_\infty$ has the same infinitesimal character as an algebraic representation of $\mathrm{RS}_{\mathbb{Q}}^{L_4}(GL_{[L_2:L_4]})$;

- $I_{L_2}^{L_4}(\chi)_y$ is supercuspidal;

- and

$$[R_l(I_{L_2}^{L_4}(\chi) \otimes (\phi' \circ \det))] = a(I_{L_2}^{L_4}(\chi) \otimes (\phi' \circ \det))$$
$$[\mathrm{Ind}_{\mathrm{Gal}(L_3^{ac}/L_2)}^{\mathrm{Gal}(L_3^{ac}/L_4)} \mathrm{rec}_{l,i}(\chi^{-1}(\phi')^{-1}| \ |^{(1-[L_2:L_4])/2})].$$

By theorem VII.1.9 and lemma VII.2.6 we see that

$$[\mathrm{rec}_l(I_{L_2}^{L_4}(\chi)_y)] = [\mathrm{Ind}_{W_{L_{2,y}}}^{W_{L_{4,y}}} \mathrm{rec}_l(\chi_y)],$$

and hence that

$$[\mathrm{rec}_l(I_{L_2}^{L_4}(\chi)_y^\sigma)] = [\mathrm{Ind}_{W_{\tilde{\sigma} L_{2,y}}}^{W_{L_{4,y}}} \mathrm{rec}_l(\chi_y^{\tilde{\sigma}})].$$

Thus

$$[\mathrm{rec}_l(I_{L_2}^{L_4}(\chi)_y)|_{W_{L_{3,y}}}] = \sum_{\tau \in \mathrm{Gal}(L_3/L_2)\backslash \mathrm{Gal}(L_3/L_4)} [\mathrm{rec}_l(\chi_y \circ N_{L_{3,y}/L_{2,y}} \circ \tau)],$$

and

$$[\mathrm{rec}_l(I_{L_2}^{L_4}(\chi)_y^\sigma)|_{W_{L_{3,y}}}] = \sum_{\tau \in \mathrm{Gal}\,(L_3/L_2)\backslash\mathrm{Gal}\,(L_3/L_4)} [\mathrm{rec}_l(\chi_y \circ N_{L_{3,y}/L_{2,y}} \circ \tau\widetilde{\sigma})].$$

In particular by our assumption on χ_y we see that

$$[\mathrm{rec}_l(I_{L_2}^{L_4}(\chi)_y)] \neq [\mathrm{rec}_l(I_{L_2}^{L_4}(\chi)_y^\sigma)]$$

and conclude that

$$I_{L_2}^{L_4}(\chi)_y \not\cong I_{L_2}^{L_4}(\chi)_y^\sigma.$$

Now set

$$I_{L_2}^{L_1}(\chi) = \mathrm{Ind}\,_{L_4}^{L_1} I_{L_2}^{L_4}(\chi).$$

By the strong multiplicity one theorem we see that

$$I_{L_2}^{L_1}(\chi)^\vee \cong I_{L_2}^{L_1}(\chi)^c.$$

If w is an infinite place of L_1 below places $x_1, \ldots, x_q$ of L_4 then $I_{L_2}^{L_1}(\chi)_x$ is a subquotient of $\mathrm{Ind}\,_{Q(L_{1,x})}^{GL_{[L_2:L_1]}(L_{1,x})}(I_{L_2}^{L_4}(\chi)_{x_1} \times \cdots \times I_{L_2}^{L_4}(\chi)_{x_q})$, where $Q \subset GL_{[L_2:L_1]}$ is a parabolic subgroup with Levi component $GL_{[L_2:L_4]}^q$. Using this one can check that $(I_{L_2}^{L_1}(\chi) \otimes (\phi \circ \det))_\infty$ has the same infinitesimal character as an algebraic representation of $\mathrm{RS}_{\mathbb{Q}}^{L_1}(GL_{[L_2:L_1]})$. Moreover

$$I_{L_2}^{L_1}(\chi)_y = \mathrm{Ind}\,_{L_{4,y}}^{L_{1,y}}(I_{L_2}^{L_4}(\chi)_y)$$

is supercuspidal by proposition 5.5 of [HH]. From theorem VII.1.9 we see that for any finite place x of L_4 not dividing l and lying below places $x_1, \ldots, x_r$ of L_2 we have

$$[\mathrm{rec}_l(I_{L_2}^{L_4}(\chi)_x \otimes (\phi_x' \circ \det))] = \sum_{i=1}^r [\mathrm{Ind}\,_{W_{L_{2,x_i}}}^{W_{L_{4,x}}} \mathrm{rec}_l(\chi_{x_i}(\phi_x' \circ N_{L_{2,x_i}/L_{4,x_i}}))].$$

By lemma VII.2.6 we conclude that for all but finitely many finite places x of L_1 lying below places $x_1, \ldots, x_r$ of L_2 we have

$$[\mathrm{rec}_l(I_{L_2}^{L_1}(\chi)_x \otimes (\phi_x \circ \det))] = \sum_{i=1}^r [\mathrm{Ind}\,_{W_{L_{2,x_i}}}^{W_{L_{1,x}}} \mathrm{rec}_l(\chi_{x_i}(\phi_x \circ N_{L_{2,x_i}/L_{1,x}}))].$$

Finally using the Cebotarev density theorem and theorem VII.1.9 we see that

$$[R_l(I_{L_2}^{L_1}(\chi) \otimes (\phi \circ \det)] =$$
$$[\mathrm{Ind}\,_{\mathrm{Gal}\,(L_3^{ac}/L_2)}^{\mathrm{Gal}\,(L_3^{ac}/L_1)} \mathrm{rec}_{l,\imath}(\chi^{-1}(\phi^{-1} \circ N_{L_2/L_1})| \;|^{(1-[L_2:L_1])/2})].$$

Now we turn to the case $L_2 \cap L_4 = L_1$. In this case by the inductive hypothesis there is a cuspidal automorphic representation $I^{L_4}_{L_2 L_4}(\chi \circ N_{L_2 L_4 / L_2})$ of $GL_{[L_2:L_1]}(\mathbb{A}_{L_4})$ such that

- $I^{L_4}_{L_2 L_4}(\chi \circ N_{L_2 L_4 / L_2})^c \cong I^{L_4}_{L_2 L_4}(\chi \circ N_{L_2 L_4 / L_2})^\vee$;

- $(I^{L_4}_{L_2 L_4}(\chi \circ N_{L_2 L_4 / L_2}) \otimes (\phi \circ N_{L_4 / L_1} \circ \det))_\infty$ has the same infinitesimal character as an algebraic representation of $\mathrm{RS}^{L_4}_{\mathbb{Q}}(GL_{[L_2:L_1]})$;

- $I^{L_4}_{L_2 L_4}(\chi \circ N_{L_2 L_4 / L_2})_y$ is supercuspidal;

- and

$$[R_l(I^{L_4}_{L_2 L_4}(\chi \circ N_{L_2 L_4 / L_2}) \otimes (\phi \circ N_{L_4/L_1} \circ \det))] =$$
$$[\mathrm{Ind}^{\mathrm{Gal}\,(L^{ac}_3/L_4)}_{\mathrm{Gal}\,(L^{ac}_3/L_2 L_4)} \mathrm{rec}_{l,i}((\chi^{-1} \circ N_{L_2 L_4 / L_2})$$
$$(\phi^{-1} \circ N_{L_4/L_1})| \ |^{(1-[L_2:L_1])/2})].$$

By theorem VII.1.9 and lemma VII.2.6 we see that for any prime x of L_4 not dividing l and lying below primes $x_1, \ldots, x_r$ of $L_2 L_4$ we have

$$[\mathrm{rec}_l(I^{L_4}_{L_2 L_4}(\chi \circ N_{L_2 L_4 / L_2})_x)] = \sum_{i=1}^{r} [\mathrm{Ind}^{W_{L_4,x}}_{W_{(L_2 L_4)_{x_i}}} \mathrm{rec}_l(\chi_x \circ N_{(L_2 L_4)_{x_i}/L_{2,x_i}})].$$

In particular we see that

$$[\mathrm{rec}_l(I^{L_4}_{L_2 L_4}(\chi \circ N_{L_2 L_4 / L_2})_x)] = [\mathrm{rec}_l(I^{L_4}_{L_2 L_4}(\chi \circ N_{L_2 L_4 / L_2})^\sigma_{\sigma x})]$$

and so by the strong multiplicity one theorem

$$I^{L_4}_{L_2 L_4}(\chi \circ N_{L_2 L_4 / L_2}) \cong I^{L_4}_{L_2 L_4}(\chi \circ N_{L_2 L_4 / L_2})^\sigma.$$

Thus by theorem 4.2 of chapter 3 of [AC] there is a cuspidal automorphic representation Π of $GL_{[L_2:L_1]}(\mathbb{A}_{L_1})$ such that $\mathrm{Res}^{L_1}_{L_4}(\Pi) = I^{L_4}_{L_2 L_4}(\chi \circ N_{L_2 L_4 / L_2})$.

Again theorem 4.2 of chapter 3 of [AC] tells us that

$$\Pi^\vee \cong \Pi^c \otimes (\eta \circ \det)$$

for some character η of $\mathbb{A}^\times_{L_1}/L^\times_1 L^\times_{1,\infty}(N_{L_4/L_1}\mathbb{A}^\times_{L_4})$. The norm map gives an isomorphism N_{L_1/L^+_1}:

$$\mathbb{A}^\times_{L_1}/L^\times_1 L^\times_{1,\infty}(N_{L_4/L_1}\mathbb{A}^\times_{L_4}) \xrightarrow{\sim} \mathbb{A}^\times_{L^+_1}/(L^+_1)^\times((L^+_{1,\infty})^\times)^0(N_{L^+_4/L^+_1}\mathbb{A}^\times_{L^+_4}).$$

On the other hand because L^+_4 is totally real we see that

$$(L^+_{1,\infty})^\times \subset (L^+_1)^\times((L^+_{1,\infty})^\times)^0(N_{L^+_4/L^+_1}\mathbb{A}^\times_{L^+_4})$$

and hence that N_{L_1/L_1^+} gives an injection

$$
\mathbb{A}_{L_1}^\times / L_1^\times L_{1,\infty}^\times (N_{L_4/L_1} \mathbb{A}_{L_4}^\times) \;\;\overset{\sim}{\to}\;\; \mathbb{A}_{L_1^+}^\times / (L_1^+)^\times (L_{1,\infty}^+)^\times (N_{L_4^+/L_1^+} \mathbb{A}_{L_4^+}^\times)
$$
$$
\hookrightarrow \;\; \mathbb{A}_{L_1}^\times / L_1^\times L_{1,\infty}^\times (N_{L_4^+/L_1^+} \mathbb{A}_{L_4^+}^\times).
$$

Thus we can find a character ψ of $\mathbb{A}_{L_1}^\times / L_1^\times L_{1,\infty}^\times$ such that $\psi \circ N_{L_1/L_1^+} = \eta$ and hence

$$
(\Pi \otimes (\psi \circ \det))^\vee = (\Pi \otimes (\psi \circ \det))^c.
$$

Note that therefore

$$
(I_{L_2 L_4}^{L_4}(\chi) \otimes (\psi \circ N_{L_4/L_2} \circ \det))^\vee = (I_{L_2 L_4}^{L_4}(\chi) \otimes (\psi \circ N_{L_4/L_2} \circ \det))^c.
$$

If x is an infinite place of L_1 lying under an infinite place $\widetilde{x}$ of L_4 we see that $\Pi_x \cong I_{L_2 L_4}^{L_4}(\chi \circ N_{L_2 L_4/L_2})_{\widetilde{x}}$. Thus $(\Pi \otimes (\phi \circ \det))_\infty$ has the same infinitesimal character as an algebraic representation of $\mathrm{RS}_{\mathbb{Q}}^{L_1}(GL_{[L_2:L_1]})$. From lemma 6.12 of chapter 1 and the discussions on pages 52/53 and 59/60 of [AC] we see that Π_y must be supercuspidal.

By lemma VII.2.6, for all but finitely many pairs $(x, \widetilde{x})$ of a place x of L_1 and a place $\widetilde{x}$ of L_4 above x we have

$$
[\mathrm{rec}_l((\Pi \otimes (\phi\psi \circ \det))_x)|_{W_{L_{4,\widetilde{x}}}}]
$$
$$
= [\mathrm{rec}_l((I_{L_2 L_4}^{L_4}(\chi) \otimes ((\phi\psi) \circ N_{L_4/L_1} \circ \det))_{\widetilde{x}})].
$$

It follows from theorem VII.1.9 and the Cebotarev density theorem that

$$
[R_l(\Pi \otimes (\phi\psi \circ \det))|_{\mathrm{Gal}\,(L_3^{\mathrm{ac}}/L_4)}] = [R_l(I_{L_2 L_4}^{L_4}(\chi) \otimes (\phi\psi \circ N_{L_4/L_1} \circ \det))]
$$

and hence that

$$
[R_l(\Pi \otimes (\phi\psi \circ \det))|_{\mathrm{Gal}\,(L_3^{\mathrm{ac}}/L_4)}] =
$$
$$
[\mathrm{Ind}_{\mathrm{Gal}\,(L_3^{\mathrm{ac}}/L_2 L_4)}^{\mathrm{Gal}\,(L_3^{\mathrm{ac}}/L_4)} \mathrm{rec}_{l,\imath}((\chi^{-1} \circ N_{L_2 L_4/L_2})(\phi^{-1}\psi^{-1} \circ N_{L_2 L_4/L_1})
$$
$$
|\;|^{(1-[L_2:L_1])/2})].
$$

As $\mathrm{Ind}_{\mathrm{Gal}\,(L_3^{\mathrm{ac}}/L_2 L_4)}^{\mathrm{Gal}\,(L_3^{\mathrm{ac}}/L_4)} \mathrm{rec}_{l,\imath}(\chi^{-1} \circ N_{L_2 L_4/L_2})|_{W_{L_{4,y}}}$ is irreducible we see that

$$
[R_l(\Pi \otimes (\phi\psi \circ \det))] = a(\Pi \otimes (\phi\psi \circ \det))
$$
$$
[\mathrm{Ind}_{\mathrm{Gal}\,(L_3^{\mathrm{ac}}/L_2)}^{\mathrm{Gal}\,(L_3^{\mathrm{ac}}/L_1)} \mathrm{rec}_{l,\imath}(\chi^{-1}(\phi^{-1}\psi^{-1}\eta' \circ N_{L_2/L_1})|\;|^{(1-[L_2:L_1])/2})],
$$

for some character η' of $\mathbb{A}_{L_1}^\times / L_1^\times L_{1,\infty}^\times (N_{L_4/L_1} \mathbb{A}_{L_4}^\times)$. Thus

$$
[(R_l(\Pi \otimes (\phi\psi \circ \det)) \otimes \mathrm{rec}_{l,\imath}(\psi \otimes (\eta')^{-1}))^c] =
$$
$$
[(R_l(\Pi \otimes (\phi\psi \circ \det)) \otimes \mathrm{rec}_{l,\imath}(\psi \otimes (\eta')^{-1}))^\vee]
$$

and hence by theorem VII.1.9, lemma VII.2.6 and the strong multiplicity one theorem we see that

$$(\Pi \otimes (\eta' \circ \det))^c = (\Pi \otimes (\eta' \circ \det))^\vee.$$

Replacing Π by $\Pi \otimes (\eta' \circ \det)$, we have that $\Pi^c = \Pi^\vee$ and

$$[R_l(\Pi \otimes (\phi\psi(\eta')^{-1} \circ \det))] =$$
$$[\text{Ind}_{\text{Gal}\,(L_3^{ac}/L_2)}^{\text{Gal}\,(L_3^{ac}/L_1)} \text{rec}_{l,i}(\chi^{-1}(\phi^{-1}\psi^{-1}\eta' \circ N_{L_2/L_1})| \ |^{(1-[L_2:L_1])/2})],$$

and hence (using theorem VII.1.9, lemma VII.2.6 and the Cebotarev density theorem)

$$[R_l(\Pi \otimes (\phi \circ \det))] =$$
$$[\text{Ind}_{\text{Gal}\,(L_3^{ac}/L_2)}^{\text{Gal}\,(L_3^{ac}/L_1)} \text{rec}_{l,i}(\chi^{-1}(\phi^{-1} \circ N_{L_2/L_1})| \ |^{(1-[L_2:L_1])/2})].$$

Thus we may set $I_{L_2}^{L_1}(\chi) = \Pi$. $\square$

Recall that if r is a continuous representation of W_K over $\mathbb{C}$ and if Ψ is a continuous additive character of K then we have the following.

- An L-factor

$$L(r, s) = \det((1 - \text{Frob}_p/(\#k(\wp_K))^s)|W_r^{I_K})^{-1},$$

where $W_r^{I_K}$ denotes the inertial invariants of r.

- An ϵ-factor

$$\epsilon(r, s, \Psi).$$

(See for instance [Tat2]. In the notation of [Tat2] we have $\epsilon(r, s, \Psi) = \epsilon(r\omega_s, \Psi, \mu_\Psi)$, where μ_Ψ is the additive Haar measure on K which is self dual with respect to Ψ.)

- A γ-factor

$$\gamma(r, s, \Psi) = L(r^\vee, 1 - s)\epsilon(r, s, \Psi)/L(r, s).$$

If moreover π_1 and π_2 are irreducible admissible representations of $GL_{g_1}(K)$ and $GL_{g_2}(K)$ then we also have the following.

- An L-factor $L(\pi_1 \times \pi_2, s)$.

- An ϵ-factor $\epsilon(\pi_1 \times \pi_2, s, \Psi)$.

- A γ-factor

$$\gamma(\pi_1 \times \pi_2, s, \Psi) = L(\pi_1^\vee \times \pi_2^\vee, 1 - s)\epsilon(\pi_1 \times \pi_2, s, \Psi)/L(\pi_1 \times \pi_2, s).$$

(See [JPSS] for the definitions. In the case $g_2 = 1$ and π_2 is trivial we will simply drop it from the notation.) Note that if π is unramified then

$$L(\pi, s) = L(\mathrm{rec}_l(\pi), s).$$

Corollary VII.2.8 *Suppose that $L_3 \supset L_2 \supset L_1$ are CM-fields such that L_3/L_1 is soluble and Galois. Suppose that χ is a character of $\mathbb{A}_{L_2}^\times / L_2^\times$ such that*

1. *$\chi^c = \chi^{-1}$;*

2. *for every embedding $\tau : L_2 \hookrightarrow \mathbb{C}$ giving rise to an infinite place x we have*

$$\chi_x : z \twoheadrightarrow (\tau z / c\tau z)^{p_\tau}$$

   where $p_\tau \in \mathbb{Z}$ and if $\tau \neq \tau'$ then $p_\tau \neq p'_\tau$;

3. *and there is a finite place y of L_1 which is inert in L_3, which does not divide l, which is unramified over L_1^+ and for which the stabiliser of the character $\chi_y \circ N_{L_3/L_2}$ of $(L_3)_y^\times$ in $\mathrm{Gal}\,(L_3/L_1)$ is $\mathrm{Gal}\,(L_3/L_2)$.*

Then for all but finitely many places x of L_1 we have

$$L(I_{L_2}^{L_1}(\chi)_x, s) = L((\mathrm{Ind}\,_{\mathrm{Gal}\,(L_3^{\mathrm{ac}}/L_2)}^{\mathrm{Gal}\,(L_3^{\mathrm{ac}}/L_1)} \mathrm{rec}_{l,i}(\chi))|_{W_{L_1,x}}, s) = \prod_{\tilde{x}|x} L(\mathrm{rec}_l(\chi_{\tilde{x}}), s).$$

Proof: We only need show that we can choose a character ϕ as in proposition VII.2.7. Let ϕ_∞ be a character of $L_{1,\infty}^\times$ of the form described in proposition VII.2.7. We have a commutative diagram with exact rows

$$
\begin{array}{ccccc}
(0) \to & L_{1,\infty}^\times \times \prod_x \mathcal{O}_{L_1,x}^\times & \to & \mathbb{A}_{L_1}^\times & \to & \mathrm{Cl}\,(L_1) \\
 & \uparrow & & \uparrow & & \uparrow \\
(0) \to & \mathcal{O}_{L_1}^\times N_{L_1/L_1^+}(L_{1,\infty}^\times \times \prod_x \mathcal{O}_{L_1,x}^\times) & \to & L_1^\times N_{L_1/L_1^+} \mathbb{A}_{L_1}^\times & \to & \mathrm{Cl}\,(L_1)^{1+c},
\end{array}
$$

where $\mathrm{Cl}\,(L_1)$ denotes the ideal class group of L_1. Thus it suffices to define ϕ on $\prod_x \mathcal{O}_{L_1,x}^\times / (\mathcal{O}_{L_1,y}^\times (N_{L_1/L_1^+} \prod_x \mathcal{O}_{L_1,x}^\times))$ so that it equals ϕ_∞^{-1} on $\mathcal{O}_{L_1}^\times$. Let $\mathcal{O}_{L_1^+}^1$ denote those elements of $\mathcal{O}_{L_1^+}^\times$ with norm down to $\mathbb{Q}$ equal to 1. Then ϕ_∞ is trivial on $\mathcal{O}_{L_1}^1$. Thus it suffices to check that

$$\mathcal{O}_{L_1}^\times \cap (\mathcal{O}_{L_1,y}^\times (N_{L_1/L_1^+} \prod_x \mathcal{O}_{L_1,x}^\times)) \subset \mathcal{O}_{L_1^+}^1.$$

So suppose $\alpha \in \mathcal{O}_{L_1}^\times \cap (\mathcal{O}_{L_1,y}^\times (N_{L_1/L_1^+} \prod_x \mathcal{O}_{L_1,x}^\times))$. Then, because y is unramified over L_1^+, we see that $\alpha \in (L_1^+)^\times \cap (N_{L_1/L_1^+}(\mathbb{A}_{L_1}^\infty)^\times)$. But we have a right exact sequence

$$(L_1^+)^\times \longrightarrow \bigoplus_x (L_1^+)_x^\times / N_{L_1/L_1^+} L_{1,x}^\times \longrightarrow \mathrm{Gal}\,(L_1/L_1^+) \longrightarrow (0).$$

Thus α must fail to be a norm at an even number of infinite places x of L_1^+, i.e. α is negative at an even number of infinite places x of L_1^+. Thus the norm down to $\mathbb{Q}$ of α is positive and hence 1. $\square$

Corollary VII.2.9 *Suppose that L_3/L_1 is a soluble Galois extension of CM-fields and suppose that L_2 and L_2' are intermediate fields between L_3 and L_1. Let $\Psi = \prod_x \Psi_x$ be a non-trivial additive character of $\mathbb{A}_{L_1}/L_1$. Suppose that χ (resp. χ') is a character of $\mathbb{A}_{L_2}^\times/L_2^\times$ (resp. $\mathbb{A}_{L_2'}^\times/(L_2')^\times$) such that*

1. *$\chi^c = \chi^{-1}$ (resp. $(\chi')^c = (\chi')^{-1}$);*

2. *for every embedding $\tau : L_2 \hookrightarrow \mathbb{C}$ (resp. $\tau' : L_2' \hookrightarrow \mathbb{C}$) giving rise to an infinite place x (resp. x') we have*

$$\chi_x : z \longmapsto (\tau z/c\tau z)^{p_\tau}$$

(resp.

$$\chi_{x'}' : z \longmapsto (\tau' z/c\tau' z)^{p_{\tau'}'})$$

where p_τ (resp. $p_{\tau'}'$) $\in \mathbb{Z}$ and if $\tau \neq \tau_1$ (resp. $\tau' \neq \tau_1'$) then $p_\tau \neq p_{\tau_1}$ (resp. $p_{\tau'}' \neq p_{\tau_1'}'$);

3. *there is a finite place y (resp. y') of L_1 which is inert in L_3, which does not divide l, which is unramified over L_1^+ and for which the stabiliser of the character $\chi_y \circ N_{L_3/L_2}$ (resp. $\chi_{y'}' \circ N_{L_3/L_2'}$) in $\mathrm{Gal}\,(L_3/L_1)$ is $\mathrm{Gal}\,(L_3/L_2)$ (resp. $\mathrm{Gal}\,(L_3/L_2')$).*

Suppose that ψ and ψ' are algebraic characters of $\mathbb{A}_{L_1}^\times$. Then for all places x of L_1 which are inert in L_3 and which do not divide l we have

$$[\mathrm{rec}_l((I_{L_2}^{L_1}(\chi))_x \otimes (\psi_x \circ \det))] = [\mathrm{Ind}\,_{W_{L_2,x}}^{W_{L_1,x}} \mathrm{rec}_l(\chi_x(\psi_x \circ N_{L_2,x/L_1,x}))]$$

and

$$[\mathrm{rec}_l((I_{L_2'}^{L_1}(\chi'))_x \otimes (\psi_x' \circ \det))] = [\mathrm{Ind}\,_{W_{L_2',x}}^{W_{L_1,x}} \mathrm{rec}_l(\chi_x'(\psi_x' \circ N_{L_2',x/L_1,x}))]$$

and

$$\gamma((I_{L_2}^{L_1}(\chi) \otimes (\psi \circ \det))_x \times (I_{L_2'}^{L_1}(\chi') \otimes (\psi' \circ \det))_x, s, \Psi_x) =$$
$$\gamma((\mathrm{Ind}\,_{W_{L_2,x}}^{W_{L_1,x}} \mathrm{rec}_l(\chi_x(\psi_x \circ N_{L_2,x/L_1,x}))) \otimes$$
$$(\mathrm{Ind}\,_{W_{L_2',x}}^{W_{L_1,x}} \mathrm{rec}_l(\chi_x'(\psi_x' \circ N_{L_2',x/L_1,x}))), s, \Psi_x).$$

Proof: This follows from the previous corollary and from theorem 4.1 of [He3]. □

Now fix a non-trivial additive character Ψ of K.

Lemma VII.2.10 *Fix a finite Galois extension K_3/K. For each pair (K_2, χ), where K_2/K is a finite subextension of K_3/K and where χ is a character of $K_2^\times$ of finite order, we can choose an irreducible admissible representation $I_{K_2}^K(\chi)$ of $GL_{[K_2:K]}(K)$ which satisfies the following properties.*

1. $[\mathrm{rec}_l(I_{K_2}^K(\chi))] = [\mathrm{Ind}\,_{W_{K_2}}^{W_K} \mathrm{rec}_l(\chi)]$.

2. *Whenever (K_2, χ) and (K_2', χ') are two such pairs (with both K_2 and K_2' inside the same fixed K_3) and ψ is a character of $K^\times$ of finite order we have*

$$\gamma((I_{K_2}^K(\chi) \otimes (\psi \circ \det)) \times I_{K_2'}^K(\chi'), s, \Psi) =$$
$$\gamma(\mathrm{Ind}\,_{W_{K_2}}^{W_K} \mathrm{rec}_l(\chi(\psi \circ N_{K_2/K})) \otimes \mathrm{Ind}\,_{W_{K_2'}}^{W_K} \mathrm{rec}_l(\chi'), s, \Psi).$$

Proof: Choose an extension of totally real fields L_3^0/L^0 and a place x_0 of L^0, which is inert in L_3^0 and for which the extension $L_{3,x_0}^0/L_{x_0}^0$ is isomorphic to K_3/K. (This may be constructed as in lemma 3.6 of [He1]. Using weak approximation one can ensure that all the number fields of that argument can be taken to be totally real). Choose an imaginary quadratic field M and a real quadratic field N such that N is disjoint from L_3^0 over $\mathbb{Q}$ and such that p splits completely in MN. Let x_1 and y_1 denote two places of MN above p which have the same restriction to M. Set $L_1 = L^0MN$ and $L_3 = L_3^0MN$. Let x (resp. y) denote the place of L_1 above x_0 and x_1 (resp. y_1). Thus x and y are inert in L_3 and the extension $L_{3,x}/L_{1,x}$ and K_3/K and $L_{3,y}/L_{1,y}$ are all isomorphic.

Fix a pair (K_2, χ) as in the lemma and let L_2/L_1 be the subfield of L_3 corresponding to $K_2 \subset K_3$ under the isomorphism of $L_{3,x}/L_{1,x}$ and K_3/K.

Choose p_τ as in corollary VII.2.9 and all divisible by the inertial degree $f_{K_2/K}$. Let χ_∞ denote the character of $L_{2,\infty}^\times$ corresponding to this choice of p_τ. Also choose a character χ_y of $K_2^\times$ for which the stabiliser of $\chi_2 \circ N_{K_3/K_2}$ in $\mathrm{Gal}\,(K_3/K_1)$ is $\mathrm{Gal}\,(K_3/K_2)$. (For this it suffices to choose a finite order character χ_y of $\mathcal{O}_{K_2}^\times$ with the same property. Again it suffices to choose a continuous homomorphism $\mathcal{O}_{K_2}^\times \to \mathbb{Z}_p$ with the same property (then compose it with a character of $\mathbb{Z}_p$ of sufficiently large order). Again it suffices to choose a $\mathbb{Q}_p$-linear map $\mathcal{O}_{K_2}^\times \otimes_{\widehat{\mathbb{Z}}} \mathbb{Q}_p \to \mathbb{Q}_p$ with the same property. But using the p-adic log and the normal basis theorem we find that there is a commutative diagram

$$
\begin{array}{ccc}
\mathcal{O}_{K_3}^\times \otimes_{\widehat{\mathbb{Z}}} \mathbb{Q}_p & \xrightarrow{\sim} & K[\mathrm{Gal}\,(K_3/K)] \\
\downarrow & & \downarrow \\
\mathcal{O}_{K_2}^\times \otimes_{\widehat{\mathbb{Z}}} \mathbb{Q}_p & \xrightarrow{\sim} & K[\mathrm{Gal}\,(K_3/K_2)\backslash\mathrm{Gal}\,(K_3/K)],
\end{array}
$$

where the top horizontal map is $\mathrm{Gal}\,(K_3/K)$-equivariant, where the left hand vertical map is induced by the norm map and where the right hand vertical map is the natural projection. The existence of the desired homomorphism is now immediate.) Now as in the proof of corollary VII.2.8 we can find a character $\widetilde{\chi}$ of $\mathbb{A}_{L_2}^\times$ such that

- $\widetilde{\chi}^{-1} = \widetilde{\chi}_c$,

- $\widetilde{\chi}|_{L_{2,\infty}^\times} = \chi_\infty$,

- $\widetilde{\chi}_x \chi^{-1}$ is unramified,

- and $\widetilde{\chi}_y \chi_y^{-1}$ is unramified.

(One must use the fact that x and y are split over the maximal totally real subfield L_2^+ of L_2. The argument is easier than in the proof of corollary VII.2.8 because $\chi_\infty|_{\mathcal{O}_{L_2^+}^\times} = 1$.)

Now set $\psi_\infty = \chi_\infty|_{L_{1,\infty}^\times}^{-1/f_{K_2/K}}$ and choose a character ψ of $\mathbb{A}_{L_1}^\times$ which is unramified at x and which restricts to ψ_∞ at ∞. One can check that $\widetilde{\chi}_x(\psi_x \circ N_{L_{2,x}/L_{1,x}})$ has finite order and hence is a twist of χ by an unramified character of finite order. Replacing ψ by a twist by a suitable character of finite order one may assume that $\chi = \widetilde{\chi}_x(\psi_x \circ N_{L_{2,x}/L_{1,x}})$.

Finally we set

$$
I_{K_2}^K(\chi) = I_{L_2}^{L_1}(\widetilde{\chi})_x \otimes (\psi_x \circ \det).
$$

The lemma follows from corollary VII.2.9. $\square$

We will let $\mathrm{Cusp}\,(GL_g(K))$ denote the set of isomorphism classes of irreducible admissible representations of $GL_g(K)$ and we will let

$$
\mathrm{Cusp}_K = \bigcup_{g=1}^\infty \mathrm{Cusp}\,(GL_g(K)).
$$

We will let $\mathbb{Z}[\mathrm{Cusp}_K]$ denote the free $\mathbb{Z}$-module with basis the elements of Cusp_K. Then we may extend the definition of $L(\pi_1 \times \pi_2, s)$, $\epsilon(\pi_1 \times \pi_2, s, \Phi)$ and $\gamma(\pi_1 \times \pi_2, s, \Phi)$ to bilinear maps from $\mathbb{Z}[\mathrm{Cusp}_K] \times \mathbb{Z}[\mathrm{Cusp}_K]$ to the multiplicative abelian group of non-zero meromorphic functions on $\mathbb{C}$. We may also extend

- rec_l to a homomorphism $\mathbb{Z}[\mathrm{Cusp}_K] \to \mathrm{Groth}_l(W_K)$,

- $\vee$ to a homomorphism $\mathbb{Z}[\mathrm{Cusp}_K] \to \mathbb{Z}[\mathrm{Cusp}_K]$,

- and $\otimes(\psi \circ \det)$ to a homomorphism $\mathbb{Z}[\mathrm{Cusp}_K] \to \mathbb{Z}[\mathrm{Cusp}_K]$, for any character ψ of $K^\times$.

(Note that to any irreducible admissible representation π of $GL_g(K)$ we can associate a class $[\pi] \in \mathbb{Z}[\mathrm{Cusp}_K]$, i.e. if π is a subquotient of n-Ind $(\pi_1 \times \cdots \times \pi_r)$ with each π_i irreducible supercuspidal then $[\pi] = [\pi_1] + \cdots + [\pi_r]$. Note however that we do not in general have $L(\pi, s) = L([\pi], s)$ etc.)

Lemma VII.2.11 *Fix a finite Galois extension K_3/K. We can associate to any irreducible g-dimensional representation r of W_K/W_{K_3}, an element $[\pi_{K_3/K}(r)]$ in $\mathbb{Z}[\mathrm{Cusp}_K]$ with the following properties.*

1. *For any such r we have $\mathrm{rec}_l[\pi_{K_3/K}(r)] = r$.*

2. *For any irreducible representations r and r' of W_K/W_{K_3} and any character ψ of $K^\times$ of finite order, we have*

$$\gamma([\pi_{K_3/K}(r) \otimes (\psi \circ \det)] \times [\pi_{K_3/K}(r')]^\vee, s, \Psi) = \\ \gamma(r \otimes \mathrm{rec}_l(\psi) \otimes (r')^\vee, s, \Psi).$$

Proof: This follows from the previous lemma and from Brauer's theorem that representations induced from characters of subgroups form a $\mathbb{Z}$-basis of the Grothendieck group of virtual representations of the finite group W_K/W_{K_3}. $\square$

Corollary VII.2.12 *Fix a finite Galois extension K_3/K. If r is an irreducible representation of W_K then $[\pi_{K_3/K}(r)]$ can be represented by a supercuspidal representation $\pi_{K_3/K}(r)$. The map*

$$r \longmapsto \pi_{K_3/K}(r)$$

is an injection from the irreducible representations of W_K/W_{K_3} to Cusp_K.

Proof: Suppose that $[\pi_{K_3/K}(r)] = \sum_i a_i[\pi_i]$ where the $[\pi_i]$ are distinct elements of Cusp_K. Then

$$\gamma([\pi_{K_3/K}(r)] \times [\pi_{K_3/K}(r)]^\vee, s, \Psi)$$

has a zero at $s = 0$ of order $\sum_i a_i^2$ (see proposition 8.1 of [JPSS]). On the other hand

$$\gamma([\pi_{K_3/K}(r)] \times [\pi_{K_3/K}(r)]^\vee, s, \Psi) = \gamma(r \otimes r^\vee, s, \Psi)$$

and so has a simple zero at $s = 1$. The corollary follows. $\square$

Corollary VII.2.13 *Fix a finite Galois extension K_3/K. If r and r' are irreducible representations of W_K/W_{K_3} and if ψ is a continuous character of $K^\times$ of finite order then*

$$L((\pi_{K_3/K}(r) \otimes (\psi \circ \det)) \times \pi_{K_3/K}(r')^\vee, s) = L(r \otimes \mathrm{rec}_l(\psi) \otimes (r')^\vee, s)$$

and

$$\epsilon((\pi_{K_3/K}(r) \otimes (\psi \circ \det)) \times \pi_{K_3/K}(r')^\vee, s, \Psi) = \epsilon(r \otimes \mathrm{rec}_l(\psi) \otimes (r')^\vee, s, \Psi).$$

In particular $\pi_{K_3/K}(r)$ and r have the same conductor.

Proof: This follows from lemma VII.2.11 and the previous corollary as in lemma 4.4 and proposition 4.5 of [He3]. $\square$

Lemma VII.2.14 *To any irreducible continuous g-dimensional representation r of W_K with finite image, we can associate an irreducible supercuspidal representation $\pi(r)$ of $GL_g(K)$ with the following properties.*

1. *For any such r we have $\mathrm{rec}_l[\pi(r)] = r$.*

2. *For any continuous irreducible representations r and r' of W_K with finite images and for any character ψ of $K^\times$ of finite order, we have*

$$L((\pi(r) \otimes (\psi \circ \det)) \times \pi(r')^\vee, s) = L(r \otimes \mathrm{rec}_l(\psi) \otimes (r')^\vee, s)$$

and

$$\epsilon((\pi(r) \otimes (\psi \circ \det)) \times \pi(r')^\vee, s, \Psi) = \epsilon(r \otimes \mathrm{rec}_l(\psi) \otimes (r')^\vee, s, \Psi).$$

Proof: As K has finitely many extensions of given degree there are countably many irreducible continuous representations r of W_K. List them $r_1, r_2, \ldots$ and set $g_i = \dim r_i$. There are only finitely many irreducible supercuspidal representations of $GL_{g_i}(K)$ with the same conductor as r_i and with central character $\det(r_i) \circ \mathrm{Art}_K$. (For instance, as there are only finitely many irreducible representations of $D_{K,g_i}^\times$ which are trivial on a given open compact subgroup and which have given central character, this can be deduced from the Jacquet-Langlands correspondence (see sections 2.5 and 2.6 of [He4]).) For any positive integer I we may find a set $\mathcal{K}_I$ of finite Galois extensions of K such that

- if K' and $K'' \in \mathcal{K}_0$ then either $K' \supset K''$ or $K'' \supset K'$;

- $\bigcup_{K' \in \mathcal{K}_I} K' = K^{\mathrm{ac}}$;

- $\mathcal{K}_I \subset \mathcal{K}_{I-1}$;

- and for each K', $K'' \in \mathcal{K}_I$ and each $i \leq I$ we have

$$\pi_{K'/K}(r_i) = \pi_{K''/K}(r_i).$$

(Argue by recursion on I.) Set $\pi(r_i) = \pi_{K'/K}(r_i)$ for any $K' \in \mathcal{K}_i$. The lemma now follows easily. $\square$

Corollary VII.2.15 *If r is an irreducible continuous representation of W_K with finite image and if ψ is a character of $K^\times$ of finite order then*

$$\pi(r \otimes \mathrm{rec}_l(\psi)) = \pi \otimes (\psi \circ \det).$$

Proof: Look at the zero at $s = 0$ of

$$\gamma([\pi(r) \otimes (\psi \circ \det)] \times [\pi(r \otimes \mathrm{rec}_l(\psi))]^\vee, s, \Psi) = \gamma(r \otimes r^\vee, s, \Psi).$$

$\square$

Corollary VII.2.16 *The map rec_l gives a bijection from the set of isomorphism classes of irreducible supercuspidal representations of $GL_g(K)$ with central character of finite order to the set of equivalence classes of g-dimensional irreducible continuous representations of W_K with finite image.*

Proof: This now follows from the previous lemma and corollary and from theorem 1.2 of [He4]. $\square$

Corollary VII.2.17 *The map rec_l gives a bijection from $\mathrm{Cusp}\,(GL_g(K))$ to the set of equivalence classes of g-dimensional irreducible continuous representations of W_K. If $\pi_1 \in \mathrm{Cusp}\,(GL_{g_1}(K))$ and $\pi_2 \in \mathrm{Cusp}\,(GL_{g_2}(K))$ then*

$$L(\pi_1 \times \pi_2, s) = L(\mathrm{rec}_l(\pi_1) \otimes \mathrm{rec}_l(\pi_2), s)$$

and

$$\epsilon(\pi_1 \times \pi_2, s, \Psi) = \epsilon(\mathrm{rec}_l(\pi_1) \otimes \mathrm{rec}_l(\pi_2), s, \Psi).$$

Proof: Any irreducible supercuspidal representation of $GL_g(K)$ is of the form $\pi \otimes (\psi \circ \det)$, where ψ_π is finite order and ψ is unramified. The corollary follows as in sections 4.2, 4.3 and 4.4 of [He5]. $\square$

Corollary VII.2.18   *1. If π is an irreducible supercuspidal representation of $GL_g(K)$ over $\overline{\mathbb{Q}}_l^{ac}$ then $r_l(\pi)$ is irreducible.*

2. *If π is an irreducible square integrable representation of $GL_g(K)$ over $\overline{\mathbb{Q}}_l^{ac}$ with $|i\psi_\pi| \equiv 1$ and if $\sigma \in W_K$, then any eigenvalue α of $r_l(\pi)(\sigma)$ satisfies*

$$|i\alpha| \in (\#k(v_K))^{\mathbb{Z}}.$$

3. *If π is an irreducible, generic, i-preunitary representation of $GL_g(K)$ with $|i\psi_\pi| \equiv 1$, then $i\pi$ is tempered if and only if for all $\sigma \in W_K$ every eigenvalue α of $r_l(\pi)(\sigma)$ satisfies*

$$|i\alpha| \in (\#k(v_K))^{\mathbb{Z}}.$$

Proof: The first part follows from corollary VII.2.17. This implies that if π is supercuspidal and $\sigma \in W_K$ then for some $N \in \mathbb{Z}_{>0}$ we have that $r_l(\pi)(\sigma)^N \in (\overline{\mathbb{Q}}_l^{ac})^\times$. The second part in the case π is supercuspidal follows from this. The rest of the second part follows from this special case (see the classification of square integrable representations recalled in the paragraph before the statement of lemma I.3.2). The third part follows from the second and from lemma I.3.8. $\square$

Corollary VII.2.19 *The bijection rec_l from $\mathrm{Cusp}\,(GL_g(K))$ to the set of equivalence classes of g-dimensional irreducible continuous representations of W_K is independent of the choice of $l \neq p$ and of the choice of isomorphism $i : \overline{\mathbb{Q}}_l^{ac} \overset{\sim}{\to} \mathbb{C}$ (which we have assumed is chosen so that $i|\ |_K^{1/2}$ is valued in $\mathbb{R}_{>0}^\times$).*

Proof: This follows from the last corollary and from theorem 4.1 of [He5]. $\square$

As described in section 4.4 of [Rod] one may naturally extend

$$\mathrm{rec}_l : \mathrm{Cusp}_K \overset{\sim}{\longrightarrow} \mathrm{Irr}_l(W_K)$$

to a series of bijections

$$\mathrm{rec}_K : \mathrm{Irr}(GL_g(K)) \longrightarrow \mathrm{WDRep}_g(W_K)$$

for all $g \in \mathbb{Z}_{>0}$. We will let $\mathrm{Sp}_g = (r, N)$ denote the g-dimensional Weil-Deligne representation of W_K on a complex vector space with basis $e_0, \ldots, e_{g-1}$ where

- $r(\sigma)e_i = |\mathrm{Art}_K^{-1}|^i e_i$ for all $\sigma \in W_K$ and all $i = 0, \ldots, g-1$

- and where $Ne_i = e_{i+1}$.

Then Rodier sets

$$\mathrm{rec}_K(\mathrm{Sp}_{s_1}(\pi_1) \boxplus \cdots \boxplus \mathrm{Sp}_{s_t}(\pi_t))$$
$$= (\mathrm{rec}_l(\pi_1) \otimes \mathrm{Sp}_{s_1}) \oplus \cdots \oplus (\mathrm{rec}_l(\pi_t) \otimes \mathrm{Sp}_{s_t}).$$

Then we have

$$\mathrm{rec}_K(\pi) = (\mathrm{rec}_l(\pi), N(\pi))$$

for some $N(\pi)$.

Theorem VII.2.20 *The map rec_K is a local Langlands correspondence. (See the introduction for the definition of a local Langlands correspondence.)*

Proof: That this follows from what we have already proved seems to be well known, but for lack of an explicit reference we sketch the argument. It follows from lemma VII.2.6 and the definition of rec_K that

- if $\pi \in \mathrm{Irr}(GL_1(K))$ then $\mathrm{rec}_K(\pi) = \pi \circ \mathrm{Art}_K^{-1}$;

- if $[\pi] \in \mathrm{Irr}(GL_g(K))$ and $\chi \in \mathrm{Irr}(GL_1(K))$ then $\mathrm{rec}_K(\pi \otimes (\chi \circ \det)) = \mathrm{rec}_K(\pi) \otimes \mathrm{rec}_K(\chi)$;

- and if $[\pi] \in \mathrm{Irr}(GL_g(K))$ then $\det \mathrm{rec}_K(\pi) = \mathrm{rec}_K(\psi_\pi)$.

Suppose that π is an irreducible admissible representation of $GL_g(K)$. Then we can write

$$\pi \cong \mathrm{Sp}_{s_1}(\pi_1) \boxplus \cdots \boxplus \mathrm{Sp}_{s_t}(\pi_t),$$

with $\pi_1, \ldots, \pi_t$ irreducible supercuspidals. Moreover we have

$$\pi^\vee \cong \mathrm{Sp}_{s_1}(\pi_1^\vee \otimes |\det|^{1-s_1}) \boxplus \cdots \boxplus \mathrm{Sp}_{s_t}(\pi_t^\vee \otimes |\det|^{1-s_t}).$$

(In the case $t = 1$ this can be deduced from proposition 1.1 (d) and proposition 2.10 of [Ze]. Then the case that π is tempered follows from another application of proposition 1.1 (d) of [Ze]. Finally the general case follows from proposition 1.1 (d) of [Ze] and corollary 2.7 of chapter XI of [BW].) Hence using lemma VII.2.6, the definition of rec_K and the isomorphism

$$\mathrm{Sp}_s^\vee \cong \mathrm{Sp}_s \otimes |\mathrm{Art}_K^{-1}|^{1-s}$$

we see that

- if $[\pi] \in \mathrm{Irr}(GL_g(K))$ then $\mathrm{rec}_K(\pi^\vee) = \mathrm{rec}_K(\pi)^\vee$.

It remains to check that if $[\pi_1] \in \mathrm{Irr}(GL_{g_1}(K))$ and $[\pi_2] \in \mathrm{Irr}(GL_{g_2}(K))$ then

$$L(\pi_1 \times \pi_2, s) = L(\mathrm{rec}_K(\pi_1) \otimes \mathrm{rec}_K(\pi_2), s)$$

and

$$\epsilon(\pi_1 \times \pi_2, s, \Psi) = \epsilon(\mathrm{rec}_K(\pi_1) \otimes \mathrm{rec}_K(\pi_2), s, \Psi).$$

Recall that in [JPSS] the factors $L(\pi_1 \times \pi_2, s)$ and $\epsilon(\pi_1 \times \pi_2, s, \Psi)$ are only defined directly for π_1 and π_2 generic. In this case if

$$\pi_i \cong \mathrm{Sp}_{s_1}(\pi_{i,1}) \boxplus \cdots \boxplus \mathrm{Sp}_{s_{t_i}}(\pi_{i,t_i})$$

with each $\pi_{i,j}$ supercuspidal, then

$$L(\pi_1 \times \pi_2, s) = \prod_{j_1=1}^{t_1} \prod_{j_2=1}^{t_2} L(\pi_{1,j_1} \times \pi_{2,j_2}, s)$$

and

$$\epsilon(\pi_1 \times \pi_2, s, \Psi) = \prod_{j_1=1}^{t_1} \prod_{j_2=1}^{t_2} \epsilon(\pi_{1,j_1} \times \pi_{2,j_2}, s, \Psi)$$

(see theorems 3.1 and 9.5 of [JPSS]). In general these formulae are used to define $L(\pi_1 \times \pi_2, s)$ and $\epsilon(\pi_1 \times \pi_2, s, \Psi)$ for any irreducible admissible π_1 and π_2. As by definition we have

$$L(r_1 \oplus r_2, s) = L(r_1, s)L(r_2, s)$$

and

$$\epsilon(r_1 \oplus r_2, s, \Psi) = \epsilon(r_1, s, \Psi)\epsilon(r_2, s, \Psi)$$

for all Weil-Deligne representations r_1 and r_2, we only need to check that

$$L(\mathrm{Sp}_{s_1}(\pi_1) \times \mathrm{Sp}_{s_2}(\pi_2), s) = L(\mathrm{rec}_K(\pi_1) \otimes \mathrm{rec}_K(\pi_2) \otimes \mathrm{Sp}_{s_1} \otimes \mathrm{Sp}_{s_2}, s)$$

and

$$\epsilon(\mathrm{Sp}_{s_1}(\pi_1) \times \mathrm{Sp}_{s_2}(\pi_2), s, \Psi)$$
$$= \epsilon(\mathrm{rec}_K(\pi_1) \otimes \mathrm{rec}_K(\pi_2) \otimes \mathrm{Sp}_{s_1} \otimes \mathrm{Sp}_{s_2}, s, \Psi)$$

for all irreducible supercuspidal representations π_1 and π_2, and for all positive integers $s_1 \geq s_2$.

By theorems 3.1 and 8.2 of [JPSS] (see also equation (14) of section 8.2 of [JPSS]), we see that if $s_1 \geq s_2$ then

$$L(\text{Sp}_{s_1}(\pi_1) \times \text{Sp}_{s_2}(\pi_2), s) = \prod_{i=0}^{s_2-1} L(\pi_1 \times (\pi_2 \otimes |\det|^{s_1+i-1}), s)$$

and

$$\gamma(\text{Sp}_{s_1}(\pi_1) \times \text{Sp}_{s_2}(\pi_2), s, \Psi)$$
$$= \prod_{i=0}^{s_1-1} \prod_{j=0}^{s_2-1} \gamma((\pi_1 \otimes |\det|^i) \times (\pi_2 \otimes |\det|^j), s, \Psi).$$

Using corollary VII.2.17, we see that it suffices to check if r_1 and r_2 are irreducible Weil-Deligne representations of W_K and if $s_1 \geq s_2$ are positive integers, then

$$L(r_1 \otimes r_2 \otimes \text{Sp}_{s_1} \otimes \text{Sp}_{s_2}, s) = \prod_{i=0}^{s_2-1} L(r_1 \otimes r_2 \otimes |\text{Art}_K^{-1}|^{s_1+i-1}), s)$$

and

$$\gamma(r_1 \otimes r_2 \otimes \text{Sp}_{s_1} \otimes \text{Sp}_{s_2}, s, \Psi) = \prod_{i=0}^{s_1-1} \prod_{j=0}^{s_2-1} \gamma(r_1 \otimes r_2 \otimes |\text{Art}_K^{-1}|^{i+j}, s, \Psi).$$

Note that if $s_1 \geq s_2$ then

$$\text{Sp}_{s_1} \otimes \text{Sp}_{s_2} \cong \sum_{i=1}^{s_2} \text{Sp}_{s_1+s_2+1-2i} \otimes |\text{Art}_K^{-1}|^{i-1}.$$

The desired equality of L-factors follows at once from the definitions in section 4.1.6 of [Tat2]. The desired equality of γ-factors follows easily if we can show that for any irreducible Weil-Deligne representation r of W_K we have

$$\gamma(r \otimes \text{Sp}_t, s, \Psi) = \prod_{i=0}^{t-1} \gamma(r \otimes |\text{Art}_K^{-1}|^i, s, \Psi).$$

To prove this we consider two cases. If r is ramified then according to section 4.1.6 of [Tat2] we have

$$\gamma(r \otimes \text{Sp}_t, s, \Psi) = \epsilon(\bigoplus_{i=0}^{t-1} r \otimes |\text{Art}_K^{-1}|^i, s, \Psi) = \prod_{i=0}^{t-1} \gamma(r \otimes |\text{Art}_K^{-1}|^i, s, \Psi).$$

Thus we may suppose that r is unramified and hence that $\dim r = 1$. Again using the formulae of section 4.1.6 of [Tat2] we see that

$$\gamma(r \otimes \mathrm{Sp}_t, s, \Psi) =$$

$$\epsilon(\bigoplus_{i=0}^{t-1} r \otimes |\mathrm{Art}_K^{-1}|^i, s, \Psi)(-1)^{t-1} \prod_{i=0}^{t-2} (r \otimes |\mathrm{Art}_K^{-1}|^i)(\mathrm{Frob}_K)$$
$$\prod_{i=0}^{t-1} L(r^\vee \otimes |\mathrm{Art}_K^{-1}|^{-i}, 1-s)(\prod_{i=0}^{t-1} L(r \otimes |\mathrm{Art}_K^{-1}|^i, s))^{-1} =$$

$$\prod_{i=0}^{t-1} \gamma(r \otimes |\mathrm{Art}_K^{-1}|^{s+i}, s, \Psi) \prod_{i=0}^{t-2} (-(r \otimes |\mathrm{Art}_K^{-1}|^{s+i})(\mathrm{Frob}_K)$$
$$L(r^\vee \otimes |\mathrm{Art}_K^{-1}|^{-i-1}, 1-s)/L(r \otimes |\mathrm{Art}_K^{-1}|^i, s)) =$$

$$\prod_{i=0}^{t-1} \gamma(r \otimes |\mathrm{Art}_K^{-1}|^{s+i}, s, \Psi).$$

The theorem follows. $\square$

Appendix

A result on vanishing cycles

by V. G. Berkovich

Let k be a complete discrete valuation field whose residue field $\tilde{k}$ is algebraically closed, and let l be a prime integer different from $\mathrm{char}(\tilde{k})$. For a special formal scheme $\mathfrak{X}$ over k° (see [Berk3]), we denote by $\mathfrak{X}(n)$ the scheme $(\mathfrak{X}, \mathcal{O}_{\mathfrak{X}}/\mathcal{I}^{n+1})$, where $\mathcal{I}$ is the maximal ideal of definition of $\mathfrak{X}$. We also denote by $\Psi_m^q(\mathfrak{X})$ the vanishing cycles sheaves $R^q\Psi_\eta(\mathbf{Z}/l^m\mathbf{Z})_{\mathfrak{X}_\eta}$ of $\mathfrak{X}$. These are étale sheaves on the closed fibre $\mathfrak{X}_s = \mathfrak{X}(0)$ of $\mathfrak{X}$.

Let $\mathfrak{X}$ be a quasi-compact special formal scheme over k° locally isomorphic to a formal scheme of the form $\widehat{\mathcal{X}}_{/\mathcal{Y}}$, where $\mathcal{X}$ is a scheme of finite type over k° whose generic fibre $\mathcal{X}_\eta$ is smooth over k and $\mathcal{Y}$ is a subscheme of the closed fibre of $\mathcal{X}$. Assume that $\mathfrak{X}$ is endowed with a continuous action of a profinite group G such that the induced action on $\mathfrak{X}_s$ is trivial. We call it the μ-action of G on $\mathfrak{X}$. Let $\{G_n\}_{n\geq 0}$ be a projective system of finite quotients of G such that G acts on $\mathfrak{X}(n)$ via G_n and $G_0 = \{1\}$.

Furthermore, assume we are given a projective system $\{\mathcal{X}_n\}_{n\geq 0}$ of finite étale Galois coverings of the closed fibre $\mathfrak{X}_s$ with the Galois group G such that the Galois group of $\mathcal{X}_n$ over $\mathfrak{X}_s$ is precisely G_n. By [Berk3], Proposition 2.1, it comes from a projective system $\{\mathfrak{X}_n\}_{n\geq 0}$ of finite étale Galois coverings of $\mathfrak{X}$. We call the corresponding action of G_n on $\mathfrak{X}_n$ the ν-action. Since the μ-action of G on $\mathfrak{X}$ induces the trivial action on $\mathfrak{X}_s$, it extends in a unique way to an action on each $\mathfrak{X}_n$ that induces the trivial action on $\mathfrak{X}_{n,s}$. Thus, for each $n \geq 0$, there are two actions of the finite group G_n

on the scheme $\mathcal{X}_n(n)$, the one induced by μ and that induced by ν, and therefore one can consider the diagonal action of G_n on $\mathcal{X}_n(n)$.

We assume that, for each $n \geq 0$, the quotient scheme $\mathfrak{Y}^{(n)}$ of $\mathcal{X}_n(n)$ by the diagonal action of G_n exists. (For example, it is always true if $\mathcal{X}_n(n)$ is quasi-projective over k°.) In this case $\{\mathfrak{Y}^{(n)}\}$ is an inductive system of schemes. We also assume that there is a special formal scheme $\mathfrak{Y}$ over k° such that $\mathfrak{Y}(n) = \mathfrak{Y}^{(n)}$ for all $n \geq 0$. Our aim is to describe the vanishing cycles sheaves $\Psi_m^q(\mathfrak{Y})$ of $\mathfrak{Y}$ in terms of the vanishing cycles sheaves $\Psi_m^q(\mathfrak{X})$ of $\mathfrak{X}$. (Notice that both sheaves are defined on the same scheme $\mathcal{X}_s = \mathfrak{Y}_s$.)

Recall that the comparison theorem 3.1 from [Berk3] implies that the sheaves $\Psi_m^q(\mathfrak{X})$ are constructible (and equal to zero for $q > \dim(\mathfrak{X}_\eta)$). Furthermore, it follows from [Berk3], Corollary 4.5, that there exists $N \geq 0$ such that any automorphism of $\mathfrak{X}$ trivial on $\mathfrak{X}(N)$ acts trivially on all the sheaves $\Psi_m^q(\mathfrak{X})$ and, in particular, the action of G on the sheaves $\Psi_m^q(\mathfrak{X})$ factors through an action of the finite quotient G_N. Let $n \geq N$. The ν-action of the group G_n on $\mathcal{X}_n$ induces an action on the sheaves $\Psi_m^q(\mathcal{X}_n) = \Psi_m^q(\mathfrak{X})|_{\mathcal{X}_{n,s}}$. On the other hand, consider the diagonal action of the group G on $\mathcal{X}_n$. By loc. cit., the induced action of G on $\Psi_m^q(\mathcal{X}_n)$ factors through the quotient G_n. We call it the diagonal action of G_n on $\Psi_m^q(\mathcal{X}_n)$. Since it is compatible with the ν-action of G_n on $\mathcal{X}_{n,s}$, $\Psi_m^q(\mathcal{X}_n)$ is the pullback of a unique sheaf Ψ_m^q on $\mathcal{X}_s = \mathfrak{Y}_s$. Notice that the construction of the sheaf Ψ_m^q does not depend on the choice of the number $n \geq N$.

Theorem. *Assume that the family of formal completions of $\mathfrak{X}$ along a closed point of $\mathfrak{X}_s$ has a finite number of isomorphism classes. Then there is a canonical system of compatible isomorphisms of sheaves $\Psi_m^q \xrightarrow{\sim} \Psi_m^q(\mathfrak{Y})$, $m \geq 0$, $q \geq 0$.*

Lemma. *Let $\mathfrak{X}$ and $\mathfrak{Y}$ be special affine formal schemes over k°, and assume that $\mathfrak{X}$ is isomorphic to the formal completion of an affine scheme of finite type over k°, whose generic fibre is smooth over k, along a subscheme of the closed fibre. Furthermore, assume we are given projective systems $\{\mathfrak{X}_n\}_{n \geq 0}$ and $\{\mathfrak{Y}_n\}_{n \geq 0}$ of finite étale coverings of $\mathfrak{X}$ and $\mathfrak{Y}$, respectively, and a compatible system of isomorphisms $\varphi_n : \mathfrak{Y}_n(n) \xrightarrow{\sim} \mathfrak{X}_n(n)$. Then for every $r \geq 1$ there exists $n_0 \geq r$ such that for any $n \geq n_0$ there exists an isomorphism of formal schemes $\psi : \mathfrak{Y}_n \xrightarrow{\sim} \mathfrak{X}_n$ whose restriction to $\mathfrak{Y}_n(r)$ coincides with that of φ_n.*

Proof. We may assume that all $\mathfrak{X}_n$ and $\mathfrak{Y}_n$ are connected. Let $\mathfrak{X} = \mathrm{Spf}(A)$ and $\mathfrak{Y} = \mathrm{Spf}(B)$, and let I and J be the maximal ideals of definition of A and B, respectively. By the first assumption, A is the $\mathbf{a}$-adic completion of a finitely generated algebra A' over k° with $A' \otimes_{k^\circ} k$ smooth over k, where $\mathbf{a}$ is an ideal of A' with $\mathbf{a}A = I$. By the second assumption, there are inductive systems $\{A_n\}_{n \geq 0}$ and $\{B_n\}_{n \geq 0}$ of finite étale algebras over A and B, respectively, and a compatible system of isomorphisms $\alpha_n : A_n/I^{n+1}A_n \tilde{\to} B_n/J^{n+1}B_n$. We set $B_\infty = \varinjlim B_n$ and $J_\infty = JB_\infty$. Since all of the pairs (B_n, JB_n) are Henselian, the pair (B_∞, J_∞) is also Henselian and, since B_∞ is faithfully flat over each B_n, one has $J_\infty \cap B_n = JB_n$. The above isomorphisms induce compatible systems of homomorphisms $f_n : A' \to B_n/J^{n+1}B_n$ and $f'_n : A' \to B_\infty/J_\infty^{n+1}$. By R. Elkik's result (*Solutions d'équations à coefficients dans un anneau hensélien*, Ann. Scient. Éc.Norm. Sup. **6** (1973), 553-604: Corollary 1 on p. 567 and Remark 2 on p. 587), there exist integers $t \geq 0$ and $m \geq t$ such that for any $n \geq m$ there exists a homomorphism $g_n : A' \to B_\infty$ which is congruent to f'_{n-t} modulo J_∞^{n-t+1}.

Suppose now we are given an integer $r \geq 1$. We may increase it and assume that $r + t \geq m$. The image of the homomorphism g_{r+t} is contained in some B_{n_0}. Increasing n_0, if necessary, we may assume that $n_0 \geq r+t$, and we claim that it satisfies the property of the lemma. Indeed, let $n \geq n_0$. By the construction, the homomorphism g_{r+t} induces a homomorphism $h_n : A' \to B_n$ congruent to f_n modulo $J^{r+1}B_n$. The latter implies that h_n uniquely extends to a continuous homomorphism $A \to B_n$ which, in its turn, extends uniquely to the finite étale A-algebra A_n, i.e., to a continuous homomorphism $\beta_n : A_n \to B_n$ congruent to α_n modulo $J^{r+1}B_n$. Since α_n is an isomorphism, it follows that β_n is an isomorphism, and the lemma follows. ∎

Proof of Theorem. By the assumption and Corollary 4.5 of [Berk3], we can find an integer $r \geq 1$ such that, for every closed point $\mathbf{x} \in \mathfrak{X}_s$, each automorphism of $\widehat{\mathfrak{X}}_{/\mathbf{x}}$ trivial on $\widehat{\mathfrak{X}}_{/\mathbf{x}}(r)$ acts trivially on $\Psi_m^q(\widehat{\mathfrak{X}}_{/\mathbf{x}})$. Notice that the comparison theorem 3.1 from [Berk3] implies that $\Psi_m^q(\widehat{\mathfrak{X}}_{/\mathbf{x}}) = \Psi_m^q(\mathfrak{X})_{\mathbf{x}}$. In particular, since all of the sheaves $\Psi_m^q(\mathfrak{X})$ are constructible, the number N defined before the formulation of the theorem can be taken equal to r. Furthermore, fix a finite covering $\{\mathfrak{X}^i\}$ of $\mathfrak{X}$ by open affine formal subschemes satisfying the first assumption of the lemma. It gives rise to a finite affine covering $\{\mathfrak{Y}^i\}$ of $\mathfrak{Y}$. Let n be the maximum of the numbers n_0 from the lemma taken for r and all of the pairs $(\mathfrak{X}^i, \mathfrak{Y}^i)$, and let ψ^i denote the isomorphisms $\mathfrak{Y}_n^i \tilde{\to} \mathfrak{X}_n^i$ constructed in the lemma. The latter isomorphisms induce isomorphisms of sheaves $\alpha^i = (\psi^i)^* : \Psi_m^q(\mathfrak{X}_n^i) \tilde{\to} \Psi_m^q(\mathfrak{Y}_n^i)$. Notice that $\mathfrak{Y}_n^i$ is an étale Galois covering of $\mathfrak{Y}^i$ with the Galois group G_n and, in particular, there is an action of G_n on the sheaves $\Psi_m^q(\mathfrak{Y}_n)$. To prove the theorem, it suffices to check the following two facts:

(1) the action of G_n on the sheaves $\Psi_m^q(\mathfrak{X}_n^i)$, which is induced via α^i by the canonical action of G_n on $\Psi_m^q(\mathfrak{Y}_n^i)$, coincides with the diagonal action;

(2) the restrictions of α^i and α^j to the intersection $\mathfrak{X}_{n,s}^i \cap \mathfrak{X}_{n,s}^j$ coincide.

(1) In the proof of (1) we can replace $\mathfrak{X}$ by $\mathfrak{X}^i$, and so we may omit the superscript i. For an element $g \in G$, let $\gamma(g)$ denote the automorphism of $\mathfrak{X}_n$ which is induced via $\psi : \mathfrak{Y}_n \tilde{\to} \mathfrak{X}_n$ by the action of g on $\mathfrak{Y}_n$.

Since all of the sheaves considered are constructible, it suffices to verify the commutativity of the following diagram for every point $\mathbf{z} \in \mathfrak{X}_{n,s}$

$$
\begin{array}{ccc}
\Psi_m^q(\mathfrak{X}_{/\mathbf{x}}) & \overset{\sim}{\to} & \Psi_m^q(\mathfrak{X}_{n/g\mathbf{z}}) \\
\downarrow \mu(g)^* & & \downarrow \gamma(g)^* \\
\Psi_m^q(\mathfrak{X}_{/\mathbf{x}}) & \overset{\sim}{\to} & \Psi_m^q(\mathfrak{X}_{n/\mathbf{z}})
\end{array}
$$

where $\mathbf{x}$ is the image of $\mathbf{z}$ in $\mathfrak{X}_s$ and the horizontal arrows are induced by the canonical isomorphisms $\mathfrak{X}_{n/\mathbf{z}} \overset{\sim}{\to} \mathfrak{X}_{/\mathbf{x}}$ and $\mathfrak{X}_{n/g\mathbf{z}} \overset{\sim}{\to} \mathfrak{X}_{/\mathbf{x}}$. But the construction of ψ gives the commutative diagram

$$
\begin{array}{ccc}
\mathfrak{X}_{n/\mathbf{z}}(r) & \overset{\sim}{\to} & \mathfrak{X}_{/\mathbf{x}}(r) \\
\downarrow \gamma(g) & & \downarrow \mu(g) \\
\mathfrak{X}_{n/g\mathbf{z}}(r) & \overset{\sim}{\to} & \mathfrak{X}_{/\mathbf{x}}(r)
\end{array}
$$

and, therefore, the required fact is true by our choice of r.

(2) Let $\mathbf{z} \in \mathfrak{X}_{n,s}^i \cap \mathfrak{X}_{n,s}^j$. The automorphism $\psi^j \circ (\psi^i)^{-1}$ of $\mathfrak{X}_{n/\mathbf{z}}$ gives rise to an automorphism of $\mathfrak{X}_{/\mathbf{x}}$ whose restriction to $\mathfrak{X}_{/\mathbf{x}}(r)$ is trivial, where $\mathbf{x}$ is the image of $\mathbf{z}$ in $\mathbf{X}_s$. It follows that the automorphism $(\alpha^i)^{-1} \circ \alpha^j$ of $\Psi_m^q(\mathfrak{X}_{n/\mathbf{z}})$ is trivial, and the required fact follows. ∎

Remark. It is very likely that the assumption on $\mathfrak{X}$ from the formulation of the theorem is true for any quasi-compact special formal scheme.

Acknowledgements. The research for this appendix was supported by the US-Israel Binational Science Foundation and by the Minerva Foundation, Germany.

Bibliography

[AC]   J.Arthur and L.Clozel, *Simple algebras, base change and the advanced theory of the trace formula*, Annals of Math. Studies 120, PUP 1989.

[Arth]   J.Arthur, *The L^2-Lefschetz numbers of Hecke operators*, Invent. Math. 97 (1991), 257-290.

[Arti]   M.Artin, *Théorèmes de représentabilité pour les espaces algébriques*, Presses de l'Université de Montréal 1973.

[BBM]   P.Berthelot, L.Breen and W.Messing, *Théorie de Dieudonné cristalline II*, LNM 930, Springer 1982.

[Berk1]   V.G.Berkovich, *Étale cohomology for non-Archimedean analytic spaces*, Pub. Math. IHES 78 (1993), 5-161.

[Berk2]   V.G.Berkovich, *Vanishing cycles for formal schemes*, Invent. Math. 115 (1994), 539-571.

[Berk3]   V.G.Berkovich, *Vanishing cycles for formal schemes II*, Invent. Math. 125 (1996), 367-390.

[Bern]   J.-N.Bernstein, *Le "centre" de Bernstein* (rédigé par P.Deligne), in J.-N.Bernstein, P.Deligne, D.Kazhdan, M.-F.Vigneras "Représentations des groupes réductifs sur un corps local", Hermann 1984.

[Bert]   P.Berthelot, *Altérations de variétés algébriques (d'après A. J. de Jong)*, Astérisque 241 (1997), 273–311.

[Bew]   J.Bewersdorff, *Eine Lefschetzsche Fixpunktformel fur Hecke Operatoren*, Bonner Math. Schriften 164.

[BHK]   C.J.Bushnell, G.Henniart and P.C.Kutzko, *Correspondance de Langlands locale pour GL_n et conducteurs de paires*, Ann. Sci. E.N.S. 31 (1998), 537-560.

[BM]   P.Berthelot and W.Messing, *Théorie de Dieudonné cristalline
            III. Théorèmes d'équivalence et de pleine fidélité.*, in "The
            Grothendieck Festschrift I", Progress in Mathematics 86,
            Birkhäuser, 1990.

[Bo]   P.Boyer, *Mauvaise reduction des varieties de Drinfeld*, Invent.
            Math. 138 (1999), 573–629.

[BR]   D.Blasius and D.Ramakrishnan, *Maass forms and Galois repre-
            sentations*, in "Galois groups over $\mathbb{Q}$", eds. Y.Ihara, K.Ribet and
            J.-P.Serre, Springer 1989.

[Bry]   J.Brylinski, *Un lemme sur les cycles évanescents en dimension
            relative 1*, Ann. Sci. ENS 19 (1986), 460-467.

[BW]   A.Borel and N.Wallach, *Continuous cohomology, discrete sub-
            groups, and representations of reductive groups*, Princeton U.P.
            1980.

[BZ]   I.N.Bernstein and A.V.Zelevinsky, *Induced representations of re-
            ductive $\mathfrak{p}$-adic groups I*, Ann. Sci. ENS (4) 10 (1977), 441-472.

[Car1]   H.Carayol, *Sur la mauvaise réduction des courbes de Shimura*,
            Comp. Math. 59 (1986), 151-230.

[Car2]   H.Carayol, *Sur les représentations l-adiques associées aux formes
            modulaires de Hilbert*, Ann. Sci. ENS 19 (1986), 409-467.

[Car3]   H.Carayol, *Non-abelian Lubin-Tate theory*, in L.Clozel and
            J.S.Milne "Automorphic forms, Shimura varieties and *L*-
            functions II", Academic Press 1990.

[Cas]   W.Casselman, *Characters and Jacquet modules*, Math. Ann. 230
            (1977), 101-105.

[Cl1]   L.Clozel, *Représentations Galoisiennes associées aux representa-
            tions automorphes autoduales de GL(n)*, Pub. Math. IHES 73
            (1991), 97-145.

[Cl2]   L.Clozel, *On the cohomology of Kottwitz's arithmetic varieties*,
            Duke Math. J. 72 (1993), 757-795.

[CL]   L.Clozel and J.-P.Labesse, *Changement de base pour les
            représentations cohomologiques des certaines groupes unitaires*,
            appendix to [Lab].

[De1]   P.Deligne, *Travaux de Shimura*, Seminaire Bourbaki in LNM 244,
            Springer 1971.

[De2]   P.Deligne, letter to Piatetski-Shapiro (1973).

[DKV]   P.Deligne, D.Kazhdan, and M.-F.Vigneras, *Représentations des algèbres centrales simples p-adiques*, in J.-N.Bernstein, P.Deligne, D.Kazhdan, M.-F.Vigneras "Représentations des groupes réductifs sur un corps local", Hermann 1984.

[Dr]   V.G.Drinfeld, *Elliptic modules*, Math. USSR Sbornik 23 (1974), 561-592.

[EGAI]   A.Grothendieck and J.Dieudonné, *Eléments de géométrie algébriques I: le langage des schémas*, Publ. Math. IHES 4 (1960).

[El]   R.Elkik, *Solutions d'équations à coefficients dans un anneau hensélian*, Ann. Sci. ENS 6 (1973), 553-604.

[En]   T.J.Enright, *Relative Lie algebra cohomology and unitary representations of complex Lie groups*, Duke Math. J. 46 (1979), 513-525.

[Fa1]   G.Faltings, *On the cohomology of locally symmetric Hermitian spaces*, in "Séminaire d'Algèbre Paul Dubreil et Marie-Paul Malliavin, 1982", LNM 1029, Springer 1983.

[Fa2]   G.Faltings, *p-adic Hodge theory*, J.A.M.S. 1 (1988), 255-299.

[Fo]   J.-M.Fontaine, *Représentations p-adiques semi-stables*, Astérisque 223 (1994), 113–184.

[Fu]   K.Fujiwara, *Rigid geometry, Lefschetz-Verdier trace formula and Deligne's conjecture*, Invent. Math. 127 (1997), 489-533.

[FK]   E.Freitag and R. Kiehl, *Étale cohomology and the Weil conjecture*, Springer 1988.

[Har1]   M.Harris, *Automorphic forms and the cohomology of vector bundles on Shimura varieties*, in "Automorphic forms, Shimura varieties, and L-functions" eds. L.Clozel and J.S.Milne, Academic Press 1990.

[Har2]   M.Harris, *Supercuspidal representations in the cohomology of Drinfel'd upper half spaces: elaboration of Carayol's program*, Invent. Math. 129 (1997), 75-120.

[Har3]   M.Harris, *The local Langlands conjecture for $GL(n)$ over a p-adic field, $n < p$*, Invent. Math. 134 (1998), 177-210.

[Has]   H.Hasse, *Die Normenresttheorie relativ-Abelscher Zahlkörper als Klassenkörpertheorie im Kleinen*, J. reine agnew. Math. 162 (1930) 145-154.

[HC]   Harish-Chandra, *Admissible invariant distributions on reductive p-adic groups*, Queen's Papers in Pure and Applied Math. 48 (1978), 281-347.

[He1]   G.Henniart, *La conjecture de Langlands locale pour $GL(3)$*, Mémoires S.M.F. 11/12 (1984).

[He2]   G.Henniart, *Le point sur la conjecture de Langlands pour $GL(N)$ sur un corps local*, in C.Goldstein "Séminaire de théorie des nombres de Paris, 1983-84" Birkhäuser 1985.

[He3]   G.Henniart, *On the local Langlands conjecture for $GL(n)$: the cyclic case*, Ann. of Math. 123 (1986), 145-203.

[He4]   G.Henniart, *La conjecture de Langlands locale numérique pour $GL(n)$*, Ann. Sci. E.N.S. (4) 21 (1988), 497-544.

[He5]   G.Henniart, *Caractérisation de la correspondence de Langlands locale par les facteurs ϵ de paires*, Invent. Math. 113 (1993), 339-350.

[He6]   G.Henniart, *Une preuve simple des conjectures de Langlands pour $GL(n)$ sur un corps p-adiques*, Invent. Math. 139 (2000), 439–455.

[HH]   G.Henniart and R.Herb, *Automorphic induction for $GL(n)$ (over local nonarchimedean fields*, Duke Math. J. 78 (1995), 131-192.

[HS]   H.Hopf and H.Samelson, *Ein Satz über die Wirkungsräume geschlossener Liescher Gruppen*, Comment. Math. Helv. 13 (1941), 240-251.

[I1]   L.Illusie, *Déformations de groupes de Barsotti-Tate, d'après A.Grothendieck* , in "Séminaire sur les pinceaux arithmétiques: la conjecture de Mordell", asterisque 127 (1985), 151-198.

[I2]   L.Illusie, *Autor du théorème de monodromie locale*, Astérisque 223 (1994), 9-57.

[JPSS] H.Jacquet, I.I.Piatetskii-Shapiro and J.Shalika, *Rankin-Selberg convolutions*, Amer. J. Math. 105 (1983), 367-483.

[JS1]   H.Jacquet and J.Shalika, *On Euler products and the classification of automorphic representations I*, Amer. J. Math. 103 (1981), 499-558.

[JS2]   H.Jacquet and J.Shalika, *The Whittaker models of induced representations*, Pacific J. of Math. 109 (1983), 107-120.

[Kat]   N.Katz, *Serre-Tate local moduli*, in "Surfaces Algébriques" LNM 868, Springer 1981.

[Ko1]   R.Kottwitz, *Stable trace formula: cuspidal tempered terms*, Duke Math. J. 51 (1984), 611-650.

[Ko2]   R.Kottwitz, *Stable trace formula: elliptic singular terms*, Math. Ann. 275 (1986), 365-399.

[Ko3]   R.Kotwitz, *Points on some Shimura varieties over finite fields*, Jour. of the AMS 5 (1992), 373-444.

[Ko4]   R.Kottwitz, *On the λ-adic representations associated to some simple Shimura varieties*, Invent. Math. 108 (1992), 653-665.

[Ko5]   R.Kottwitz, *Tamagawa Numbers*, Annals of Math. 127 (1988), 629-646.

[Ku]   P.Kutzko, *The Langlands conjecture for GL(2) of a local field*, Ann. of Math. 112 (1980), 381-412.

[KM]   N.Katz and B.Mazur, *Arithmetic moduli of elliptic curves*, Princeton University Press 1985.

[KS]   M.Kuga and J.Sampson, *A coincidence formula for locally symmetric spaces*, Amer. J. Math. 94 (1972), 486-500.

[Lab]   J.-P.Labesse, *Cohomologie, stabilisation et changement de base*, Astérisque 257 (1999).

[Lan]   R.P.Langlands, *Problems in the theory of automorphic forms*, in "Lectures in modern analysis and applications III", LNM 170, Springer 1970.

[LRS]   G.Laumon, M.Rapoport and U.Stuhler, *D-elliptic sheaves and the Langlands correspondence*, Invent. Math. 113 (1993), 217-338.

[LT]   J.Lubin and J.Tate, *Formal complex multiplication in local fields*, Ann. of Math. 81 (1965), 380-387.

[Ma]   H.Matsumura, *Commutative algebra*, (second edition) Benjamin/Cummings 1980.

[Me]   W.Messing, *The crystals associated to Barsotti-Tate groups: with applications to abelian schemes*, LNM 264, Springer 1972.

[Mu1]   D.Mumford, *Abelian varieties*, OUP 1974.

[Mu2]   D.Mumford, *The red book of varieties and schemes*, LNM 1358, Springer 1988.

[MW]   C.Moeglin and J.-L.Waldspurger, *Le spectre résiduel de GL(n)*, Ann. Sci. ENS 22 (1989), 605-674.

[O]   F.Oort, *Subvarieties of moduli space*, Invent. Math. 24 (1974), 95-119.

[P]   R.Pink, *On the calculation of local terms in the Lefschetz-Verdier trace formula and its application to a conjecture of Deligne*, Annals of Math. 135 (1992), 483-525.

[Rod]   F.Rodier, *Représentations de GL(n, k) où k est un corps p-adiques*, Astérisque 92-93 (1982), 201-218.

[Rog1]   J.Rogawski, *An application of the building to orbital integrals*, Comp. Math. 42 (1981), 417-423.

[Rog2]   J.Rogawski, *Representations of GL(n) and division algebras over a p-adic field*, Duke Math. J. 50 (1983), 161-196.

[RZ1]   M.Rapoport and Th.Zink, *Uber die lokale Zetafunktion von von Shimuravarietäten, Monodromiefiltration und verschwindende Zyklen in ungleicher Characteristik*, Invent. Math. 68 (1982), 21-201.

[RZ2]   M.Rapoport and Th.Zink, *Period spaces for p-divisible groups*, Annals of Mathematics Studies 141, PUP 1996.

[Sc]   W.Scharlau, *Quadratic and hermitian forms*, Springer-Verlag (1985).

[Sen]   S.Sen, *Lie algebras of Galois groups arising from Hodge-Tate modules*, Ann. of Math. 97 (1973), 160–170.

[Ser]   J.-P.Serre, *Cohomologie des groupes discrets* in Prospects in Mathematics, Annals of Math. Studies 70 (1971), 77-169.

[SGA4$\frac{1}{2}$] P.Deligne et al, *Cohomologie Etale*, LNM 569, Springer 1977.

[SGA5]   A.Grothendieck et al, *Cohomologie l-adique et fonctions L*, LNM 589, Springer 1977.

[SGA7]   A.Grothendieck et al, *Groupes de monodromie en géométrie algébrique I*, LNM 288, Springer 1972.

[Tad]   M.Tadic, *Classification of unitary representations in irreducible representations of general linear group (non-archimedean case)*, Ann. Sci. E.N.S. 19 (1986), 335-382.

[Tat1]   J.Tate, *Classes d'isogénie des variétés abéliennes sur un corps fini*, Séminaire Bourbaki 352, LNM 179, Springer 1971.

[Tat2]   J.Tate, *Number theoretic background*, in A.Borel and W.Casselman "Automorphic forms, representations and L-functions", Proc. Symposia in Pure Math. 33 (2), AMS 1979.

[Tay1]   R.Taylor, letter to Clozel dated December 11, 1991.

[Tay2]   R.Taylor, *l-adic representations associated to modular forms over imaginary quadratic fields II*, Invent. Math. 116 (1994), 619-643.

[V1]   M.-F.Vigneras, *Caractérisation des intégrales orbitales sur un groupe réductif p-adique*, J. Fac. Sci. Univ. Tokyo 28 (1982), 945-961.

[V2]   M.-F.Vigneras, *On the global correspondence between GL(n) and division algebras*, lecture notes from IAS, Princeton, 1984.

[VZ]   D.Vogan and G.Zuckerman, *Unitary representations with nonzero cohomology*, Compositio Math. 53 (1984), 51–90.

[Ze]   A.V.Zelevinsky, *Induced representations of reductive p-adic groups II: on irreducible representations of GL(n)*, Ann. Sci. E.N.S. (4) 13 (1980), 165-210.

[Zi]   Th.Zink, *Isogenieklassen von Punkten von Shimuramannigfaltigkeiten mit Werten in einem endlichen Körper*, Math. Nachr. 112 (1983), 103–124.

Index

Roman symbols

$A_\flat$, abelian variety, 158

$(A_\flat, i_\flat)$, abelian variety with endomorphisms, 159

$A_{K,g}$, 19

Art $_L$, the Artin map, 19, 20

Aut ^{1}, 66

$\mathbb{A}_L$, adele ring, 20

$\quad \mathbb{A}_L^S$, adeles away from S, 20

$\quad \overline{\mathbb{A}}^S$, 20

$\mathcal{A}$, universal abelian variety, 98, 111

L^{ac}, algebraic closure, 17

$\widehat{K}^{\mathrm{ac}}$, completion of algebraic closure of p-adic field, 19

B, specific division algebra centre F, 51

BC (π), base change, 199, 200

$C_{(A,\lambda,i)}$, endomorphism algebra, 152

C_z, endomorphism algebra, 152

C_0, endomorphism algebra, 173

$C_c^\infty(G)$, smooth compactly supported functions, 22

$C_c^\infty(G, \psi)$, smooth functions compactly supported mod centre, 22

Cusp $(GL_g(K))$, 247

Cusp $_K$, 247

c, complex conjugation, 20

$D_{K,g}$, division algebra with centre K and Hasse invariant $1/g$, 19

D_0, division algebra, 173

D_j, division algebra, 176

D_y^{LA}, division algebra, 176

$D_{\mathrm{DR},\sigma}(R)$, 99

d, specific integer $[F^+ : \mathbb{Q}]$, 51

$\det_{C/L}$, 45

E, specific imaginary quadratic field, 51

End $^0(A)$, 18

$\mathcal{E}_\mu$, coherent sheaf, 101

$e[\rho]$, 87

$e_K = e_{v_K}$, 19

H^{et}, ind-etale part of H, 60, 62

F, specific CM-field EF^+, 51

F^+, specific totally real field, 51

F, $F_{Y/X}$, relative Frobenius, 18

F^m, filtration, 101

$\overline{F}^m$, filtration, 101

Fix$([U^p(m,0)gU^p(m,0)])$, fixed point set, 153

$\mathrm{FP}_{\mathrm{AV}}^{(h)}$, 166

$\mathrm{FP}_{\tau,\mathrm{LA}}^{(h)}$, 176

$\mathrm{FP}_\tau^{(h)}$, 180

Fr, absolute Frobenius on rings in characteristic p, 18

$\widetilde{\mathrm{Fr}^*}^{f_1}$, 123

$\widetilde{\mathrm{Fr}_\varpi^*}^{f_1(n-h)}$, "canonical" Frobenius lift, 126

Frob$_k$, geometric Frobenius, 19

$\mathcal{F}_\rho$, specific lisse sheaf, 136

$\mathcal{F}(\Psi_{F_w,l,n-h,t}^j)$, specific lisse sheaf, 137

$\mathcal{F}(\Psi_{F_w,l,n-h,t}^j)_{x_\infty}$, direct limit of stalks, 140

$\mathcal{F}(\Psi_{F_w,l,n-h,t}^j[\rho])$, specific lisse sheaf, 137

$f_K = f_{v_K}$, 19

f_i, specific integer f_{w_i}, 51

G, specific unitary similitude group over $\mathbb{A}^\infty$, 55

G_β, specific unitary similitude group over $\mathbb{Q}$, 52

G_τ, specific unitary similitude group over $\mathbb{Q}$, 55

G_1, specific unitary group over $\mathbb{A}^\infty$, 55